# HEINEMANN MATHEMATICS 6

## Teacher's Notes

Heinemann Mathematics 6 is intended for use with children

- working in Key Stage 2, mainly at Level 4, of the National Curriculum (England and Wales)
- working at Level 4 of the Common Curriculum (Northern Ireland) and
- working towards Level D of Mathematics 5–14 (Scotland).

Heinemann Educational,
a division of Heinemann Publishers (Oxford) Ltd,
Halley Court, Jordan Hill, Oxford OX2 8EJ

OXFORD LONDON EDINBURGH
MADRID ATHENS BOLOGNA PARIS
MELBOURNE SYDNEY AUCKLAND SINGAPORE TOKYO
IBADAN NAIROBI HARARE GABORONE
PORTSMOUTH NH (USA)


First published 1995

99 98 97 96 95
10 9 8 7 6 5 4 3 2 1

**Writing team**
John T Blair
Ian K Clark
Aileen P Duncan
Percy W Farren
Archie MacCallum
John Mackinlay
Myra A Pearson
Catherine D J Preston
Dorothy S Simpson
John W Thayers
David K Thomson

Designed by Miller, Craig & Cocking
Produced by Gecko Limited, Bicester, Oxon

Printed in the UK by Scotprint Ltd, Edinburgh.

ISBN 0 435 02228 8

# Contents

# Preface

Heinemann Mathematics is a balanced course which provides progression and continuity from nursery through the infant and junior stages to secondary level. It aims to help children to apply mathematics in a variety of situations. The course draws on feedback from practising teachers and experience gained through classroom use of the SPMG materials 'Primary Mathematics – A development through activity'.

## A practical approach

The course is based on the belief that mathematics is best learned through practical experience and discussion. The use of materials, diagrams, and pictures is encouraged throughout the Heinemann Mathematics course to help pupils acquire concepts and understand techniques.

## Problem solving

There is emphasis on problem solving and investigative activity. Pupils are encouraged to apply their mathematics to solve practical, simulated and real problems. In addition to the problem solving work in the Textbook and Workbook, there are further ideas in the Problem Solving Activities booklet in the Assessment and Resources Pack.

## Contexts

Mathematics is presented in a variety of contexts, some related to the world outside the classroom and others to the world of the children's imagination. Such contexts are more likely to stimulate an interest in mathematics and encourage positive attitudes.

## Calculation

Heinemann Mathematics encourages mental, paper and pencil, and calculator methods of calculation. All these methods are used in the Textbook, Workbook, Extension Textbook and Reinforcement Sheets.

## Approach to learning and teaching

The approach adopted in Heinemann Mathematics fits well with the guidelines provided by the National Curriculum (England and Wales), Mathematics 5–14 (Scotland) and the Northern Ireland Common Curriculum. The course has been designed to provide teachers with resources and a structure to meet the requirements of each curriculum.

## Assessment

There is an Assessment booklet in the Assessment and Resources Pack. It contains Check-ups and Round-ups related to the sections of work in the Textbook and Workbook.

Teachers may also wish to select specific questions from the Textbook and Workbook for use in assessment.

# Introduction

Heinemann Mathematics 6 is intended for use with children

- working in Key Stage 2 mainly at Level 4 of the National Curriculum (England and Wales)
- working at Level 4 of the Common Curriculum (Northern Ireland) and
- working towards Level D of Mathematics 5–14 (Scotland).

## FORMAT OF THE COURSE

The Heinemann Mathematics 6 materials consist of

- Textbook
- Workbook
- Extension Textbook
- Reinforcement Sheets
- Teacher's Notes
- Answer Book
- Assessment and Resources Pack

The Teacher's Notes suggest activities for introducing mathematical concepts and techniques. Written work for children is then provided in the Textbook and Workbook.

The Teacher's Notes also contain suggestions for further consolidation and additional activities. The Extension Textbook and Reinforcement Sheets supply activities to enrich and consolidate the children's learning. Assessment material related to the content of Heinemann Mathematics 6 is included in the Assessment and Resources Pack.

## TEXTBOOK AND WORKBOOK

The Textbook contains sections of work as outlined below. The Workbook is linked to the Textbook and referenced from it.

### Textbook

- addition and subtraction
- numbers to millions
- multiplication and division
- fractions, decimals and percentages
- pattern
- length, weight, area, volume, time
- co-ordinates, 2D shape, symmetry, 3D shape, angles
- handling data, probability
- other activities

### Workbook

The pages support topics which appear in the Textbook.

Charts giving details of the mathematical content of Heinemann Mathematics 6 are given on pages 8 to 13.

**Number and money work** should be used in more or less the order in which it appears, to provide a continuous development. **Measure, Shape and Handling Data sections should be interspersed with number work** to give balanced coverage of the various aspects of mathematics. The order in which the Measure, Shape and Handling Data sections are used is fairly flexible, and at the teacher's discretion. The end of each section of work is indicated by the instruction

**Ask your teacher what to do next.**

Some of the pages in the Textbook contain references to a Workbook page where the development of the section is continued. For example on Textbook page 2 there is the instruction

**Go to Workbook page 1.**

Some of the pages in the Workbook contain references back to the Textbook. For example, the following appears on Workbook page 3.

**Go to Textbook page 3.**

# EXTENSION TEXTBOOK

The Extension Textbook contains work to extend some topics featured in the Textbook and Workbook of Heinemann Mathematics 6. It also provides other activities to enrich the children's experience of mathematics.

References to Extension Textbook pages which are appropriate to particular sections of work are included in the Overviews for those sections in the Teacher's Notes. The Record of Work grids at the back of the Workbook also show how the Extension Textbook relates to the other materials comprising Heinemann Mathematics 6.

# REINFORCEMENT SHEETS

This booklet contains photocopiable sheets designed to provide further practice for children, to consolidate selected topics from Heinemann Mathematics 6. Each page is referenced from a specific page of the Textbook or Workbook. For example,

**R7** appears on Textbook page 24 to indicate that reinforcement sheet 7 contains further work appropriate to that section of the Textbook.

This symbol also appears beside the note for Textbook page 24 in the Teacher's Notes.

# SPECIAL FEATURES

## Other activity pages

These pages are identified in the Textbook, Workbook and Extension Textbook by a **purple** box containing the teacher's heading and the page number. They are intended to give further opportunities for the children to apply the mathematics they have been learning, or to extend their experience. The Record of Work grids at the back of the Workbook show where these pages are located in the Textbook and Extension Textbook.

These pages can be used in any order as long as the related concepts have been met by the children. The teaching notes for these pages give appropriate advice about this. The teacher should choose when one of the activities is to be used, and the children for whom it is appropriate.

## Problem solving

**Problem solving**

Throughout the Textbook and Workbook, 'flags' have been used to indicate that a page or a question contains this type of work. These flags also appear in the Teacher's Notes. The problem solving flag indicates work of a non-routine nature that will require mathematical thinking if the children are to find a solution. Some of this work may also be described as investigative as the children are involved in finding out about a mathematical situation for themselves.

Problem solving flags do not appear on Extension Textbook pages as a great deal of the work is of this nature.

## Calculators

This symbol is used to indicate that a calculator is specifically recommended. However calculators may be used throughout to introduce concepts, to extend the work in some aspects and in problem solving.

## Contexts

Most of the activities in the Textbook and Workbook are set in a context. For example, the sections of work on addition and subtraction, numbers to millions and rounding are related to the theme of 'Shipwrecked' where a group of children have a series of adventures. The sections on multiplication and division involve two characters, Superhero and Mog.

Further details about contexts are given on page 16 in the section 'Using the Course'.

# TEACHER'S NOTES

The Teacher's Notes are the central element of the course. They suggest activities and teaching methods for the mathematical content. Ways of introducing and developing contexts are discussed. The notes also include advice about how the Textbook, Workbook and other materials in Heinemann Mathematics 6 might be used to provide a well-balanced course for a school's programme of study.

For each section of work, an 'Overview' describes the purpose and content of the Textbook and/or Workbook pages. 'Resources' for the section are also listed.

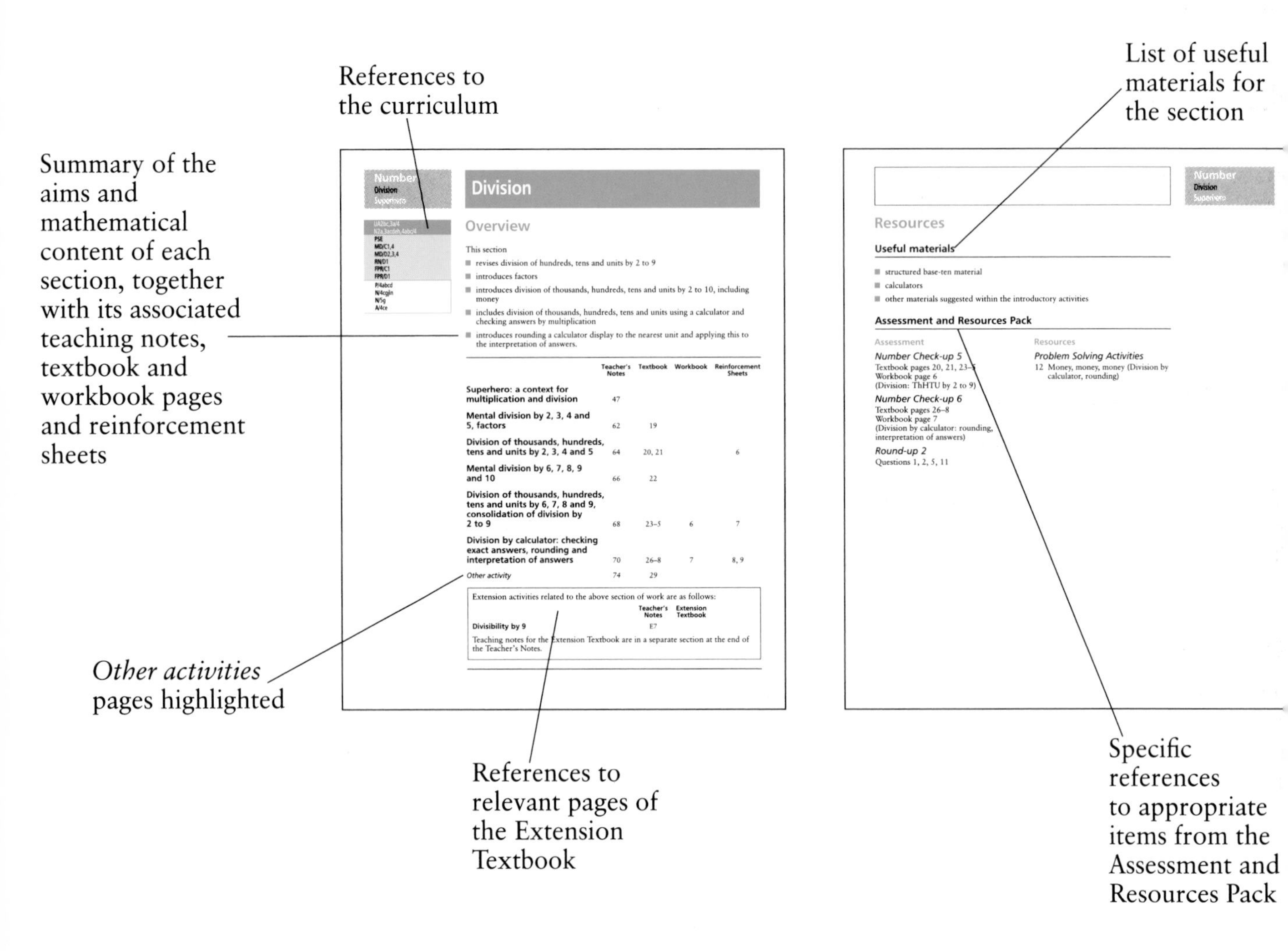


These pages are supported by teaching notes which provide

- advice about introducing any extended context used in the section of work
- references to the curriculum
- a variety of activities from which the teacher can select to introduce each new section of the work
- notes about pages in the Textbook or Workbook which relate to that section of work
- occasional additional activities to give further practice or enrichment
- notes about any 'other activity' pages which appear in this part of the Textbook.

**Teaching notes for the Extension Textbook appear in a separate section at the end of the main Teacher's Notes.**

# ANSWER BOOK

The Answerbook contains lists of answers for the Textbook and Extension Textbook pages and photo reductions of the Workbook pages with the answers inserted.

# ASSESSMENT AND RESOURCES PACK

The pack contains assessment and resource material, including photocopiable record sheets. The teacher should select materials from the pack to suit the needs of particular children.

## Assessment

The Assessment booklet contains twenty-five Check-ups. Thirteen are related to number, six to measure, three to shape and three to handling data. There are also three short Round-ups, each containing a number of questions related to the same theme and covering a range of mathematical topics. The work is presented as photocopiable sheets for use with children.

The booklet provides advice to teachers, answers and curriculum references.

## Problem Solving Activities

This booklet contains thirty-two activities in the form of photocopiable sheets for children. These are intended to supplement the problem solving activities in the Textbook, Workbook and Extension Textbook. They could be used for assessment purposes. Some of the activities are specifically designed to make use of a calculator. The booklet offers advice to teachers and provides answers.

## Resource Cards

The Resource Cards contain games and activities for children. They are linked to the work in the Textbook, Workbook and Teacher's Notes and, in some cases, use the same contexts. A booklet of notes for teachers gives details about preparing and using the cards.

# References to the Curriculum

Detailed references to the National Curriculum (England and Wales), Mathematics 5–14 (Scotland) and Common Curriculum (Northern Ireland) are included in the Teacher's Notes in the Overview to each section of work and in the Assessment and Resources Pack. These references are provided as an aid to planning and recording.

## National Curriculum (England and Wales)

- The Teacher's Notes contain detailed references to the Programme of Study statements within the four Sections at Key Stage 2. The following code is used to identify the Sections.

  **UA** Using and **A**pplying Mathematics **N** Number **SSM** **S**hape, **S**pace and **M**easures **HD** **H**andling **D**ata

The referencing shows each Section, Subsection, Part and Level. For example,

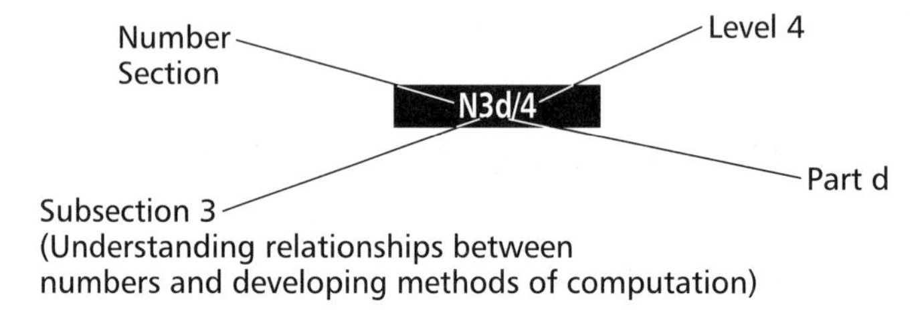


- Some of the outcomes specified in the Programme of Study statements relate to skills and concepts which children develop over an extended period. In these circumstances the Level is sometimes indicated by an arrow. For example,

  **N3d/4→5**

  refers to Number work appropriate to children at Level 4 'working towards' a statement which is at Level 5 in the Level Descriptions.

- The work in a section of Heinemann Mathematics 6, or even on a single page, frequently relates to more than one Subsection or Part. On these occasions, for economy of presentation, statements at the same Level in a Section are combined. For example,

  **N3eg,4a/4**

- Within each Section, Subsection 1 overarches all the other Subsections. As the Heinemann Mathematics course includes coverage of all the statements in these other Sections, no references to Subsection 1 statements are included.

## Mathematics 5–14 (Scotland)

In the Guidelines, the strands within each attainment outcome are not numbered or lettered. The following code is used in Heinemann Mathematics:

PSE Problem Solving and Enquiry
C Collect
O Organise
D Display
I Interpret
RTN Range and Type of Numbers
M Money
AS Add and Subtract
MD Multiply and Divide
RN Round Numbers
FPR Fractions, Percentages and Ratio
PS Patterns and Sequences
FE Functions and Equations
ME Measure and Estimate
T Time
PFS Perimeter, Formulae, Scales
RS Range of Shapes
PM Position and Movement
S Symmetry
A Angle

Each reference then consists of the *strand* code followed by the appropriate *level* and the *target*. For example.

**MD/D2**
second target
'Multiplication and Division' strand
Level D

## Common Curriculum (Northern Ireland)

The referencing for the Common Curriculum of Northern Ireland shows the attainment target, level and statement of attainment. For example,

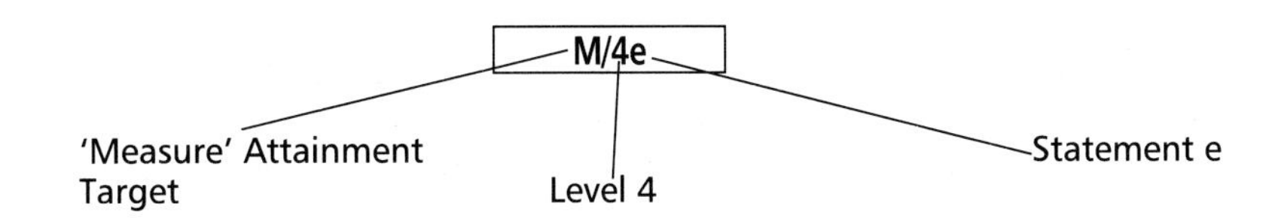


# DEVELOPMENT CHARTS

The following charts outline the mathematical content of Heinemann Mathematics 5 and 6.

One of the charts relates to Mathematics in the National Curriculum (England and Wales), one to Mathematics 5–14 (Scotland), and a third to the Common Curriculum of Northern Ireland.

# Mathematics in the Nation

| | | Number: Knowledge and use of numbers | Algebra |
|---|---|---|---|
| HEINEMANN MATHEMATICS 5 | Practical tasks, mathematical and real-life problems and investigations occur throughout | Mental addition and subtraction<br>Place value to 9999<br>Adding and subtracting three- and four-digit whole numbers with/without a calculator<br>Rounding three-digit whole numbers to the nearest 100 and 10<br>Estimating and approximating to check addition and subtraction calculations<br>Multiplying three- and four-digit whole numbers by 2 to 10 with/without a calculator<br>Understanding and using the effect of multiplying whole numbers by 10 and by 100<br>Multiplying by a two-digit whole number with/without a calculator (introduction)<br>Dividing two- and three-digit whole numbers by 2 to 10 with/without a calculator<br>Money: +, −, ×, ÷, with/without a calculator<br>Recognizing and understanding simple fractions, finding fractions of a quantity, eg '$\frac{1}{8}$ of'<br>One-place decimals:<br>– adding and subtracting<br>– multiplying and dividing by 2 to 10 with/without a calculator<br>Solving +, −, ×, ÷, problems using a calculator | Generalizing in words about patterns arising in various topics and in prob solving |
| HEINEMANN MATHEMATICS 6 | Practical tasks, mathematical and real-life problems and investigations occur throughout | Mental addition and subtraction<br>Place value to millions<br>Rounding:<br>– two- and three-digit whole numbers to the nearest 10 and 100<br>– four-digit numbers to the nearest 100 and 1000<br>Adding and subtracting three- and four-digit whole numbers with/without a calculator<br>Checking addition and subtraction when using a calculator<br>Mental multiplication and division using tables<br>Multiplying mentally two- and three-digit whole numbers by 10 and 100<br>Multiplying three- and four-digit whole numbers by 2 to 9 with/without a calculator<br>Dividing three- and four-digit whole numbers by 2 to 10 with/without a calculator, checking answers, interpreting answers, rounding a calculator display to the nearest appropriate unit<br>Money: +, −, ×, ÷, with/without a calculator<br>Multiplying and dividing by a two-digit whole number with/without a calculator<br>Fractions:<br>– simple equivalences<br>– fractions of quantities, eg '$\frac{1}{10}$ of'<br>One- and two-place decimals:<br>– adding and subtracting with/without a calculator<br>– multiplying mentally by 10<br>– multiplying and dividing with a calculator<br>Percentages:<br>– concept, 100% is 1 whole<br>– simple percentages of quantities, eg '10% of'<br>Negative numbers (temperatures) | Patterns and relationships in multiplication tables and other topics<br>Recognizing and continuing sequences including square and triangular numbers<br>Use of simple function machines for operations<br>Using and identifying simple word formulae in context |

## evelopment Chart

# urriculum (England and Wales)

| Shape, Space and Measures | | Handling Data |
|---|---|---|
| **Shape, position and movement** | **Measures** | |
| nstructing simple 2D and 3D shapes and sing associated language:<br>naking nets of cubes and cuboids;<br>keleton models<br>onstructing circles<br>ling patterns<br><br>flecting shapes in a mirror line; more than wo lines of symmetry<br><br>ecifying location using references such as ·, 2<br><br>ing and understanding language ssociated with angle: degrees, omparing angles, compass directions<br><br>eating paths on squared paper using ommands such as FD 10, RT 90, etc | Using a wider range of metric units of length, area, weight, capacity, volume and standard units of time:<br>length – tenth of a metre;<br>weight – 20g, 10g, 5g;<br>area – $m^2$;<br>capacity – $\frac{1}{4}$ litre; volume – $cm^3$<br>time – durations, counting on and back in hours and minutes, am/pm<br><br>Choosing and using appropriate units and instruments; interpreting numbers on a range of measuring instruments with appropriate accuracy<br><br>Making sensible estimates of a range of measures in relation to everyday objects<br><br>Understanding relationships between units<br><br>Finding perimeters of simple shapes in m and cm<br><br>Finding areas in $cm^2$ by counting squares; areas of irregular shapes<br><br>Finding volumes in $cm^3$ by counting cubes | Constructing and interpreting bar graphs<br><br>Conducting a survey, using grouped tallies<br><br>Understanding the idea of the mode, ie the most frequent item<br><br>Understanding and calculating the mean of a set of data<br><br>Interrogating and interpreting data in a computer database<br><br>Probability:<br>– placing events in order of likelihood and using words to identify the chance |
| ing co-ordinates to locate a point<br><br>llecting, discussing and making:<br>2D shapes including parallelograms, equilateral and isosceles triangles and circles, using side, angle, diagonal, congruent, radius, diameter, circumference<br>3D shapes including cubes, cuboids, triangular prisms and pyramids, using face, edge, vertex<br><br>rning through angles given in degrees, eight point compass<br><br>eating paths, shapes and rotating patterns using computer turtle graphics<br><br>mpleting the missing half of a symmetrical shape or pattern<br>cognizing and drawing up to 4 lines of symmetry in shapes and patterns<br><br>omparing angles – right, acute, obtuse<br>easuring angles accurately within 5 degrees | Choosing and using standard units<br>– length, m, cm<br>– weight, kg, g<br>– area, $cm^2$, $m^2$<br>– volume, litre, ml, $cm^3$<br><br>Relationship between units<br>– 1km = 1000m<br>– 1l = 1000ml = $1000cm^3$<br><br>Imperial units in common use<br><br>Choosing and using appropriate units and instruments, interpreting numbers on a range of measuring instruments with appropriate accuracy<br><br>Calculating perimeters of simple shapes in m and cm<br>Using simple scales to calculate true lengths<br><br>Areas of irregular shapes, rectangles and composite shapes in $cm^2$ by counting squares<br><br>Volumes of cuboids and prisms in $cm^3$ by counting cubes<br><br>Time: reading and writing times to the nearest minute, am/pm, durations including counting on and back in hours and minutes, 24-hour notation including simple durations, timing in seconds | Constructing and interpreting pictograms, bar, bar-line and trend graphs<br><br>Interpreting data<br>– by identifying the mode and median<br>– by calculating the mean<br><br>Constructing and interpreting bar graphs for grouped data<br><br>Organizing and interpreting data using a computer database or spreadsheet<br><br>Conducting a survey<br><br>Probability:<br>– understanding and using ideas of 'likelihood'<br>– distinguishing between 'fair' and 'unfair'<br>– listing possible outcomes of an event |

# Mathematics 5–1

| | Problem Solving and Enquiry | Information Handling | Number, Money<br>Range and Type of Numbers | Money | Add and Subtract | Multiply and Divide | Round Numbers | Fractions Percent-ages and Ratio |
|---|---|---|---|---|---|---|---|---|
| HEINEMANN MATHEMATICS 5 | Problem solving and investigations occur throughout | Constructing and interpreting bar graphs<br>Conducting a survey; using grouped tallies<br>Identifying the most and least frequent items from displays and databases<br>Organizing and interpreting data in a computer database<br>Understanding and calculating the average or 'mean' of a set of data | Whole numbers to 10 000 (count, order, read/write)<br>Place value to 9999<br>Fractions: thirds, fifths, sixths, sevenths, eighths, ninths, tenths<br>Simple equivalences<br>Working with first place decimals | Using coins / notes to £20 worth<br>+, −, ×, ÷ of money with / without a calculator | Mental addition and subtraction<br>Adding and subtracting three- and four -digit numbers with / without a calculator<br>First place decimals: adding and subtracting, multiplying and dividing by 2 to 10, with / without a calculator | Multiplying and dividing mentally using tables<br>Multiplying mentally two- and three-digit numbers by 10 and 100<br>Multiplying two-, three- and four -digit numbers by 2 to 9 with / without a calculator<br>Multiplying by a two-digit whole number with a calculator<br>Dividing two- and three-digit numbers by 2 to 9 with / without a calculator | Rounding three-digit whole numbers to the nearest 100 and 10; checking addition and subtraction calculations | Finding simple fractions quantities involving two-digit whole numbers '$\frac{1}{8}$ of' 48 |
| HEINEMANN MATHEMATICS 6 | Problem solving and investigations occur throughout | Constructing and interpreting pictograms, bar, bar-line and trend graphs<br>Interpreting data<br>– by identifying the mode (most frequent) and median (middle)<br>– by calculating the average or 'mean'<br>Constructing and interpreting bar graphs for grouped data<br>Organizing and interpreting data using a database and spreadsheet<br>Conducting a survey | Whole numbers:<br>– to 100 000 (count, order, read/ write)<br>– to millions (read/write)<br>Fractions:<br>– to hundredths<br>– simple equivalences<br>– mixed numbers with halves and quarters<br>Decimals: working with one- and two-place decimals<br>Percentages:<br>– concept<br>– 100% is 1 whole<br>Negative numbers (temperature) | +, −, ×, ÷ of money with / without a calculator | Mental addition and subtraction<br>Adding and subtracting three- and four-digit whole numbers with/without a calculator<br>Checking addition and subtraction when using a calculator<br>Adding and subtracting one- and two-place decimals with/without a calculator | Mental multiplication and division using all tables to 10<br>Multiplying mentally two- and three-digit numbers by 10 and 100<br>Multiplying three- and four-digit whole numbers by 2 to 9 with/ without a calculator<br>Mutliplying and dividing by a two-digit whole number with/ without a calculator<br>Dividing three- and four-digit whole numbers by 2 to 10 with/without a calculator; checking and interpreting answers<br>Multiplying mentally one- and two-place decimals by 10<br>Multiplying and dividing one- and two-place decimals with a calculator | Rounding:<br>– two- and three-digit numbers to the nearest 10 and 100<br>– four-digit numbers to the nearest 100 and 1000<br>Checking addition and subtraction calculations<br>Rounding a calculator display to the nearest appropriate unit | Finding simple fractions of quantities involving two- and three-digit numbers<br>Finding simple percentag of quantiti |

# evelopment Chart

# cotland)

| d Measurement | | | | | Shape, Position and Movement | | | |
|---|---|---|---|---|---|---|---|---|
| atterns nd equences | Functions and Equations | Measure and Estimate | Time | Perimeter Formulae Scales | Range of Shapes | Position and Movement | Symmetry | Angle |
| atterns nd relation- ips in ultiplica- on tables nd other pics<br><br>ontinuing nd escribing equences | Use of simple 'function machines' for operations | Measuring in standard units:<br>length - $\frac{1}{10}$ of a metre<br>weight - 20g, 10g, 5g<br>area - $m^2$<br>areas of irregular shapes in $cm^2$<br>volume - $\frac{1}{4}$ litre<br>volumes in $cm^3$ by counting cubes<br><br>Reading scales on measuring devices for weight<br><br>Conservation of area<br><br>Estimating lengths, weights and volumes in standard units<br><br>Selecting appropriate measuring devices for length, weight | Reading and writing times to the nearest minute; am / pm<br><br>Using 12-hour clock times in simple timetables<br><br>Durations in hours / minutes<br><br>Counting on and back in hours and minutes<br><br>Using calendars | Calculating perimeters of simple shapes in m and cm | Collecting, discussing and making 3D and 2D shapes:<br>– identifying 2D shapes within 3D shapes<br>– recognizing 3D shapes from 2D drawings<br>– constructing circles<br>– making nets of cubes and cuboids, skeleton models, tiling patterns | Using grid references such as 4,2<br><br>Creating paths on squared paper using commands such as FD 10, RT 90, etc<br><br>Describing and giving directions for routes and journeys<br><br>Turning through right angles and angles given in degrees, eight point compass | Completing the missing half of a symmetrical shape or pattern<br><br>Finding lines of symmetry of shapes drawn on grids<br><br>Recognizing and drawing one, two, more than two lines of symmetry of a shape<br><br>Creating symmetrical shapes | Right angle as 90°, angles in degrees related to position and movement<br><br>Comparing angles |
| atterns and lationships multipli– ation tables nd other pics<br><br>ecognizing nd ontinuing equences cluding uare and iangular umbers | Use of simple 'function machines' for operations<br><br>Using and identifying simple word formulae in context | Measuring in standard units:<br>– length, m, cm<br>– weight, kg, g<br>– area, $cm^2$, $m^2$<br>– volume, litres, ml, $cm^3$<br><br>Relationships between units:<br>– 1 kilometre = 1000m<br>– 1l = 1000ml = $1000cm^3$<br>– Imperial units in common use<br><br>Estimating lengths, weights, volumes in standard units<br><br>Reading scales to the nearest mark on measuring devices in weight and volume<br><br>Areas of irregular shapes, rectangles and composite shapes by counting squares<br><br>Volumes of cuboids and prisms in $cm^3$ by counting cubes | Reading and writing times to the nearest minute, am/pm<br><br>Durations including counting on and back in hours and minutes<br><br>24-hour notation including simple durations<br><br>Timing in seconds | Calculating perimeters of simple shapes in m and cm<br><br>Using simple scales to calculate true lengths | Collecting, discussing and making:<br>– 2D shapes including parallelo- grams, equilateral and isosceles triangles and circles, using side, angle, diagonal, congruent, radius, diameter, circum- ference<br>– 3D shapes including cubes, cuboids, triangular prisms and pyramids, using face, edge, vertex | Using co- ordinates to locate a point<br><br>Turning through angles given in degrees, eight point compass<br><br>Creating paths, shapes and rotating patterns using computer turtle graphics | Completing the missing half of a symmetrical shape or pattern<br><br>Recognizing and drawing up to 4 lines of symmetry in a shape<br><br>Creating symmetrical shapes<br><br>Recognizing rotational symmetry in shapes and patterns | Comparing angles – right, acute, obtuse<br><br>Measuring angles accurately within 5 degrees |

coverage at Level C | coverage at Level D | coverage at Level E

# Mathematics in the Commo

| | Process in Mathematics | Number | Algebra |
|---|---|---|---|
| HEINEMANN MATHEMATICS 5 | Practical tasks, mathematical and real-life problems and investigations occur throughout | Mental addition and subtraction<br>Adding and subtracting three- and four-digit whole numbers with / without a calculator<br>Place value to 9999<br>Multiplying three- and four-digit numbers by 2 to 10 with / without a calculator<br>Understanding the effect of multiplying whole numbers by 10 and by 100<br>Multiplying by a two-digit whole number with a calculator/<br>without a calculator (introduction)<br>Dividing two- and three-digit numbers by 2 to 9 with / without a calculator, including remainders<br>+, −, ×, ÷ of money with a calculator<br>Recognizing and understanding simple fractions;<br>finding fractions of quantities eg '$\frac{1}{8}$ of'; simple equivalences<br>Using and understanding first place decimals; adding and subtracting; multiplying and dividing by 2 to 10, with / without a calculator<br>Approximating three-digit whole numbers to the nearest 100 and 10<br>Estimating in addition and subtraction calculations to obtain approximate answers | Generalizing, in words, patterns arising in various situations |
| HEINEMANN MATHEMATICS 6 | Practical tasks, mathematical and real-life problems and investigations occur throughout | Mental addition and subtraction, estimating to obtain approximate answers<br>Place value to millions<br>Rounding: two- and three-digit numbers to nearest 10 and 100, four-digit numbers to nearest 100 and 1000<br>Adding and subtracting three- and four-digit whole numbers, with / without a calculator.<br>Mental multiplication and division using tables<br>Multiplying and dividing:<br>– three- and four-digit numbers by 2 to 9<br>– by a two-digit whole number with / without a calculator<br>Rounding a calculator display to the nearest appropriate unit<br>Money: +, −, ×, ÷ with / without a calculator<br>Fractions: understanding and using simple equivalences, fractions of quantities<br>Decimals: understanding and using decimals to two places, solving addition and subtraction problems with / without a calculator, multiplying mentally by 10, mutliplying and dividing with a calculator<br>Percentages: recognizing and understanding simple percentages, percentages of quantities<br>Negative numbers: in the context of temperature | Generalizing, in words, patterns arising in various situations<br>Understanding and using multiple and facto<br>Understanding and using simple formulae expressed in words<br>Understanding that addition and subtractio are inverse operations and using this to che calculations, with a calculator when necessary<br>Recognizing and continuing sequences including square and triangular numbers |

# evelopment Chart

# urriculum (Northern Ireland)

| Measures | Shape and Space | Handling Data |
|---|---|---|
| ng a wider range of metric units:<br>gth – tenth of a metre<br>ight – 20g, 10g, 5g<br>a – $m^2$<br>pacity – $\frac{1}{4}$ litre<br>ume – $cm^3$<br>e – durations: counting on and back in hours and minutes; am / pm<br><br>oosing and using appropriate units and truments in a variety of situations, erpreting numbers on a range of asuring instruments<br><br>aking sensible estimates of a range of asures in relation to everyday objects<br><br>derstanding relationships between units<br><br>derstanding the concept of perimeter; ding perimeters of simple shapes in m d cm<br><br>ding areas in $cm^2$ by counting squares; eas of irregular shapes<br><br>nding volumes in $cm^3$ by counting cubes | Recognizing reflective symmetry in 2D shapes; more than two lines of symmetry<br><br>Making simple 2D and 3D shapes and knowing associated language:<br>– making nets of cubes and cuboids; skeleton models<br>– constructing circles<br><br>Using and understanding language associated with angle: degrees, comparing angles, compass directions<br><br>Specifying location using references such as 4,2 | Constructing and interpreting bar graphs<br><br>Conducting a survey, using grouped tallies<br><br>Understanding and calculating the mean of a set of data<br><br>Interrogating data in a computer database<br><br>Placing events in order of 'likelihood' and using appropriate words to identify the chance |
| nderstanding the relationship between its<br>ength, m and cm, m and km<br>weight, kg and g<br>area, $m^2$ and $cm^2$<br>volume, litres, ml and $cm^3$<br>time, h, min, s<br><br>aking sensible estimates of a range of easures<br><br>nderstanding the concept of perimeter, lculating perimeters of simple shapes in m d cm<br><br>nding areas, in $cm^2$, by counting squares d volumes, in $cm^3$, by counting cubes<br><br>nderstanding and using the 12-hour and 4-hour clock, durations<br><br>nderstanding the notion of scale<br><br>nowing imperial units in daily use and eir rough metric equivalent<br><br>easuring angles to the nearest 5° | Specifying location by means of co-ordinates (in first quadrant)<br><br>2D Shape: making shapes such as the parallelogram, isosceles and equilateral triangles, circles, from given information; using associated language such as side, angle, diagonal, congruent, radius and diameter<br><br>3D Shape: making shapes such as triangular prisms and pyramids from given information; using associated language such as face, edge and vertex<br><br>Understanding the eight points of the compass, using clockwise and anti-clockwise appropriately<br>Understanding and using language associated with angle–right, acute, obtuse<br><br>Identifying the reflective symmetries in various shapes<br><br>Recognizing rotational symmetry | Collecting, grouping and ordering data using tallying methods with given equal class intervals and creating frequency tables and diagrams for grouped data<br><br>Understanding and using range and mode, and calculating the mean of a set of data<br><br>Organizing and interpreting data in a computer database or spreadsheet<br><br>Constructing and interpreting<br>– bar, bar-line graphs<br>– trend graphs, knowing that intermediate values may or may not have meaning<br><br>Conducting a survey<br><br>Placing events in order of 'likelihood' and using appropriate words to identify the chance<br><br>Distinguishing between 'fair' and 'unfair'<br><br>Listing possible outcomes of an event |

| Key | |
|---|---|
| ☐ | coverage at Level 3 |
| ☐ | coverage at Level 4 |
| ☐ | coverage at Level 5 |

# Using the course

## APPROACHES TO LEARNING AND TEACHING

- The teaching approach on which the course is based is one where new ideas are introduced through the involvement of the teacher and the children in activities and discussion. This is usually followed by further activities for the children to do by themselves. Written work involving the use of the Textbook and Workbook is only attempted after sufficient practical work and discussion have taken place.
- Given this approach, the Teacher's Notes are a crucial component of Heinemann Mathematics 6, as they provide suggestions for teaching activities and guidance on the use of other resources.
- Teaching by the teacher is essential and cannot be replaced by the use of the Textbook, Workbook or other course materials. The function of such materials is to check on what has already been taught, to provide a record of work completed and to set new challenges where the children can apply the mathematics they have learned. The course materials are not designed to teach new concepts to children working through them on their own, without prior teaching and discussion.
- Heinemann Mathematics 6 provides more ideas for pupil activities, more Textbook and Workbook pages, and more assessment material than should be attempted by any one child. The teacher should *select* the most appropriate activities from the materials to suit the programme of work for particular groups or individual children.
- Teaching suggestions for each section of mathematics are found in the Teacher's Notes. Some of the activities are structured with initial teacher direction and follow-up tasks for the children. Other tasks are designed to be investigative and to encourage the children to explore new mathematical ideas through a problem solving approach. It is likely that children will complete these tasks in a variety of ways and so follow-up discussion between teacher and children will be required to ensure meaningful learning. The children should also have opportunities to discuss their work among themselves and to report on work completed.

## PLANNING

### Starting points

- It is likely that the teacher will identify the mathematical content by referring to national curricular guidelines and the school's programme of study for mathematics. The charts starting on page 297 summarize the curriculum coverage provided by Heinemann Mathematics 6.
- Information about the children's previous experiences in mathematics can be found by consulting
  - the children's records of achievement in mathematics
  - the development charts on pages 8–13.
- Information about the work contained in Heinemann Mathematics 6 is found by consulting
  - the contents list in the Textbook
  - the record grids at the back of the Workbook or in the Assessment and Resources Pack 'Overview' booklet
  - the overview at the beginning of each new section in the Teacher's Notes.

- The introductory pages at the beginning of each section of mathematics in the Teacher's Notes give
  - the aims for the section of work
  - a list of the relevant pages in the Teacher's Notes, the Textbook, Workbook and Reinforcement Sheets
  - a note of any resources the teacher may wish to collect and have available
  - references to the Extension Textbook and to the assessment, problem solving and other resource materials in the Assessment and Resources Pack.
- The teacher should read the Teacher's Notes for the chosen section of work, *selecting* the introductory activities, the Textbook and/or Workbook pages, and additional activities which are the most appropriate to the children's needs. These activities can be supplemented, where desired, by ideas from other sources.

## Routes through the material

- Heinemann Mathematics 6 has been designed to be used in a flexible manner. There are many routes through the material. The important principle is to use sections on Number mostly in the given order and to 'slot in' sections from Measure, Shape and Handling data, thus providing a balanced programme of study to meet the needs of the children.
- All the work of a particular section, for example, Decimals, need not necessarily be tackled through *one* sequence of work. Where sections are quite long, the teacher may find it profitable to break these up at a suitable point with other work.
- The sections on Measure, Shape and Handling data, as well as the 'Other activity' pages, need not be tackled in the order given.
- The notes for Extension Textbook and 'Other activity' pages give details, where appropriate, of units of work from the Textbook or Workbook which should be completed before these are attempted.
- The Record of Work grids at the back of the Workbook and in the Assessment and Resources Pack 'Overview' booklet are useful in planning routes through the material.

## Differentiation

- There will be times when the teacher will wish to work with the whole class or with individual children. For most activities, however, working with groups of children will be appropriate to allow the teacher to teach and to differentiate work. The teaching suggestions in the Teacher's Notes contain activities for group teaching which might be adapted for use with the whole class or individuals.
- The following features of Heinemann Mathematics 6 can help the teacher to plan differentiated programmes:
  - suggestions in the Teacher's Notes for introductory activities and additional activities
  - the Extension Textbook
  - Reinforcement Sheets
  - 'Other activity' pages in the Textbook and Extension Textbook
  - additional Problem Solving Activities including calculator activities, and Resource Cards in the Assessment and Resources Pack.
- The teacher should omit pages, parts of pages or questions in the Textbook and the Workbook which are not appropriate for specific children. However, all the children should experience some problem solving work.
- Reinforcement sheets should be used to provide consolidation work for those children thought to require it. While the Extension Textbook contains some activities which are designed to challenge the more able children, others should be accessible to most children and can be used for enrichment purposes.

- The children's progress through the materials is determined by their understanding of the mathematical ideas and should *not* be planned on the basis of a 'page a day'. Children should have suitable experience of activities such as those suggested in the sections on introductory activities in the Teacher's Notes, before attempting to complete the Textbook or Workbook pages.

## Contexts

- An important consideration when planning any piece of mathematics teaching is the context or setting through which the relevant mathematical concepts, knowledge and skills can be developed. This is true whether the mathematics is to be taught through a single lesson, a block of work, or a more extended mathematical theme or topic. These contexts, real, imaginary, simulated or purely mathematical, can be used to
  - present mathematics in an interesting and motivating way
  - encourage the children to draw on and talk about experiences relevant to the context
  - help to develop a positive attitude towards mathematics
  - provide opportunities for other related mathematics
  - provide opportunities for cross-curricular work
  - provide a stimulus for year or stage collaboration.
- Heinemann Mathematics 6 contains a range of contexts through which various sections of mathematics are developed. The more extended contexts used are as follows:
  - Shipwrecked
  - Superhero
  - Merlin Castle
  - Alltmouth
  - Lynchester Airport
  - Animal Protection Education Centre (APEC)
  - Lands Beyond
  - Channel 6TV
  - Greenwatch Campaign
- The 'Shipwrecked' context is used for Place Value to millions, Addition and Subtraction to 9999, Rounding and Estimation. Four children have been shipwrecked and land on a desert island. They forage for food and supplies, meet another castaway, discover an abandoned submarine and a sunken galleon laden with treasure. They also save some fish threatened by polluted water before finally setting a course for home.
- 'Superhero' provides a context for work on Multiplication and Division by a single-digit number. The context follows Superhero and his cat-like computer, Mog, as they attempt to help people after an earthquake, sort out the chaos at Medic-Aid, foil some jewel thieves, enjoy a Superparty and a Superholiday.
- 'Merlin Castle' is the context used to develop work on Fractions and Percentages. Aspects of life in the mediaeval castle feature in the Fraction work. The Percentage work is set in the present time when the ruins of the castle have become a sightseeing location for tourists.
- 'Alltmouth' is the area around the estuary of the River Allt and provides the setting for the work on Decimals. The context involves the children in visiting different locations including the fishing harbour at Allton, Point of Allt, the lighthouse and the suspension bridge.
- 'Lynchester Airport' provides the context for Multiplication and Division by a two-digit number. The context introduces the children to the features and facilities of this international airport.

- 'APEC' is an organization concerned with animal welfare and conservation. The context is used to develop the work on Length, Weight and Volume.
- 'Lands Beyond' is a fantasy world where new locations, such as the Guard Room and the Groovy Gallery are entered by walking through doors scattered across the strange landscape. The context is used for the work on Area and Angles.
- 'Channel 6TV' is the setting for the work on Time. This context features Class 6, who, having watched the mathematics programme '*Timewarp*', on schools' television, find out more about how programmes are made by visiting the 6TV studios.
- The 'Greenwatch Campaign' involves children in Topperton Primary school in a campaign to improve the quality of their environment – the classrooms, the school, the playground and the waste ground beside the school. The context is used for the work on Handling Data.
- Contexts in Heinemann Mathematics 6 are used primarily to support the learning and teaching of mathematics. They are not meant to dominate or be restrictive. Suggestions on how each context might be used are included in the appropriate section in these Teacher's Notes. Introductory and additional activities are often contextualized – sometimes in the same context as the Textbook and Workbook pages.

# ORGANIZING AND IMPLEMENTING

## Organizing the materials

- At the beginning of each section of work in the Teacher's Notes, the materials required for the Textbook and/or Workbook pages are listed under the heading 'Resources'. Materials from this list, for a particular page or group of pages, should be easily accessible to the children. Materials required for an introductory or additional activity are detailed in the notes for the activity.
- A complete listing of the materials needed to resource Heinemann Mathematics 6 is given on page 314.
- Teachers should take particular note of the calculator symbol which appears on certain Textbook and Workbook pages. The symbol indicates that a calculator is necessary for particular questions.
- Exercise books should be available to the children for the work of the Textbook pages. Sometimes plain, grid or dotty paper is also required. This will usually be listed under 'Resources' for the section.
- Some Textbook and Workbook pages contain a symbol such as **R6** . This indicates that a Reinforcement Sheet is available to provide extra practice in the work of the section from which it is referenced. Sheets should only be photocopied for those children who require additional practice.
- The Teacher's Notes also contain the notes for the Extension Textbook. In general this work is for more able children, but pages which will be of value and interest to most children are also included.
- The Teacher's Notes contain references to Resource Cards from the Assessment and Resources Pack which are relevant to the section. These cards should be prepared in advance. They mainly provide supplementary activities for individuals or small groups. The notes which accompany these cards describe how they relate to the section of work and how they could be used.

## Using the materials

- Introductory teaching, including related practical work, should be carried out before asking the children to attempt the Textbook and Workbook pages. Suggestions for teaching activities are given in the Teacher's Notes. Teachers should select which ones to use or, where appropriate, replace them by activities of their own choosing.
- The activities in the Teacher's Notes have, in most instances, been designed for group teaching and should be adapted for whole class teaching where this is thought to be more appropriate for a particular activity or a specific class. Activities suggested in the Teacher's Notes for children to attempt by themselves, individually or in small groups, should also precede written work.
- Having undertaken appropriate activities, when the children are about to attempt a Textbook or a Workbook page, it may be necessary for the teacher to discuss some of the following:
  - what they have to do, focusing on any difficulties with vocabulary or interpretation of instructions
  - where to find any materials they may need
  - the meaning of any symbols, for example, for the calculator
  - whether they are to use an exercise book
  - how to set out their work and record their answers
  - which questions they should do or omit. For example, some more testing questions, highlighted in the Teacher's Notes, should be omitted by certain children.
- Teachers may wish, on occasions, to ask the children to
  - do questions from a Textbook or a Workbook page with no written record being kept by the children
  - read a question aloud and then express what they have to do in their own words
  - work in pairs or small groups with only one child recording the answers.
- Problem solving questions in the Textbook and Workbook are intended for most children. Such work might be tackled by small groups where the children discuss ways of *starting* the problem. When the children are *doing* the problem, the teacher's role is to observe their progress and offer help if this is requested or the children have misunderstood or cannot progress. Such help should be limited so that the children are left to do their own 'mathematical thinking'. *Reporting* by a group of what they did and found out can often be done orally to the other groups or to the whole class. The reporting might be delayed until several groups have attempted the same problem solving question and discussion can focus on their different methods and, possibly, answers.
- The Problem Solving Activities booklet in the Assessment and Resources Pack provides 'stand alone' problems for use with individuals or small groups. These problems can be attempted at any time as long as the children have met the mathematics involved. The problems are intended to give the children experience in developing simple strategies and mathematical thinking skills. Some of these problems are designed to involve the use of a calculator.
- Calculator activities are an integral part of the course. The children are expected to use a calculator whenever the symbol appears on a page. There will be other occasions when it is sensible to use a calculator, for example, in problem solving.
- The Reinforcement Sheets should be used selectively with those children who need extra practice in specific topics. They should not be used with children who have already mastered a topic. Also, they will not help children whose understanding of the topic is so poor that they require further teaching rather than extra written work. Answers are provided at the back of the Reinforcement Sheets booklet.

- The Extension Textbook provides extra material to enrich and extend the children's experience. Some of the pages are related to topics in the Textbook and Workbook. Information about this is given:
  – in the Overview for these topics in the Teacher's Notes
  – in the notes for the Extension Textbook pages
  – in the Record of Work grids at the back of the Workbook.

  These pages can be attempted when the related work in the Textbook and Workbook has been completed.
- The remaining pages in the Extension Textbook contain 'Other activities' which can be attempted at any time, provided that the children have sufficient understanding of the mathematics involved.

  Teachers should choose the pages which are appropriate to particular children. It is important that the completed work is discussed with the children as much of it is of a problem solving or investigative nature.
- 'Other activity' pages can be attempted by individuals, pairs or groups. The teacher should choose which ones are appropriate for particular children. Some of these pages can be tackled by the children without introductory activities or preceding teacher discussion. The teacher should, however, discuss the work with the children when they have completed the activities.

# ASSESSING AND RECORDING

## Day-to-day assessment

- Much of the assessment of children's learning in the primary classroom is informal and continuous. Many everyday tasks that the children are involved in, such as practical work, Textbook and Workbook activities, problem solving and investigations, games, and using a calculator and computer, provide evidence which helps the teacher make informal judgements on a number of important learning and teaching issues. These include establishing the level or stage at which a particular child is working. This evidence may be gathered over a period of time and in different ways. For example, by
  – studying the children's written work
  – talking with them, posing questions and noting responses
  – listening to their explanations and reports on work carried out
  – observing them working, noting individual strengths and weaknesses (including personal qualities such as perseverance and the ability to work co-operatively)
  – pupil self-assessment.

  However, some more formal assessment, related to set objectives or targets is also required. The Assessment booklet, containing Check-ups and Round-ups in the Assessment and Resources Pack helps to meet this need.

## Assessment booklet

- When the teacher wishes to use a more objective specific task to check on an individual's or group's understanding of a particular target in mathematics, one of the twenty-five Check-ups provided in the Assessment booklet in the Assessment and Resources Pack could be used. A Check-up might be used after a unit of work has been completed. Each Check-up is linked to a section of the Textbook and/or Workbook of Heinemann Mathematics 6.
- The notes for the teacher in the Assessment booklet give details of the mathematical topics and relevant Textbook and Workbook pages for each Check-up.

- Three Round-ups, each based on a single theme, are included in the Assessment booklet. Each covers several aspects of mathematics and a range of assessment targets.
- The Check-ups provide a valuable record of achievement that can be
  - discussed with individuals or groups of children
  - shared with parents
  - transferred along with other information to the next class teacher.
- Each Check-up is referenced to the curriculum. The evidence gathered from the Check-ups, combined with information from other sources, including continuous assessment, can help establish an accurate picture of pupil performance. The evidence can also assist in the planning of future programmes of work.
- A class record guide is included with the notes in the Assessment booklet in the Assessment and Resources Pack.

## Problem Solving Activities

- The Problem Solving Activities booklet in the Assessment and Resources Pack may be used to assess the children's problem solving skills with particular reference to Using and Applying Mathematics in the National Curriculum (England and Wales), the Processes Attainment Target (Northern Ireland) and the Problem Solving and Enquiry Outcome in Mathematics 5–14 (Scotland).

## Record keeping

- There are record keeping grids at the back of the Workbook. A more detailed version of these, which also includes boxes for the Problem Solving Activities and Resource Cards, appears in the Overview booklet in the Assessment and Resources Pack.

  The grids could be used to show when work has been completed or, in a more qualitative way, to show how well a child has performed, for example, by using a code.

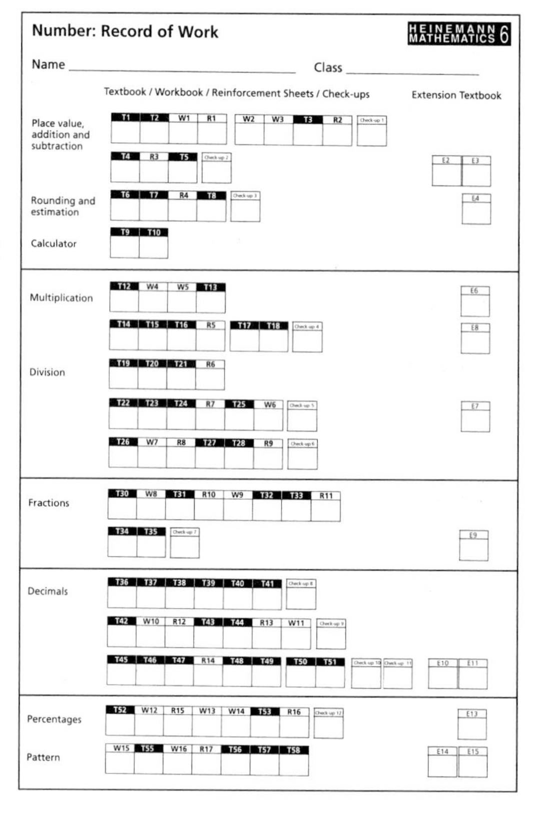

**Number: Record of Work** — HEINEMANN MATHEMATICS 6

Name ______________ Class ______________

| | Textbook / Workbook / Reinforcement Sheets / Check-ups | Extension Textbook |
|---|---|---|
| Place value, addition and subtraction | T1, T2, W1, R1 / W2, W3, T3, R2 / Check-up 1 | |
| | T4, R3, T5 / Check-up 2 | E2, E3 |
| Rounding and estimation | T6, T7, R4, T8 / Check-up 3 | E4 |
| Calculator | T9, T10 | |
| Multiplication | T12, W4, W5, T13 | E6 |
| | T14, T15, T16, R5 / T17, T18 / Check-up 4 | E8 |
| Division | T19, T20, T21, R6 | |
| | T22, T23, T24, R7, T25, W6 / Check-up 5 | E7 |
| | T26, W7, R8, T27, T28, R9 / Check-up 6 | |
| Fractions | T30, W8, T31, R10, W9, T32, T33, R11 | |
| | T34, T35 / Check-up 7 | E9 |
| Decimals | T36, T37, T38, T39, T40, T41 / Check-up 8 | |
| | T42, W10, R12, T43, T44, R13, W11 / Check-up 9 | |
| | T45, T46, T47, R14, T48, T49 / T50, T51 / Check-up 10 / Check-up 11 | E10, E11 |
| Percentages | T52, W12, R15, W13, W14, T53, R16 / Check-up 12 | E13 |
| Pattern | W15, T55, W16, R17, T56, T57, T58 | E14, E15 |

- The Assessment and Resources Pack 'Overview' booklet contains simple check lists for the Key Stage 2 programme of study for England and Wales as well as Northern Ireland and mainly at Level D for Scotland. Space is provided to note work which has been attempted or the quality of the performance of the individuals or groups.
- The Assessment and Problem Solving Activities booklets each contain their own charts and record grids.

The Number part of Heinemann Mathematics 6 has nine sections, each with an Overview and accompanying notes.

# Shipwrecked

## A context for place value, addition, subtraction, rounding and estimation

The work on Textbook pages 1–10 and Workbook pages 1–3 is set in the context of the imaginary adventures of four children who have been shipwrecked off a desert island. The pages record their experiences as they forage for food and supplies. They meet Simon, a fellow castaway, with whom they discover an underground cavern containing an abandoned submarine, the *Nemo* and a sunken galleon laden with treasure. The children and Simon use the *Nemo* to rescue fish threatened by polluted water before finally setting a course for home.

### Introducing the context

This context uses a number of ideas which should be familiar to most children through books, television and films (*Robinson Crusoe, Swiss Family Robinson,* etc.). An initial discussion could focus on the most urgent priorities facing a shipwrecked person. For example,

— finding sources of food and fresh water

— making fire for warmth and cooking

— building a shelter

— making simple tools

— salvaging materials (if wreck is accessible)

— attracting attention by making a beacon or signal

— building a raft or canoe

— finding a way of recording the passage of time.

The following activities could be used in an on-going way as the children proceed through the work.

### 1 Desert island choice

Ask the children to consider carefully and then to list the 10 items they would most like to have with them in the event of being shipwrecked. Remind them that utilities such as electric power and plumbing would not be available.

## 2 Desert island maps

The children could use their imaginations to draw their own desert island maps with various named features. The maps could, however, include locations from the pages of the Textbook and Workbook such as the ruins and the underground cavern.

Completed maps could provide the stimulus for some imaginative writing. If the maps were drawn on squared grids, 'treasure hunt' activities could be undertaken using grid references and/or compass directions.

## 3 Pollution

The issue of pollution of the sea and of the seashore could provide the focus for a discussion. Some children might be interested enough to do a little research into the causes of sea pollution and its effects, possibly giving a report to the class.

# Place value, addition, subtraction, rounding and estimation

UA2bcd,3ab,4d/4
N2a,3cdh,4ac/4
PSE
AS/C1, 2
AS/D1, 2
AS/E3
RN/C1
RN/D1
RTN/D1, 2
P/4a
N/3ah
N/4abhjm

## Overview

This section

- revises and extends work on mental calculation: for example,

  135 + 9    401 – 6
  42 + 57    96 – 25

- introduces mental methods for examples of the types

  37 + 40    66 – 50
  24 + 48    78 – 49

- extends place value to 999 999 and then introduces millions
- revises and extends pencil-and-paper methods for adding and subtracting thousands, hundreds, tens and units, and applies them to problem solving situations
- revises rounding of two- and three-digit numbers to the nearest 10 or 100
- revises and extends estimation to obtain approximate answers to additions and subtractions within 999
- introduces rounding of numbers greater than 999, to the nearest 1000 or 100
- includes addition and subtraction of large numbers using a calculator
- introduces a method of checking addition by subtraction and vice versa, when using a calculator.

| | Teacher's Notes | Textbook | Workbook | Reinforcement Sheets |
|---|---|---|---|---|
| Shipwrecked: a context for place value, addition, subtraction and estimation | 22 | | | |
| Mental addition and subtraction | 26 | 1, 2 | 1 | 1 |
| Place value: numbers to 999 999, millions | 31 | 3 | 2, 3 | 2 |
| Addition and subtraction of ThHTU | 35 | 4, 5 | | 3 |
| Place value: rounding to the nearest hundred and to the nearest ten | 38 | 6 | | |
| Addition and subtraction of HTU, estimation | 40 | 6, 7 | | 4 |

| | Teacher's Notes | Textbook | Workbook | Reinforcement Sheets |
|---|---|---|---|---|
| Place value: rounding to the nearest thousand | 42 | 8 | | |
| Place value, addition and subtraction, calculator | 44 | 9, 10 | | |
| *Other activities* | *46* | *11* | | |

| Extension activities: | Teacher's Notes | Extension Textbook |
|---|---|---|
| Number: place value, +, – | 278 | E2, 3 |
| Rounding and estimating | 280 | E4 |

# Resources

## Useful materials

- 0–100 number line
- 1–100 number square
- other materials suggested within the introductory activities

## Assessment and Resources Pack

### Assessment

*Number Check-up 1*
Textbook page 3
Workbook pages 2, 3
(Place value to millions)

*Number Check-up 2*
Textbook pages 1, 2, 4, 5
Workbook 1
(Mental addition and subtraction,
Addition and subtraction of ThHTU)

*Number Check-up 3*
Textbook pages 6–8
(Rounding to the nearest 10, 100, 1000
Addition and subtraction of HTU,
estimation)

*Round-up 1*
Questions **1(a, b), 2(a, b)**

### Resources

*Problem Solving Activities*
2 Rows and columns (Addition of 1- and 2-digit numbers)
3 Five across (Addition: magic square)
4 Finding the right route (Addition)
10 Change (Money: change)
11 Puzzles prices (Money: addition)

*Resource Cards*
1–3 All in the mind (Mental calculation)
4–6 Escape from the island (Rounding, addition and subtraction)

# Teaching notes

## MENTAL ADDITION AND SUBTRACTION

UA2bc/4 N2a,3cd/4
AS/C1,2 AS/D1
N/4h

Heinemann Mathematics 5 included mental methods for adding and subtracting

— single-digit numbers to/from two-digit or three-digit numbers, for example,

87 + 5  512 – 9

— two-digit numbers **without** 'bridging a ten', for example,

21 + 37  69 – 32

— multiples of ten, 9: for example,

120 + 70  250 – 40

This work is revised and then extended to include

— multiples of ten to/from two-digit numbers, for example,

28 + 20  85 – 30

— more difficult examples involving two-digit numbers, with 'bridging a ten', for example,

54 + 37  82 – 53

The 'Shipwrecked' context begins with the four children exploring the beach.

## Introductory activities

### 1 Pebbles *(mental addition/subtraction of a single-digit number to/from a two-digit number)*

- Use an example such as

  'Dorothy has collected 36 round pebbles and 7 flat pebbles. How many pebbles has she altogether?'

  Revise methods such as

  36 + 7 thought of as 30 + 6 + 7
  leading to 30 + 13 ⟶ **43**

  36 + 7 thought of as 36 + 4 + 3
  leading to 40 + 3 ⟶ **43**

- Subtraction can be dealt with in a similar way. For example,

  'How many more round pebbles than flat pebbles has Dorothy collected?'

  One possible method is as follows:

  36 – 7 thought of as 36 – 6 – 1
  leading to 30 – 1 ⟶ **29**

## 2 Shells *(mental addition/subtraction of a single-digit number to/from a three-digit number)*

- Examples involving three-digit numbers can be introduced in a similar way. For example,

  'Roy has collected 125 white shells and only 8 pink shells. How many shells altogether does he have?'

  Two possible methods are as follows:

  **125 + 8** thought of as 100 + 25 + 8
  leading to 100 + 33 ⟶ **133**

  (This is the method illustrated on Textbook page 1.)

  **125 + 8** thought of as 125 + 5 + 3
  leading to 130 + 3 ⟶ **133**

- Similarly for subtraction you can use an example such as

  'How many more white shells than pink shells has Roy collected?'

  Two possible methods are as follows:

  = **125 – 8**

  = 125 – 5 – 3

  = 120 – 3

  = **117**

  (This is the method illustrated on Textbook page 1.)

  **125 – 8**  25 – 8 = 17

  so 125 – 8 = **117**

  The children's own mental methods can be discussed and used in preference to teaching a specific method.

**At this point the children could try Textbook page 1.**

## 3 Beach bonfire *(mental addition/subtraction of two-digit numbers without 'bridging a ten')*

- Introduce the method of 'exploding' the numbers and dealing with the tens and units separately, using an example such as

  'Sean collected 26 pieces of driftwood for the bonfire and Elaine collected 12 pieces. How many pieces of wood were collected altogether?'

  **26 add 12** thought of as (20 + 6) + (10 + 2)

  Add the tens ⟶ 20 + 10 = 30

  Add the units ⟶ 6 + 2 = 8

  30 + 8 = **38**

- Now deal with subtraction by asking, for example,

  'How many more pieces of driftwood did Sean collect than Elaine?'

  A similar method to that shown above for addition is illustrated on Textbook page 2.

  **26 subtract 12**

  Subtract the tens ⟶ 20 – 10 = 10

  Subtract the units ⟶ 6 – 2 = 4

  10 + 4 = **14**

## 4 Roasted nuts *(mental addition/subtraction of a multiple of ten to/from a two-digit number)*

- 'Dorothy gathered 34 nuts to roast on the fire and Elaine found another 20. How many nuts did they find altogether?'

  Children may be helped to visualize the mental process if they are first given experiences with materials such as:

  — a large 0–100 number line and strips of paper or card cut to an appropriate length to represent intervals of 10. The children could place these against the number line to illustrate the addition.

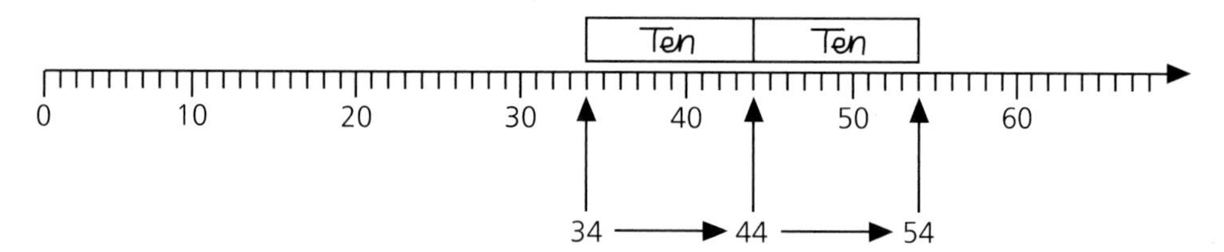


  'Jumps' drawn in chalk could be used instead of paper strips.

  — a 1–100 number square with the children sliding a ruler down to see the pattern of answers.

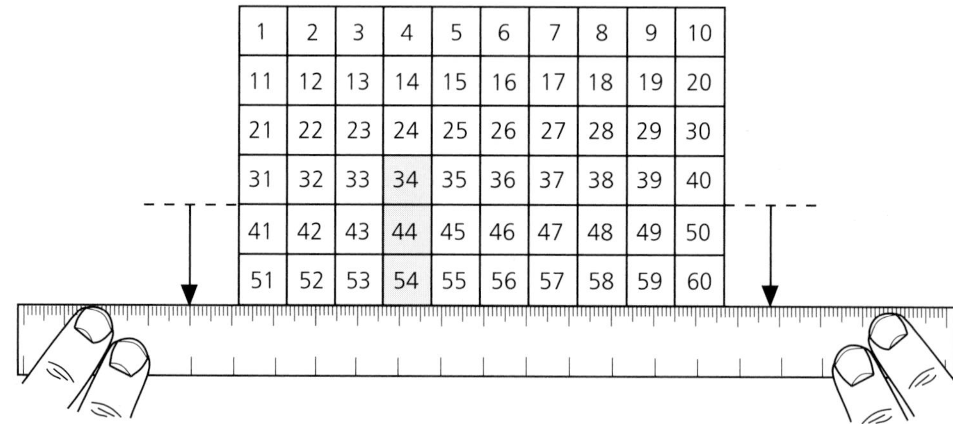

| 1 | 2 | 3 | 4 | 5 | 6 | 7 | 8 | 9 | 10 |
|---|---|---|---|---|---|---|---|---|---|
| 11 | 12 | 13 | 14 | 15 | 16 | 17 | 18 | 19 | 20 |
| 21 | 22 | 23 | 24 | 25 | 26 | 27 | 28 | 29 | 30 |
| 31 | 32 | 33 | 34 | 35 | 36 | 37 | 38 | 39 | 40 |
| 41 | 42 | 43 | 44 | 45 | 46 | 47 | 48 | 49 | 50 |
| 51 | 52 | 53 | 54 | 55 | 56 | 57 | 58 | 59 | 60 |

  — base-ten materials.

- The method illustrated on Textbook page 2 should be discussed to highlight the thinking process.

**34 + 20**

Add the tens

34 + 20 = **54**

- Subtraction examples can be tackled in a similar way.

  The children's own mental methods could be discussed and used in preference to teaching a specific method.

**At this point the children could try Textbook page 2.**

## 5 Beach wildlife *(mental addition of two-digit numbers involving 'bridging a ten')*

- Use a question such as

  'Dorothy counted 27 birds along the beach and Sean counted 26 in the trees. How many birds were counted altogether?'

  The method suggested on Workbook page 1 involves adding the **tens** of the second number to the first number and then its **units**. For example,

  **27 + 26** thought of as 27 + 20 + 6
  leading to 47 + 6 = **53**

- An alternative method would be to add the **units** of the second number to the first number and then its **tens**. For example,

  **27 + 26** thought of as 27 + 6 + 20
  leading to 33 + 20 = **53**

  This method makes use of a basic fact, 7 + 6 = 13, giving 27 + 6 = 33.

## 6 Crab count *(mental subtraction of two-digit numbers involving 'bridging a ten')*

- Use a question such as

  'Roy counted 42 crabs in the pool. 27 disappear under rocks. How many crabs can Roy still see?

  Two possible methods are as follows:

— **method 1** (illustrated on Workbook page 1)

  **42 – 27** thought of as 42 – 20 – 7
  leading to 22 – 7 = **15**

— **method 2**

  **42 – 27** thought of as 42 – 7 – 20
  leading to 35 – 20 = **15**

- Other strategies are also possible. For example, some children may think of a **complementary addition** method, counting on from the smaller number to make the larger number.

**42 – 27**

27 add what gives 42?

27 + [3] = 30
↓
30 + [10] = 40
↓
40 + [2] = 42
↓
**15**

N2a,3cd/4
AS/C1 AS/D1
N/4h

## Textbook pages 1 and 2 *Mental addition and subtraction*
## Workbook page 1

Textbook page 1 revises work introduced in Heinemann Mathematics 5.

Where additions are written with the smaller number first, as in questions 3(f) and (h), the children should be aware that they can change the order when calculating mentally.

Some children may have more difficulty with questions 3(j) and 4(f), (i) and (j), where 'bridging' of a hundred is required.

On Textbook page 2, it may be necessary to clarify the meaning of 'flares'.

None of the examples in questions 1 and 2 requires 'bridging a ten'.

On Workbook page 1, in examples which have a '9' in the units place, some children might use these methods:

— **48 + 29** ⟶ Add 30 and **take away 1**

— **78 – 49** ⟶ Subtract 50 and **add 1**

# PLACE VALUE: NUMBERS TO 999 999, MILLIONS

UA3a/4 N2a/4
RTN/, D1, 2
N/4ab

In Heinemann Mathematics 5, work on place value was restricted to numbers to 9999.

This work is now extended by dealing with the recognition, naming, writing and ordering of numbers to 999 999, and then by introducing millions.

The Shipwrecked context continues with the four children meeting a castaway, Simon, and helping him to discover a hidden submarine.

## Introductory activities

### 1 Number sequences

It is important for the children to investigate the sequences of numbers around 10 000 and 100 000.

- Write the number 9995 on the chalkboard and ask the children to say its name. Ask them to count on in ones, saying each new number. Write each new number on the chalkboard.

Explore other sequences of numbers around 10 000 in a similar way. For example,

— start with 9950 and **add on 10** each time:
9950, 9960, 9970, 9980, 9990, 10 000, 10 010, 10 020 . . .

— start with 9500 and **add on 100** each time:
9500, 9600, 9700, 9800, 9900, 10 000, 10 100, 10 200 . . .

— start with 5000 and **add on 1000** each time:
5000, 6000, 7000, 8000, 9000, 10 000, 11 000, 12 000 . . .

The children could use calculators to investigate these sequences.

- The sequences of numbers around 100 000 can be explored in a similar way, possibly using a calculator.

— start with 99 995 and **add on 1** each time:
99 995, 99 996, 99 997, 99 998, 99 999, 100 000, 100 001, 100 002 . . .

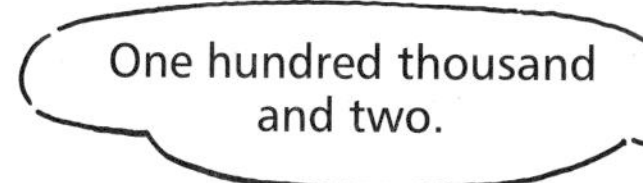

— start with 99 950 and **add on 10** each time:
99 950, 99 960, 99 970, 99 980, 99 990, 100 000, 100 010, 100 020 . . .

— start with 99 500 and **add on 100** each time:
99 500, 99 600, 99 700, 99 800, 99 900, 100 000, 100 100, 100 200 . . .

— start with 95 000 and **add on 1000** each time:
95 000, 96 000, 97 000, 98 000, 99 000, 100 000, 101 000, 102 000 . . .

- Other sequences can be explored, possibly using a calculator, to establish the patterns of intermediate numbers:
  - — start with 12 000: 12 001, 12 002, 12 003 . . . 12 020
  - — start with 25 090: 25 091, 25 092, 25 093 . . . 25 101
  - — start with 101 000: 101 001, 101 002, 101 003 . . . 101 011
  - — start with 879 295: 879 296, 879 297, 879 298 . . . 879 302
  - — start with 10 000: 20 000, 30 000, 40 000 . . . 90 000, 100 000, 110 000

## 2 Number names

- The children will require practice in associating numerals with the corresponding number names, particularly the way in which the number of thousands is said first followed by the other part of a number.

For example,

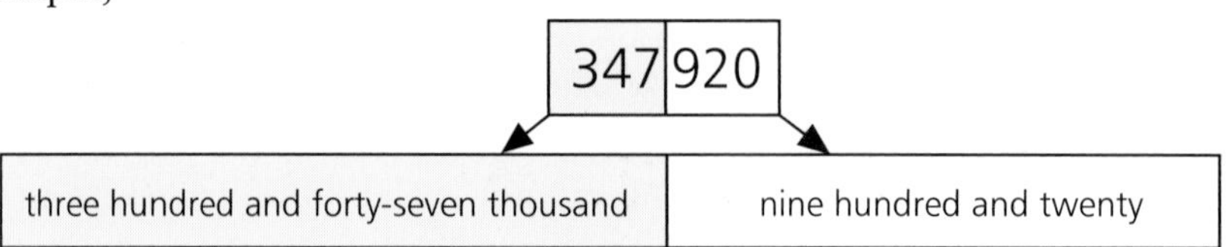


- Write large numbers on card or on the chalkboard and ask the children to read out the number names.

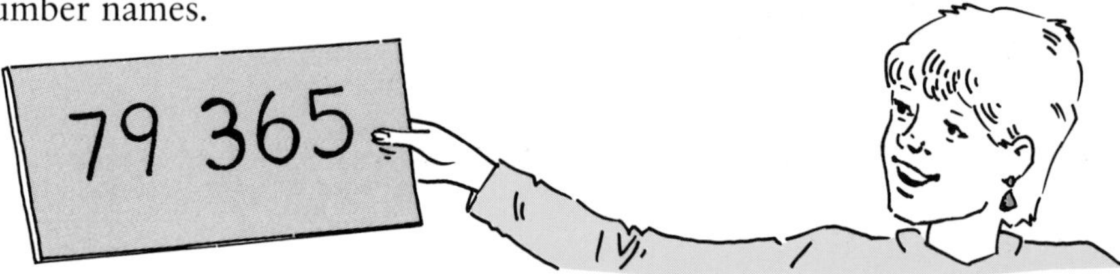


- Say large numbers and ask the children to write them.

## 3 Writing large numbers

Make sure that the children understand the conventions for writing large numbers:

— when a number has four or fewer digits, these are written together with no comma or space, for example,

9527

— when a number consists of more than four digits, they are grouped in threes from the right-hand side, and each group is separated by a small space, for example,

950 270    95 270

— when writing a number in words, hyphens and commas are used, for example, seven hundred and sixty-two thousand, eight hundred and four.

## 4 Place value

'he relative values of the digits in different columns should be explored for large .umbers. For example,

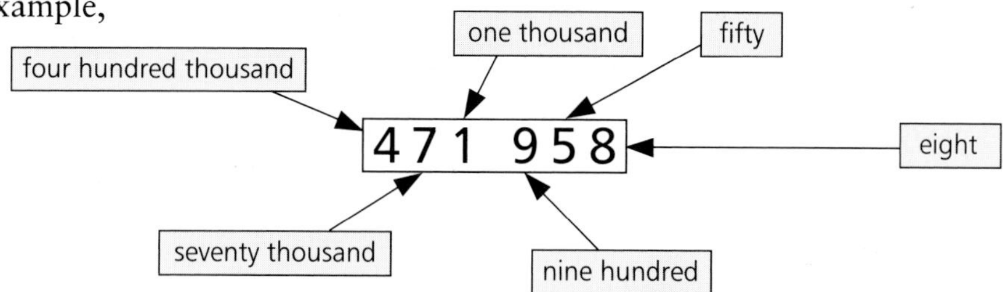


. wall display, similar to the illustration shown, could be ıade and used to help children answer questions about ıe values of digits in different place value columns. or example,

| Thousands | | | | | |
|---|---|---|---|---|---|
| H | T | U | H | T | U |
| 4 | 7 | 1 | 9 | 5 | 8 |

**.t this point the children could try Workbook pages 2 and 3.**

## 5 Introducing millions

- Introduce another sequence of numbers in a similar way to that described above in activity 1.

  Start with 999 995 and **add on 1** each time with the children saying and writing each new number. When the sequence reaches 999 999, ask the children to predict the next number name and to suggest how it could be written. Some might suggest 'one thousand thousand', which is quite correct in place value terms. Others will already be aware of millions and know that 'one million' is used instead of 'one thousand thousand'.

  Show the children how one million is written, emphasizing the grouping and spacing of the digits.

  1 000 000

- Repeat some of the sequencing, number names and place value activities described in activities 1, 2 and 4 for numbers greater than one million.

## 6 Large numbers in the environment

sk the children to suggest situations in which large numbers occur. Examples from ewspapers or magazines could be collected and displayed. For example,

**Number**
**Place value**
**Shipwrecked**

UA2b,3a/4 N2a/4
PSE RTN/D1,2
P/4a N/4ab

**Problem solving**

## Workbook pages 2 and 3 *Place value to 999 999*

On Workbook page 2, in questions 1 and 2, the children should observe the conventions for writing large numbers in figures – with digits grouped and spaced but **no** commas – and in words – with hyphens and commas where appropriate.

In questions 3 and 4, the children should realize that it is possible to increase or decrease the numbers by changing only one digit each time.

On Workbook page 3, in question 1, several digits in each of the given numbers need to be changed to make the new numbers.

Some children may find questions 2 and 3 quite challenging. In question 2, the easiest way to start is to write the two 2s in the units and hundreds columns.

A 4 can now be entered in the thousands column as the thousands digit is double the units digit.

|   |   |   |   |   |
|---|---|---|---|---|
|   | 4 | 2 |   | 2 |

Since the final number has to be less than 70 000, the last two digits can now be entered.

|   |   |   |   |   |
|---|---|---|---|---|
| 6 | 4 | 2 | 8 | 2 |

In question 3, some children may try to answer the first part by counting on in ones 25 387, 25 388, 25 389 . . . until they reach 25 395, which satisfies the condition.

This strategy is not appropriate for the second and third parts of the question, where place value knowledge will be required to give 25 450 and 31 111 respectively.

In question 4(c), the number which opens the door is 25 395.

UA3a/4 N2a/4
RTN/D1, 2
N/4ab

## Textbook page 3 *Place value: millions*

In questions 1 and 2, remind the children about the conventions for writing large numbers in figures or in words.

Some children may find the numbers 27 027 and 600 006 in question 2 more difficult to express in words.

In question 3, the specific values of the red digits should be given. For example, the 4 in question 3(b), ④6 959, has the value forty thousand as opposed to the more general 'tens of thousands'.

**R2**

# ADDITION AND SUBTRACTION OF THOUSANDS, HUNDREDS, TENS AND UNITS

N3d,4a/4
AS/D2
N/4k

Addition and subtraction of thousands, hundreds, tens and units were introduced in Heinemann Mathematics 5. All addition examples had totals less than 10 000.

This work is revised and then extended by

- introducing addition examples with totals greater than 10 000
- emphasizing subtraction involving several exchanges
- applying written methods of adding and subtracting in problem solving situations.

The Shipwrecked context continues with the four children helping their new friend, Simon, to rescue fish from polluted water.

## Introductory activities

The following activities emphasize written algorithms for addition and subtraction, and their associated language. Throughout Heinemann Mathematics written methods of calculation have been developed from prior use of structured base-ten material, and this is reflected in the language used. However, the larger numbers now involved mean that the use of such material is no longer appropriate. If it is considered necessary to provided further practice in the use of material, the relevant sections of the Teacher's Notes for Heinemann Mathematics 5 should be consulted.

### Long fish and round fish *(addition of ThHTU: totals greater than 10 000)*

Consider a problem such as

'In the lagoon, there are 7421 long fish and 5823 round fish. How many fish are there altogether?'

Discuss the appropriate language and the written technique. Emphasize the new step of adding and recording where the sum is greater than ten thousand.

| **Recording** | **Language** |
|---|---|
| 7421<br>+ 5823<br>244<br>1 | 'Add the thousands.<br>1 and 5 and 7 give 13 thousands.' |
| 7421<br>+ 5823<br>13244<br>1 | 'There are 13 244 fish altogether.' |

- Throughout Heinemann Mathematics the 'carrying' figure is placed below the answer line as shown above. Depending on school policy, the 'carrying' figure may be placed in another position.
- Provide further practice in adding two or more four-digit numbers giving totals greater than 10 000. Include examples which involve zero digits in the answers. For example,

```
  2705
  4134
 +3164
 10003
```

- Present some examples in a horizontal format to give the children practice in correct setting down. For example,

  4758 + 405 + 6093 + 97

$$\begin{array}{r} 4758 \\ 405 \\ 6093 \\ +\quad 97 \\ \hline \\ \hline \end{array}$$

## 2 Striped fish and spotted fish *(subtraction of ThHTU)*

- The decomposition method of subtraction and the associated language can be revised using a question such as

  'In the cove there are 3002 striped fish and 1769 spotted fish. How many more striped fish than spotted fish are there?'

| **Recording** | **Language** |
|---|---|
| $\begin{array}{r} 3\,0\,0\,2 \\ -\ 1\,7\,6\,9 \\ \hline \\ \hline \end{array}$ | 'Subtract the units.<br>2 take away 9, I cannot.<br>There are no tens and no hundreds to exchange.<br>I need to exchange one thousand.' |
| $\begin{array}{r} {}^{2}\cancel{3}{}^{1}0\,0\,2 \\ -\ 1\,7\,6\,9 \\ \hline \\ \hline \end{array}$ | 'Exchange 1 thousand for 10 hundreds, giving 2 thousands and 10 hundreds.' |
| $\begin{array}{r} {}^{2}\cancel{3}{}^{1}\cancel{0}^{9}\,{}^{1}0\,2 \\ -\ 1\,7\,6\,9 \\ \hline \\ \hline \end{array}$ | 'Exchange 1 hundred for 10 tens giving 9 hundreds and 10 tens.' |
| $\begin{array}{r} {}^{2}\cancel{3}{}^{1}\cancel{0}^{9}\,{}^{1}\cancel{0}^{9}\,{}^{1}2 \\ -\ 1\,7\,6\,9 \\ \hline \\ \hline \end{array}$ | 'Exchange 1 ten for 10 units giving 9 tens and 12 units.' |
| $\begin{array}{r} {}^{2}\cancel{3}{}^{1}\cancel{0}^{9}\,{}^{1}\cancel{0}^{9}\,{}^{1}2 \\ -\ 1\,7\,6\,9 \\ \hline 1\,2\,3\,3 \\ \hline \end{array}$ | 'Subtract the units.<br>12 take away 9 leaves 3 units.<br>Subtract the tens.<br>9 take away 6 leaves 3 tens.<br>Subtract the hundreds.<br>9 take away 7 leaves 2 hundreds.<br>Subtract the thousands.<br>2 take away 1 leaves 1 thousand.' |

'There are 1233 more striped fish than spotted fish.'

Some children may be able to move directly from the first step to the last step by regarding 3 thousands as 300 tens then exchanging 1 ten for 10 units giving 299 tens and 12 units.

$$\begin{array}{r} 3\,0\,0\,2 \\ -\ 1\,7\,6\,9 \\ \hline \\ \hline \end{array} \qquad \begin{array}{r} \overset{2\ 9\ 9}{\cancel{3\,0\,0}}{}^{1}2 \\ -\,1\,7\,6\,9 \\ \hline 1\,2\,3\,3 \\ \hline \end{array}$$

- Repeat for other examples involving several exchanges and/or zero digits, such as

| 9521 | 4030 | 6000 |
|---|---|---|
| − 4916 | − 2689 | − 2456 |
| ____ | ____ | ____ |

## Textbook pages 4 and 5 *Addition and subtraction ThHTU*

UA2bc/4 N3d/4
PSE AS/D2
P/4a N/4k

Throughout Textbook pages 4 and 5, the emphasis is on addition leading to totals greater than 10 000 and subtractions involving several exchanges and/or zero digits.

**Problem solving**

On page 4, in question 5, the children should first find the number of white fish by halving the total, 6000, to give 3000 white fish.

This means that the other 3000 fish are either red or purple. The number of each colour can be found in different ways. For example, since there are 800 more red fish than purple fish, 800 could be subtracted from 3000 to give 2200, which is then divided by 2 (halved) to give 1100. This is the number of purple fish. The number of red fish is then 1100 + 800 = 1900 fish. Encourage the children to check these figures. For example,

| | |
|---|---|
| 1900 | red fish |
| − 1100 | purple fish |
| 800 | more red fish |

| | |
|---|---|
| 1900 | red fish |
| 1100 | purple fish |
| + 3000 | white fish |
| 6000 | fish altogether |

R3

On Textbook page 5, in question 1(b), the children should realize from their answers to 1(a) that only tank D can accommodate the 384 Zens which cannot be held in tank B.

**Problem solving**

In question 2, a possible strategy is first to find how many fish could **still** be put into tanks E and F. For example,

| **tank E** | | **tank F** | |
|---|---|---|---|
| 4300 | | 4200 | |
| − 3987 | Goldies | − 2748 | Jets |
| 313 | | 1452 | |

The Zens and Gliders can then be dealt with by trying to put the largest number of fish (6429 Zens) into the largest available tank (tank G) and subtracting to find the number of Zens that cannot be accommodated in tank G. For example,

| | |
|---|---|
| 6429 | Zens |
| − 5000 | |
| 1429 | |

The Gliders can then be matched with tank H in a similar way. For example,

| | |
|---|---|
| 5196 | Gliders |
| − 4900 | |
| 296 | |

The number of Zens and Gliders left over can then be matched with the amount of room still left in tanks E and F. For example,

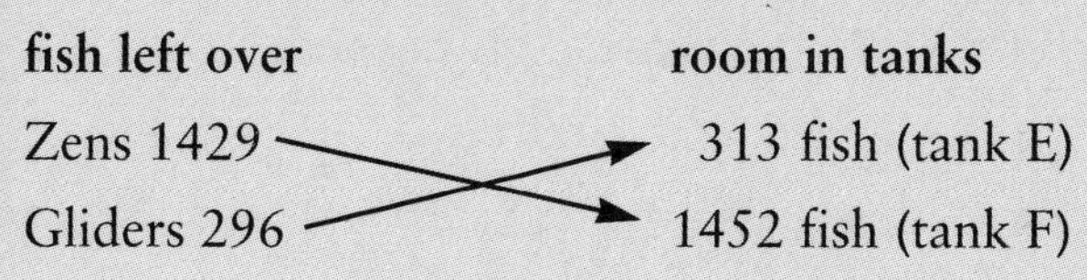

Problem solving

In question 4, the children need to realize that if the two tanks are each to hold the same amount, this must be 4000 litres (half of the 8000 litres needed altogether). The volume of fuel in each of tanks P and Q should therefore be subtracted from 4000 litres to find the amount that needs to be added to each tank.

| tank P | tank Q |
|---|---|
| 4300 | 4000 |
| − 3679 | − 2943 |
| 321 litres | 1057 litres |

## PLACE VALUE: ROUNDING TO THE NEAREST HUNDRED AND TO THE NEAREST TEN

UA3b,4d/4 N2a/4
RN/C1 RN/D1
N/3ah

In Heinemann Mathematics 5, the rounding of two-digit numbers to the nearest ten and three-digit numbers to the nearest hundred was revised. Rounding of three-digit number to the nearest ten was introduced, including examples where the units digit was 5.

This work is now consolidated.

The Shipwrecked context continues with preparations to use Nemo's mini-submarine

### Introductory activities

#### 1 Suit yourself *(rounding to the nearest 100 and to the nearest 10)*

■ Tell the children that the Nemo has a mini-sub for exploring the seabed. The castaways practise putting on diving gear before using the sub. Show the children a table giving the times taken to put on the diving gear.

| | **Time in seconds** |
|---|---|
| Roy | 487 |
| Elaine | 443 |
| Sean | 639 |
| Dorothy | 524 |
| Simon | 565 |

Draw a scale on the chalkboard and invite the children to indicate the approximate position of each time on the scale.

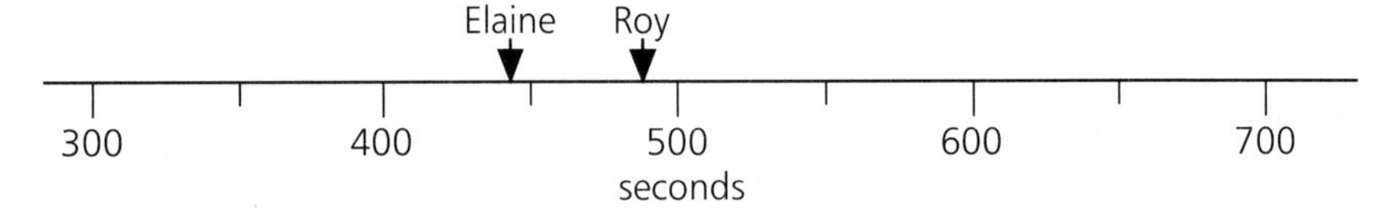


Discuss a time such as Elaine's which is

— between 400 and 500

— less than 450, and so nearer to 400 than to 500.

Ask the children to give each time to the nearest hundred seconds. Create a new column in the table to record the results.

| | **Time in seconds** | **To the nearest 100 seconds** |
|---|---|---|
| Roy | 487 | 500 |
| Elaine | 443 | 400 |
| Sean | 639 | 600 |
| Dorothy | 524 | 500 |
| Simon | 565 | 600 |

- Discuss rounding the times to the nearest **ten** seconds, emphasizing that
  - — a units digit greater than 5 means 'rounding up' to the next ten:
    487 is 490 to the nearest ten
  - — a units digit less than 5 indicates 'rounding down':
    443 is 440 to the nearest ten
  - — a number whose units digit is 5 can be rounded up or down:
    565 is 560 or 570 to the nearest ten.

At this stage the children might continue to use an informal 'rule' of always rounding up. Later, they will judge from the context whether to round up or down. If there is an existing school policy on rounding, this should of course be followed.

The table can now be altered to show times to the nearest 10 seconds.

| | **Time in seconds** | **To the nearest 10 seconds** |
|---|---|---|
| Roy | 487 | 490 |
| Elaine | 443 | 440 |
| Sean | 639 | 640 |
| Dorothy | 524 | 520 |
| Simon | 565 | 560 **or** 570 |

## 2 In the bin

A simple group activity could be used to provide practice in rounding. Write headings on a chalkboard as shown and draw a 'bin' for unallocated numbers.

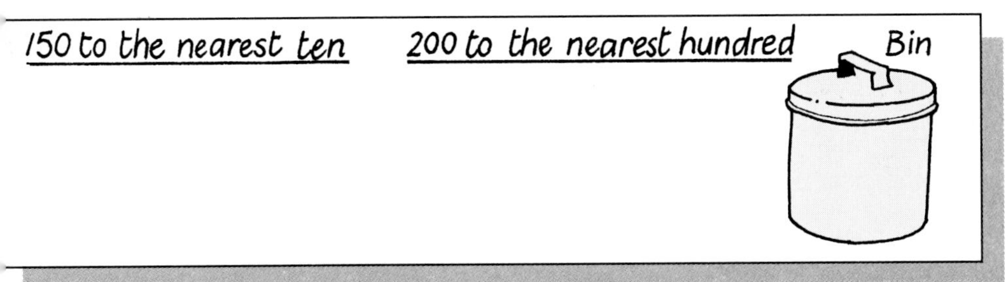


Show the children a set of numbers on card, one at a time, and ask them to list the numbers in the appropriate columns or put them in the bin.

This should produce the following:

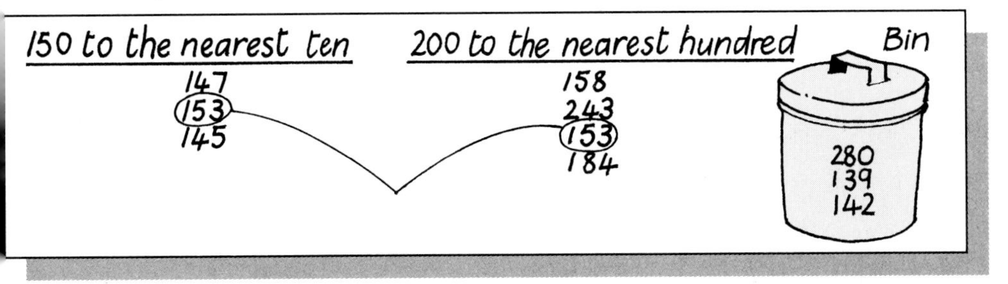


Discuss the appearance of 153 in both columns.

**At this point the children could try Textbook page 6, questions 1 to 4.**

# ADDITION AND SUBTRACTION OF HTU, ESTIMATION

N2a,3d/4
AS/C1 AS/D1 RN/C1 RN/D1
N/3ah N/4m

In Heinemann Mathematics 5, the children rounded two- and three-digit numbers to the nearest ten to obtain estimated answers to additions and subtractions such as 123 + 69 and 257 – 35.

This work is now revised and then extended to include estimates for examples such as 237 + 456 and 876 – 209, by rounding each number to the nearest hundred.

The context continues with the children using a mini-sub to visit a sunken galleon.

## Introductory activities

### 1 Fuel *(estimating after rounding to the nearest ten)*

- Tell the children that there are two tanks on board the *Nemo* which hold fuel for the mini-sub.

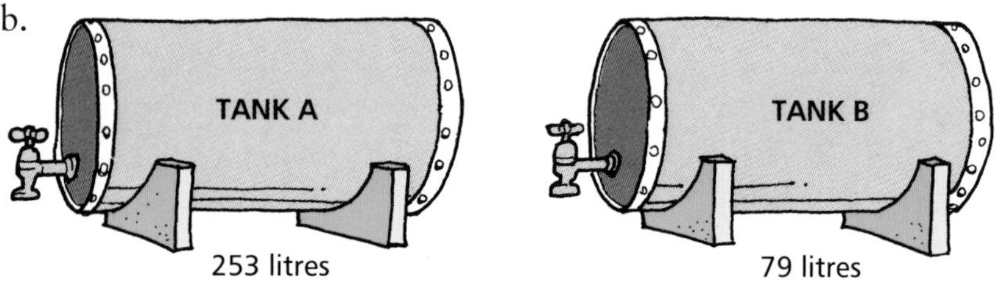


Discuss the **mental** process of estimating how much fuel there is altogether by

— rounding each volume to the nearest 10 litres:

253 litres is about 250 litres

79 litres is about 80 litres

— and adding 250 + 80 to give a total volume of fuel of **about 330 litres.**

- In a similar way, find the approximate difference between the volumes in the two tanks,

250 – 80 = 170

Tank A contains **about 170 litres** more than Tank B. Repeat for other subtraction examples such as 181 – 34.

N2a,3d/4
AS/C1 AS/D1 RN/C1 RN/D1
N/3ah N/4m

### Textbook page 6 *Place value: rounding to the nearest 100, 10*

Discuss the context and the readings on the dials for Tanks 1 and 2.

In question 2, part (f) will require discussion because 350 can be rounded to either 300 or 400, to the nearest hundred. The children might 'round up' as a rule at this stage, although later they will learn to consider the context when deciding whether to round up or down.

Question 4 contains examples where the units digit is 5 and where rounding can be either up or down. Part (j) may cause some difficulty as 797 rounded to the nearest ten is 800.

The children should tackle question 5 **mentally** and record the estimated answers only. In part (h), 265 can be rounded to either 260 or 270, giving possible estimated answers for 265 - 47 of 210 **or** 220. Similarly, two answers are possible in part (l).

## 2 Coins *(estimating after rounding to the nearest hundred)*

- Tell the children that the castaways will be using the mini-sub to explore a sunken galleon. Ask them about the sort of things they might find on the wreck.
- Coins might be found on an old wreck.

215 coins

493 coins

Discuss a **mental** method of estimating how many coins there are altogether.

> 215 is 200 to the nearest hundred.
> 493 is 500 to the nearest hundred.
> Altogether there are **about 700 coins.**

The estimate can then be compared with the exact total (708) to show that the estimate is reasonable.

Repeat for other addition examples such as 374 + 193.

- In a similar way, find how many more silver coins than gold coins there are.

> 493 is about 500.
> 215 is about 200.
> 500 – 200 is 300.
> There are **about 300 more silver coins.**

Repeat for other subtraction examples such as 622 – 175.

## 3 What can you buy?

TV
£538

VCR
£325

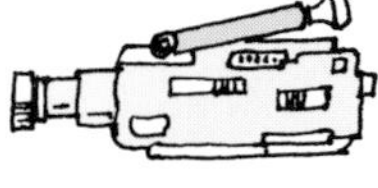

Camcorder
£385

CD player
£279

Use a price list like this to provide further practice in estimating answers, after rounding to the nearest hundred pounds. For example,

- Estimate the total cost of
  - — the TV and the camcorder (500 + 400 → about £900)
  - — the VCR and the CD player (300 + 300 → about £600)
- Your family has won £1000. After buying the TV and VCR, could they also afford the CD player? (No)

. . . and so on.

N2a,3d/4
AS/D1 RN/D1
N/3ah N/4m

### Textbook page 7 *Addition and subtraction of HTU: estimation*

Discuss the worked examples to make clear that the children should find estimated answers mentally by rounding to the nearest hundred.

In question 1, examples (g) and (i) may cause difficulty as each gives an estimated answer of '10 hundreds' or 1000.

In question 3, the children may need to write the numbers of hooks in each box in **rounded** form

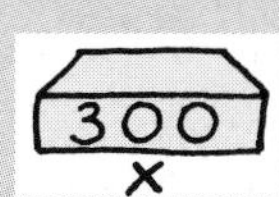

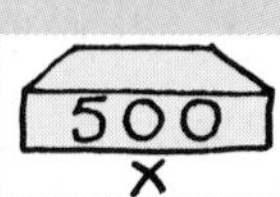


before deciding that Nipper is talking about the yellow (about 500) and green (about 300) boxes.

R4

# PLACE VALUE: ROUNDING TO THE NEAREST THOUSAND

Rounding of numbers greater than 999 to the nearest thousand, or to the nearest hundred, is now introduced.

The children continue their exploration of the sunken galleon.

## Introductory activities

### 1 Fish *(rounding to the nearest thousand)*

- Tell the children that the fish are to be released from some of the tanks inside the *Nemo*.

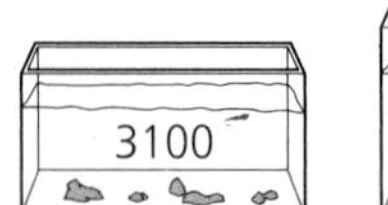

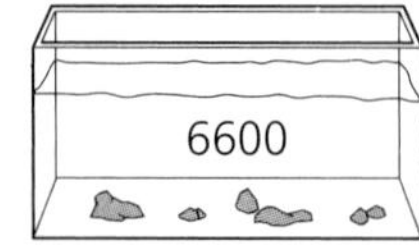

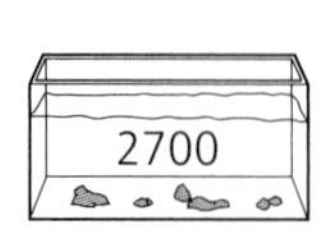

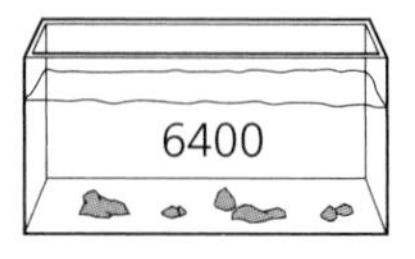

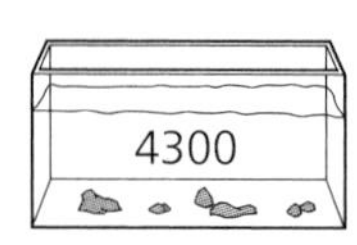


Use a scale drawn on the chalkboard and ask the children about the values of the intermediate markings, 500, 1500, 2500 . . .

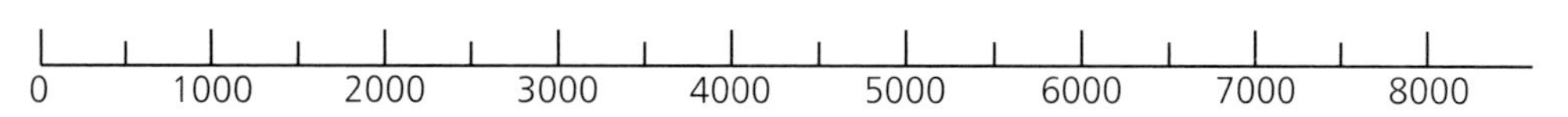


Ask them to indicate the numbers of fish on the scale.

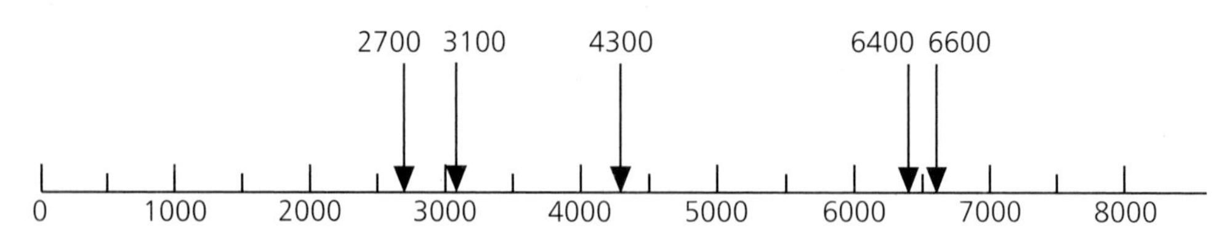


Discuss each number. For example, 4300 is

— between 4000 and 5000

— less than 4500, and so 4300 is **4000 to the nearest thousand.**

- Repeat for other numbers of fish such as

| | | **To the nearest thousand** |
|---|---|---|
| Jets | 6371 | 6000 |
| Goldies | 3987 | 4000 |
| Gliders | 5142 | 5000 |
| Zens | 3550 | 4000 |

Ask the children to read the numbers. For example,

'Zens – three thousand, five hundred and fifty.'

This helps to make clear that

— the choice lies between 3000 and 4000

— the children should consider the 550 – it is greater than 500 and so they should round up.

3550 is **4000 to the nearest thousand.**

- Extend the idea to numbers greater than 9999. For example,

    Cavern fish 29 614

  Ask the children to read the number:

    'Twenty-nine thousand, six hundred and fourteen.'

  It may help them to think in this way:

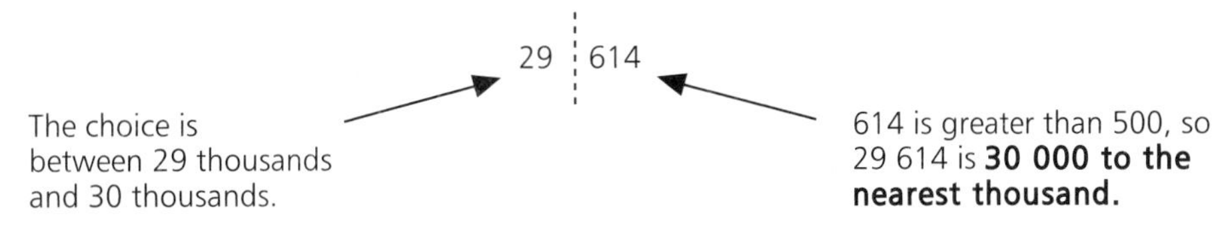


**At this point the children could try Textbook page 8, questions 1 and 2.**

## 2 Medallions *(rounding to the nearest hundred)*

- Draw a scale like this on the chalkboard.

  Discuss values for the intermediate markings, 2600, 2700 . . .

  Ask a child to indicate the number of medallions found in the wreck, 2873.

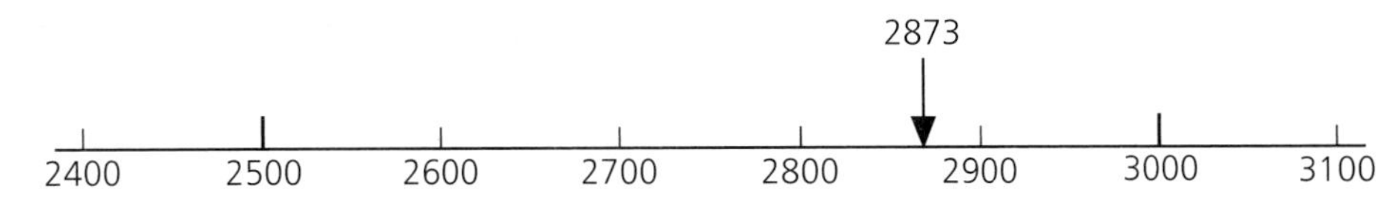


  Discuss rounding 2873 **to the nearest hundred.**

  It is nearer to 2900 than to 2800, so

    2873 is 2900 **to the nearest hundred.**

- Consider other numbers such as

    4126 → 4100 **to the nearest hundred.**

  It may help some children to think in this way:

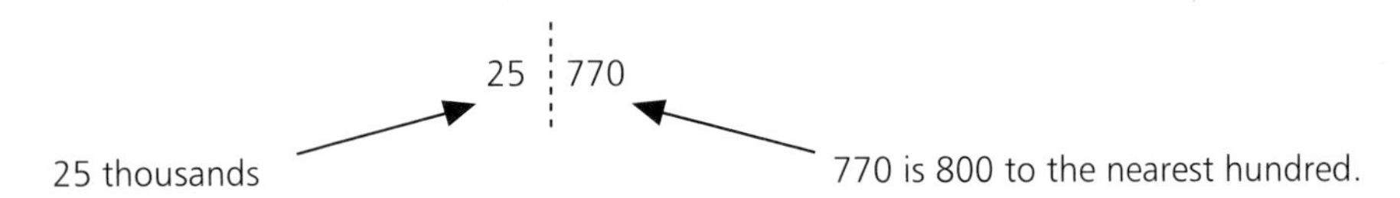


  So 25 770 is 25 800 **to the nearest hundred.**

N2a/4
RN/D1
N/4ab

### Textbook page 8
***Place value: rounding to the nearest 1000, 100***

In question 2, part (o), the number 22 500 can be rounded to either 22 000 or 23 000, to the nearest thousand.

# PLACE VALUE, ADDITION AND SUBTRACTION, CALCULATOR

UA2d/4 N3h,4c/4
AS/D3 AS/E3 RTN/D2
N/4abj

In Heinemann Mathematics 5, the children used a calculator for the four operations with numbers less than 10 000. They checked calculations by repeating them.

Heinemann Mathematics 6 introduces

— the use of a calculator to add and subtract large numbers up to 1 million

— a method of checking addition by subtraction and vice versa.

The castaways are now on their way home.

## Introductory activities

### 1 If only . . .

- Ask the children to imagine that they are able to buy something very, very expensive . . . what would it be?

Invent a price for one of the items suggested, say it to the children and ask them to enter it in their calculators. Check that they have entered it correctly.

Repeat for other large numbers up to 1 million.

- Use some of the fantasy prices to give practice in adding and subtracting using a calculator. Encourage the children to check each example by **repeating the calculation.**

  In the case of addition, they can enter the prices in a different order when checking.

**At this point the children could try Textbook pages 9 and 10, questions 1 to 7.**

### 2 A new way of checking

- This new method of checking depends or the sophisticated idea of inverse operations and may not be appropriate for some pupils at this stage.
- Show the children the statement ?? + 78 = 123.

  Ask them to enter 123 in their calculators.

Ask what they can do to the 123 to find the missing number . . . **subtract** 78 to give →

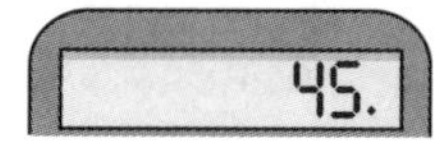


Stress that if 45 **add** 78 gives 123, then 123 **subtract** 78 gives the original number 45.

- Write an addition such as 13 852 + 8605 on the chalkboard.

  Show the answer on a calculator display and ask the children how they could check it.

They could check by repeating the addition, but there is another way.

Ask them to enter 22 457 in their calculators. Remind them that 8605 was **added** to 13 852 to give this answer. What should be left if they now **subtract** the 8605 from this answer?

**Subtracting gives the original number 13 852** again, thus checking that the **addition** was correct.

Repeat this method for other examples where the children enter and add the numbers and, **without clearing the calculator**, subtract to check.

- Where appropriate, extend the idea to checking subtractions by adding. For example,

  8709 – 2517 → 6192.

The **subtraction** of 2517 can be checked by **adding** 2517 to the answer, which should give the original number 8709 again.

## Textbook pages 9 and 10

UA2cd/4 N3h,4c/4
AS/D3 AS/E3 RTN/D2
N/4abj

### *Calculator: place value, addition and subtraction*

Discuss the scenario where a puzzle is being solved to pass the time on the journey home. The idea is to use a calculator to find Simon's 'funny' second name.

All the work in questions 1 to 7 relates to the puzzle grid in question 1. This should be carefully copied, perhaps on squared paper, with the five squares shaded, as it is used to record specified answers from questions 1 to 7.

In question 3, it is intended that the answers are found mentally, by looking at the place value structure of the given numbers, and then confirmed by using a calculator.

Similarly, in question 4, the children could round to the nearest thousand and estimate to see which pair of numbers might add to 100 000, before using a calculator to check.

In question 6, the children should try repeatedly adding 50 to 126 and then look at the **pattern** of their answers:

126 176 226 276 326 376 . . .

They should realize that, if they kept doing this for a very long time, 10 676 would appear 'at some time' in the display.

The completed grid in question 7 should be as shown. When the number on the leading diagonal (376 616) is entered in the display and the calculator is turned upside down, Simon's second name is revealed as 'Giggle'.

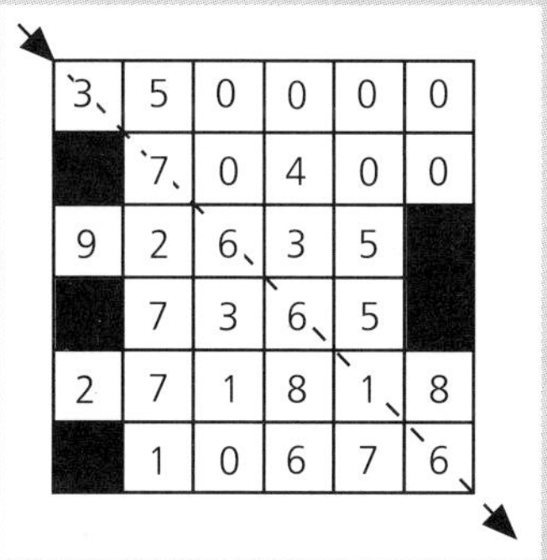


The processes involved in questions 8 to 10 should be carefully explained before the children attempt this work. The ideas may be inappropriate for some children at this stage.

## Textbook page 11 *Other activity: game, mental calculation*

- This game involves mental calculation and can be attempted at any time after Textbook pages 1 and 2 and Workbook page 1 have been completed.
- The children should read and discuss the rules. Check their understanding before they start to play.

The essential points are:

— choosing the pool and the black number for your **partner** by placing a counter on a yellow square. The number should be chosen to make it as difficult as possible for the partner.

— the calculation must be done **mentally**. Complementary addition can be used: for example,

'27 and what gives 50?'

Subtracting 50 – 27 gives 23. The player should check that this number (23) is in the pool. Instead of subtracting, the children may scan the pool to look for a likely number and **add** it to 27 to see if the total is 50.

— the player who placed the counter should check the answer with a calculator.

- The children could be asked to record correct pairs as they go along:

  14 + 36 = 50      47 + 53 = 100 . . . and so on.

There are two 'distractors' in each pool which do **not** pair up with black numbers. The distractors are

'50' pool: 25, 27

'100' pool: 47, 75

# Superhero

## A context for multiplication and division

The work on Textbook pages 12–28 and Workbook pages 4–7 is set within a context which features Superhero and Mog. Although Superhero has some extraordinary powers (he can fly, he is very strong . . .), he is not completely superhuman and so is helped by a cat-like computer called Mog. As the children proceed through the work, they visit Superhero's headquarters and follow his attempts to help after an earthquake, sort out the chaos at Medic-Aid, foil some jewel thieves and enjoy a Superparty and a Superholiday.

### Introducing the context

The context could be introduced by discussing some fictional heroes the children may have read about or seen in films or on television, such as Batman, Superman and Thunderbirds.

The activities below can be used

- to introduce Superhero and Mog
- in an on-going way as the children progress through the work.

### 1 A hero is . . .

The children could collect articles, stories and cartoons from magazines and newspapers, and use these to create a 'Hero montage'. This montage could show real heroes, fictional heroes and so on.

### 2 Superhero and Mog

The Superhero character could be introduced by telling the children that one day someone, not much older than themselves, who was feeling bored, found a mystery Superhero outfit. When it was put on, the outfit gave the wearer special powers. However, one of the armbands was missing, so the Superhero was not quite perfect. Mog, a cat-like computer with human abilities, appeared as a helper.

The children could create large pictures of Superhero and Mog, and talk about the special strengths and weaknesses each character possesses.

## 3 Create your own Super Character

The children could be asked to create a Super Character – a sportsperson, a musician, a dancer or someone else of their own choosing – by cutting out pictures from magazines and catalogues and joining them together.

The children could then label the character using eight words to describe their creation's strengths.

Alternatively, they could create a badge to show four of these strengths.

## 4 Headlines

If possible, use a set of headlines cut out of newspapers or magazines and ask the children to tell the story of how Superhero and Mog helped.

## 5 Year of the Child

The children could be asked to help Superhero and Mog to organize a two-day festival to celebrate the Year of the Child.

The children would have to consider

- who to invite to the festival
- what entertainment and events would be held
- what food and drink to have
- the timing of events
- the cost
- what special things Superhero and Mog might do.

An illustrated programme of events could be created and displayed.

UA2bc,3a,4b/4
N2a,3acd,4ac/4
PSE
MD/C1,4
MD/D1,2,3,4
PS/C1
P/4ab
N/4cgjl
A/4bc

# Multiplication

## Overview

This section

- revises multiplication facts from the 2 to 10 times tables, including the rules for multiplication by 10 and 100
- explores patterns, including multiples, digital roots and links between different tables
- deals with associativity
- revises and extends the multiplication of thousands, hundreds, tens and units by 2 to 9 to include products greater than ten thousand and money
- includes multiplication by a two-digit number using a calculator.

| | Teacher's Notes | Textbook | Workbook | Reinforcement Sheets |
|---|---|---|---|---|
| **Superhero: a context for multiplication and division** | 47 | | | |
| **Multiplication: table facts, patterns, multiples, digital roots** | 52 | 12 | 4, 5 | |
| **Multiplication: by 10, 100, associativity** | 55 | 13 | | |
| **Multiplication: thousands, hundreds, tens and units by 2 to 9, money** | 56 | 14–16 | | 5 |
| **Multiplication: by one- and two-digit numbers, using a calculator, money** | 58 | 17, 18 | | |

Extension activities related to the above section of work are as follows:

| | Teacher's Notes | Extension Textbook |
|---|---|---|
| **Table patterns** | 281 | E6 |

Teaching notes for the Extension Textbook are in a separate section at the end of the Teacher's Notes.

# Resources

## Useful materials

- calculators
- other materials suggested within the introductory activities

## Assessment and Resources Pack

### Assessment

*Number Check-up 4*
Textbook pages 14–16
(Multiplication: ThHTU by 2 to 9, money)

*Round-up 1*
Questions 5, 8

### Resources

*Problem Solving Activities*
1 Corner conquest (Multiplication by calculator of two-digit numbers)
6 Number cells (Multiples)

# Teaching notes

In Heinemann Mathematics 5, multiplication of thousands, hundreds, tens and units by 2 to 9 (products to 9999) and rules for multiplying by 10 and 100 were introduced. Using a calculator to multiply by a two-digit number was also included. A pencil-and-paper method for multiplying by selected two-digit numbers was suggested.

In Heinemann Mathematics 6, multiplication tables are consolidated through the exploration of multiples, links between tables, digital roots, multiplication by 10 and 100, and the concept of associativity.

Thereafter, multiplication of thousands, hundreds, tens and units by 2 to 9 is extended to include products greater than 10 000.

The use of a calculator for multiplication by two-digit numbers is consolidated.

The children are introduced to Superhero and Mog, and follow some of their adventures.

## MULTIPLICATION: TABLE FACTS, PATTERNS, MULTIPLES, DIGITAL ROOTS

UA3a,4b/4 N3acd/4
MD/C1 PS/C1
N/4g A/4bc

### Introductory activities

#### 1 Super security *(2 to 10 times tables)*

- On the chalkboard, draw two of the security buttons that appear on Superhero's door.

  Write a random selection of numbers from 1 to 10 on each button.

  Ask the children to write or say as many multiplication facts as they can using one number from each button.

  6 7 9 2 8 | 1 9 4 8 10

  6 X 1 = 6
  7 X 9 = 63
  2 X 4 = 8
  8 X 10 = 80

  This activity could be timed and a record kept to try to encourage the children to improve the time taken to recall tables facts.

- The numbers on the security buttons could be changed to give other multiplication facts.

#### 2 Mog's multiplication *(2 to 10 times tables)*

- Make up a set of large cards, each with a multiplication drawn from the tables.

  The teacher, or one of the children, holds up each card in turn and the group or class say what the product is.

9×8

72

- This activity could be extended to include examples such as

| $(8\times9)-2$ | $(5\times6)+4$ |
|---|---|

## 3 Games (*2 to 10 times tables*)

Many games, designed to reinforce table facts, are available from commercial sources. For example,

— Multiplication bingo

— Multiplication snap

— Dominoes.

**At this point the children could try Textbook page 12.**

## 4 Hundred squares (*introduction of multiples*)

- State that 2, 4, 6, 8 . . . are called multiples of 2.
  5, 10, 15, 20 . . . are called multiples of 5.

  Explain that, to find a **multiple** of any number, you multiply it by any other number. For example, multiples of 7 are 7, 14, 21, 28, 35 . . .

- Ask for some multiples of 3, 4, 6, 8, 9, 10.
- Ask 'Is 12 a multiple of 3?'
  'Is 36 a multiple of 9?'
  'Is 36 a multiple of 5? . . . Why not?'
- Using the '100 square', ask the children to colour the multiples of a number between 1 and 10, such as 3, and look at the pattern produced. Ask the children to colour the multiples of two different numbers on the same '100 square', such as 5 and 10, and look for common multiples.

| 1 | 2 | 3 | 4 | 5 | 6 | 7 | 8 | 9 | 10 |
|---|---|---|---|---|---|---|---|---|---|
| 11 | 12 | 13 | 14 | 15 | 16 | 17 | 18 | 19 | 20 |
| 21 | 22 | 23 | 24 | 25 | 26 | 27 | 28 | 29 | 30 |
| 31 | 32 | 33 | 34 | 35 | 36 | 37 | 38 | 39 | 40 |
| 41 | 42 | 43 | 44 | 45 | 46 | 47 | 48 | 49 | 50 |
| 51 | 52 | 53 | 54 | 55 | 56 | 57 | 58 | 59 | 60 |
| 61 | 62 | 63 | 64 | 65 | 66 | 67 | 68 | 69 | 70 |
| 71 | 72 | 73 | 74 | 75 | 76 | 77 | 78 | 79 | 80 |
| 81 | 82 | 83 | 84 | 85 | 86 | 87 | 88 | 89 | 90 |
| 91 | 92 | 93 | 94 | 95 | 96 | 97 | 98 | 99 | 100 |

**Multiples of 3**

| 1 | 2 | 3 | 4 | 5 | 6 | 7 | 8 | 9 | 10 |
|---|---|---|---|---|---|---|---|---|---|
| 11 | 12 | 13 | 14 | 15 | 16 | 17 | 18 | 19 | 20 |
| 21 | 22 | 23 | 24 | 25 | 26 | 27 | 28 | 29 | 30 |
| 31 | 32 | 33 | 34 | 35 | 36 | 37 | 38 | 39 | 40 |
| 41 | 42 | 43 | 44 | 45 | 46 | 47 | 48 | 49 | 50 |
| 51 | 52 | 53 | 54 | 55 | 56 | 57 | 58 | 59 | 60 |
| 61 | 62 | 63 | 64 | 65 | 66 | 67 | 68 | 69 | 70 |
| 71 | 72 | 73 | 74 | 75 | 76 | 77 | 78 | 79 | 80 |
| 81 | 82 | 83 | 84 | 85 | 86 | 87 | 88 | 89 | 90 |
| 91 | 92 | 93 | 94 | 95 | 96 | 97 | 98 | 99 | 100 |

**Multiples of 5 and 10**

**At this point the children could try Workbook page 4.**

## 5 Digital roots

- Explain that the digital root of a number is found by adding the digits together until a **single-digit** number is produced.

$25 \rightarrow 2 + 5 = 7$ The digital root of 25 is 7.

$76 \rightarrow 7 + 6 = 13 \rightarrow 1 + 3 = 4$ The digital root of 76 is 4.

■ Explore the digital roots of some of the multiplication tables.

9 times table

| Multiplication | 9 | 18 | 27 | 36 | 45 |
|---|---|---|---|---|---|
| Digital root | 9 | 9 | 9 | 9 | 9 |

3 times table

| Multiplication | 3 | 6 | 9 | 12 | 15 | 18 |
|---|---|---|---|---|---|---|
| Digital root | 3 | 6 | 9 | 3 | 6 | 9 |

UA3a,4b/4 N3acd/4
MD/C1 PS/C1
N/4g A/4bc

## Textbook page 12 *Multiplication: table facts, patterns, multiples,*
## Workbook page 4 *table links*

The Superhero context is introduced with the children exploring Superhero's headquarters.

On Textbook page 12, in question 1, some children may need to be reminded of the meaning of the word 'product'. Although the children will be able to find each product mentally, they will have to record the products so that they can add them up and identify the order in which to press the buttons.

In question 2, the children should record the multiplications that Superhero steps on, i.e. $8 \times 7$, $7 \times 4$, $8 \times 6$, $6 \times 6$, $8 \times 5$, $5 \times 6$, $9 \times 8$.

In question 3(d), some children may have difficulty in identifying a pattern in its entirety, and may need to look at the simpler patterns in each part. For example, in (a)

$$(8 \times 1) + 1 = 9$$
$$(8 \times 2) + 2 = 18$$
$$(8 \times 3) + 3 = 27$$

↑ **8 times table** ↑ **adding 1 more each time** ↑ **adding 9 more each time *or* multiples of 9**

On Workbook page 4, the children may need to be reminded of the meaning of the word 'multiple'. Some children may need to be told to colour **all** of the multiples of a number **including the ones outside the tables.**

UA3a,4b/4 N3acd/4
MD/C1 PS/C1
N/4g A/4bc

## Workbook page 5 *Multiplication: tables, digital roots*

The work is concerned with an investigation of the 'digital roots' of the stations of the multiplication tables, and the patterns produced when the sequences of digital roots are joined up on circles whose circumferences have been divided into 9 equal parts.

It may be necessary to remind some children that the digital root of a number is found by adding the digits repeatedly until a **single-digit** number is produced. For example, $85 \rightarrow 8 + 5 = 13 \rightarrow 1 + 3 = 4$.

In question 1, the children should note that

— the digital root pattern for multiples of 3 is 3, 6, 9, 3, 6, 9 . . .

— the digital root pattern for multiples of 6 is 6, 3, 9, 6, 3, 9 . . .

— the design on each dial is the same – a triangle.

In question 2, the children should find that the designs on the dials look like this:

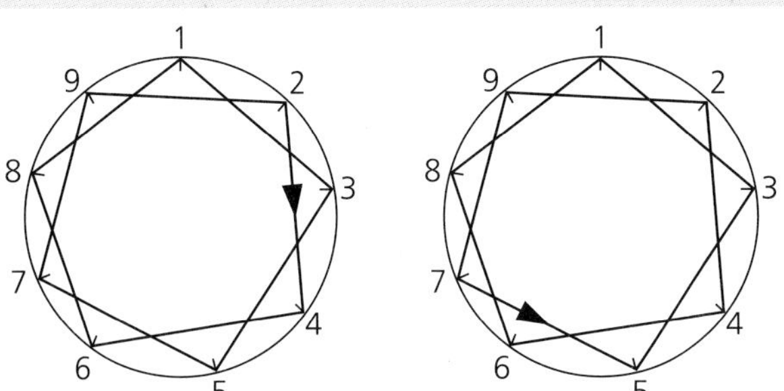


## MULTIPLICATION: BY 10, 100, ASSOCIATIVITY

N2a,3d,4c/4
MD/D1,2
N/4cg

In Heinemann Mathematics 5, the children discovered rules for multiplying two-digit numbers by 10 and 100 mentally.

This work is now revised and extended to include multiplying three-digit numbers by 10 and 100.

The concept of associativity to simplify mental multiplication is introduced.

The context continues in Superhero's headquarters.

## Introductory activities

### 1 Multiplying by 10 and 100

- Display a number of examples like these on the chalkboard.

$\begin{array}{r} 73 \\ \times 10 \\ \hline \end{array}$ $\begin{array}{r} 86 \\ \times 10 \\ \hline \end{array}$ $\begin{array}{r} 123 \\ \times 10 \\ \hline \end{array}$ $\begin{array}{r} 103 \\ \times 10 \\ \hline \end{array}$

Ask different children to complete the multiplications using the 10 times table and saying their thinking aloud.

Discuss the pattern linking starting number and product.

Revise the rule for multiplying by 10 which emphasizes the digit shift. For example,

'To multiply by 10, move each digit one place to the left. Put a 0 in the units place.'

- Repeat for examples in which the multiplication is by 100.

Revise the rule for multiplying by 100 which emphasizes the digit shift. For example,

'To multiply by 100, move each digit two places to the left. Put a 0 in the units place **and** in the tens place.'

**Number**
**Multiplication**
**Superhero**

## 2 All change *(associativity)*

- Use a set of number cards, each displaying one of the numbers 2 to 10.
- Select 3 cards, such as 2, 4 and 5, and discuss with the children how to find the product by multiplying the numbers **in different orders.**

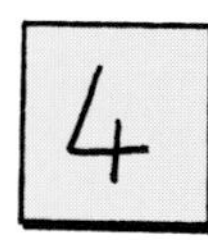


5

$$\overbrace{2 \times 4} \times 5 = 8 \times 5 = 40$$
$$2 \times \overbrace{4 \times 5} = 2 \times 20 = 40$$
$$\overbrace{2 \times 4 \times 5} = 10 \times 4 = 40$$

Emphasize that, regardless of the order in which the numbers are multiplied, **the answer each time is the same.**

- Repeat for other examples.

N2a,3d,4c/4
MD/D1, 2
N/4cg

### Textbook page 13 *Multiplication by 10, 100, associativity*

In question 1, children who have not completed the introductory activities may need to be reminded of the rules for multiplying by 10 and 100.

In question 3, while many children will be able to do most of the calculations mentally, they should record how they found the answer. For example,

3(d) $8 \times 5 \times 2 \rightarrow \overbrace{8 \times 5} \times 2 = 40 \times 2 = 80$ or $8 \times \overbrace{5 \times 2} = 8 \times 10 = 80$ or $\overbrace{8 \times 5 \times 2} = 16 \times 5 = 80$

In question 4, the children should be encouraged to look initially for pairs of numbers which give a product of 10 or 100 to help make the second multiplication easier. For example,

(a) $\overbrace{2 \times 76 \times 5} = 10 \times 76 = 760$

(f) $\overbrace{2 \times 50} \times 6 = 100 \times 6 = 600$

## MULTIPLICATION: THOUSANDS, HUNDREDS, TENS AND UNITS BY 2 TO 9, MONEY

Heinemann Mathematics 5 introduced the algorithm for multiplication of thousands, hundreds, tens and units by a single digit with products less than 10 000.

This work is now consolidated and extended to include some products greater than 10 000. Work with money, in pounds only, is also included.

The context is set in the aftermath of an earthquake in which Superhero and Mog help to pack supplies, rebuild damaged villages and raise money for the Earthquake Fund.

# Introductory activity

## Food Aid

This activity is intended to revise the algorithm for multiplication of thousands, hundreds, tens and units, and extend it to include products greater than 10 000.

For some children, it may be appropriate to work with products less than 10 000 or to revise the algorithm for hundreds, tens and units only.

- Write an example on the chalkboard related to the earthquake context.

  'Superhero puts 3891 food parcels in each of the 4 relief trucks. How many parcels does he load altogether?'

  Discuss the pencil-and-paper algorithm and the accompanying language.

  Multiplying the units by 4, then the tens by 4 leads to → 3891 ×4 = 64 (carry 3)
  which is then extended as follows:

| Recording | Language |
|---|---|
| 3891 × 4 = 564 (carries 3 3) | 'Multiply the hundreds. 4 times 8 is 32 and 3 is 35 hundreds. Write the 5 in the hundreds column and carry 3 thousands.' |
| 3891 × 4 = 15564 (carries 3 3) | 'Multiply the thousands. 4 times 3 is 12 and 3 is 15 thousands. Write the 5 in the thousands column. Write the 1 in the ten thousands column.' |

  Superhero loads **15 564** parcels.

- Repeat for other examples, including money. For example 6 × £1732.

## Textbook pages 14, 15 and 16

*Multiplication: thousands, hundreds, tens and units by 2 to 9*

N3cd,4a/4
MD/D3
N/4I

Discuss the tasks Superhero and Mog have to undertake within the earthquake context.

On Textbook page 14, in questions 1 and 2 the products are less than 10 000.

In question 3, the products are greater than 10 000.

On Textbook page 15, the progression is repeated for multiplication by 6 to 9.

In question 5, the children should explain that the people need 7644 kg of rice and 11 466 litres of water, and Superhero has brought enough. He has not brought enough beans as 10 192 kg are needed.

On Textbook page 16, questions 1 to 3 involve money. Examples are of the type £898 × 6 or £1568 × 9 with the emphasis on the latter.

For question 4, 1 cm squared paper should be sufficiently large for the number search grid.

R5

# MULTIPLICATION: BY ONE- AND TWO-DIGIT NUMBERS USING A CALCULATOR, MONEY

UA2bc,3a/4 N3cd,4a/4
PSE MD/C4 MD/D4
P/4ab N/4jl

The work on Textbook pages 17 and 18 provides practice in the use of a calculator to

— multiply numbers with up to four digits by 2 to 9

— multiply numbers with two or three digits by a two-digit number.

There is an emphasis on examples involving money.

The context is a 'thankyou' party for Superhero.

## Textbook page 17 *Multiplication: ThHTU by 2 to 9 by calculator, money*

## Textbook page 18 *Multiplication: by TU, calculator, money*

As there is very little completely new work on these pages, there is an emphasis on using and applying mathematics to solve problems. Any introductory work should be confined to a discussion of the context.

On Textbook page 17, Superhero is exhausted from repairing the houses damaged in the earthquake. Everyone is grateful for his efforts and Mog takes a major role in planning a surprise 'thankyou' party for Superhero. The questions are concerned with items – paper hats, whistles, cakes, ice-cream and so on – which Mog is ordering for the party.

**Problem solving**

On Textbook page 17, questions 1, 2 and 3 are numerical examples whereas questions 4 and 5 involve money. It is possible to interchange questions 3 and 4 so that the work ends with both problem solving examples. In question 3, encourage the children to adopt a systematic approach. For example, they could list all the multiples of 9 between 1000 and 1050 and check which of these are also multiples of 5:

'999 is a multiple of 9.
So suitable multiples of 9 are 1008, 1017, 1026, 1035 and 1044.
1035 is a multiple of 5.
Hence Mog orders 1035 candles.'

Other approaches are, of course, possible.

**Problem solving**

The problem in question 5 is challenging and many children are likely to adopt a guess and check strategy. A systematic approach might involve listing the cost of 1, 2, 3 . . . large boxes and 1, 2, 3 . . . small boxes, then looking for a pair of costs which total £12. For example,

| | 1 | 2 | 3 | 4 | 5 | 6 |
|---|---|---|---|---|---|---|
| large boxes | £2·20 | £4·40 | (£6·60) | £8·80 | £11·00 | £13·20 |
| small boxes | £1·08 | £2·16 | £3·24 | £4·32 | (£5·40) | £6·48 |

Hence Mog orders 3 large and 5 small boxes.

On Textbook page 18, the party is in full swing. The illustrations are of a fancy dress party and it is clear that everyone is having fun. However, Mog is a little worried about the cost of the party and, rather anxiously, is using a calculator to work out some costs.

**Problem solving**

Question 4 could be tackled using division, but it is intended that children adopt a 'trial and improvement' strategy based on multiplication. The fundamental questions they must ask themselves are

'26 times what gives £832?' and 'What times £18 gives £666?'

For example,

$26 \times 20 = 520$ (Too small.)

$26 \times 30 = 780$ (Too small.)

$26 \times 40 = 1040$ (Too large.)

$26 \times 35 = 910$ (Too large.)

and so on until the cost for an alien costume is found to be £32.

Similarly, the number of robot costumes is found to be 37.

**Problem solving**

As the problem in question 5 is again a challenging one, many children are likely to adopt a 'guess and check' strategy. They must realize that they have to find two numbers which multiply together to give 493 and at the same time have a difference of 12. If any guidance is needed, suggest that the children try numbers, with a difference of 12, which when multiplied end in a 3. For example,

$11 \times 23 = 253$ ✗

$21 \times 33 = 693$ ✗

$7 \times 19 = 193$ ✗

$17 \times 29 = 493$ ✓

Hence 29 people are dressed as dinosaurs and 17 as dragons.

UA2bc,3a/4
N2a,3acdeh,4ab/4
N4c/4→5
PSE
MD/C1,4
MD/D2,3,4
RN/D1
FPR/C1
FPR/D1
P/4abcd
N/4cgjln
N/5g
A/4ce

# Division

## Overview

This section

- revises division of hundreds, tens and units by 2 to 9
- introduces factors
- introduces division of thousands, hundreds, tens and units by 2 to 10, including money
- includes division of thousands, hundreds, tens and units using a calculator and checking answers by multiplication
- introduces rounding a calculator display to the nearest unit and applying this to the interpretation of answers.

| | Teacher's Notes | Textbook | Workbook | Reinforcement Sheets |
|---|---|---|---|---|
| **Superhero: a context for multiplication and division** | 47 | | | |
| **Mental division by 2, 3, 4 and 5, factors** | 62 | 19 | | |
| **Division of thousands, hundreds, tens and units by 2, 3, 4 and 5** | 64 | 20, 21 | | 6 |
| **Mental division by 6, 7, 8, 9 and 10** | 66 | 22 | | |
| **Division of thousands, hundreds, tens and units by 6, 7, 8 and 9, consolidation of division by 2 to 9** | 68 | 23–5 | 6 | 7 |
| **Division by calculator: checking exact answers, rounding and interpretation of answers** | 70 | 26–8 | 7 | 8, 9 |
| *Other activity* | *74* | *29* | | |

Extension activities related to the above section of work are as follows:

| | Teacher's Notes | Extension Textbook |
|---|---|---|
| **Divisibility by 9** | 282 | E7 |

Teaching notes for the Extension Textbook are in a separate section at the end of the Teacher's Notes.

# Resources

## Useful materials

- structured base-ten material
- calculators
- other materials suggested within the introductory activities

## Assessment and Resources Pack

### Assessment

*Number Check-up 5*
Textbook pages 20, 21, 23–5
Workbook page 6
(Division: ThHTU by 2 to 9)

*Number Check-up 6*
Textbook pages 26–8
Workbook page 7
(Division by calculator: rounding, interpretation of answers)

*Round-up 2*
Questions 1, 2, 5, 11

### Resources

*Problem Solving Activities*
12 Money, money, money (Division by calculator, rounding)

# Teaching notes

In Heinemann Mathematics 5, division of hundreds, tens and units by 2 to 10 was introduced. The work covered

— remainders and their interpretation in context

— a rule for division by 10

— using a calculator for divisions where there were no remainders.

The work in Heinemann Mathematics 6 begins with revision of mental division by 2, 3, 4 and 5 and the introduction of the term 'factor'.

The Superhero context continues with Superhero and Mog sorting out the chaos at the factory, Medic-Aid, which supplies a range of medical equipment.

## MENTAL DIVISION BY 2, 3, 4 AND 5, FACTORS

### Introductory activities

**1 Medical items** *(2, 3, 4 and 5 times tables facts and remainders)*

- Items have to be sorted into sets of 2, 3, 4 and 5. Draw a set of boxes as shown.

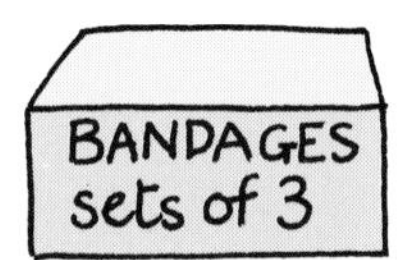

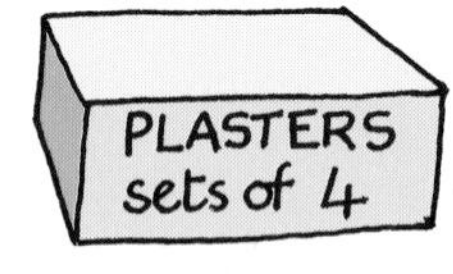

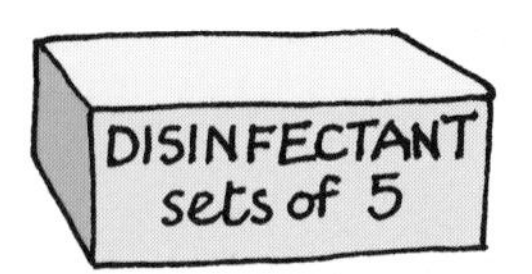


Give the children a number (of sponges), such as 19, and ask how many sets of sponges there are and how many are left over. Repeat for the other items and for different starting numbers within the limits of the respective multiplication tables.

- Include'sharing' language. For example,

  'Share 35 packets of cotton wool equally among 4 boxes.'

  The children can find the answer by asking,

  '4 times what is 35?'

  4 times eight is 32 and 3 left over.

  Each box has 8 packets and there are 3 packets left over.

- Revise examples involving fractions. For example,

  'One fifth of 45 bottles have lost their tops. How many tops are lost?'

  Associate finding $\frac{1}{5}$ of 45 with dividing 45 by 5.

  $\frac{1}{5}$ of 45 $= 45 \div 5 = 9$

  '9 bottle tops are lost.'

## 2 Soap boxes (*introduction of factors*)

Number
**Division**
Superhero

- Use 6 bars of soap and 6 boxes. Discuss whether or not the bars of soap can be divided equally among 6 boxes (yes), 5 boxes (no), 4 boxes (no), 3 boxes (yes), 2 boxes (yes), and put into 1 box (yes). Discuss each successful outcome.

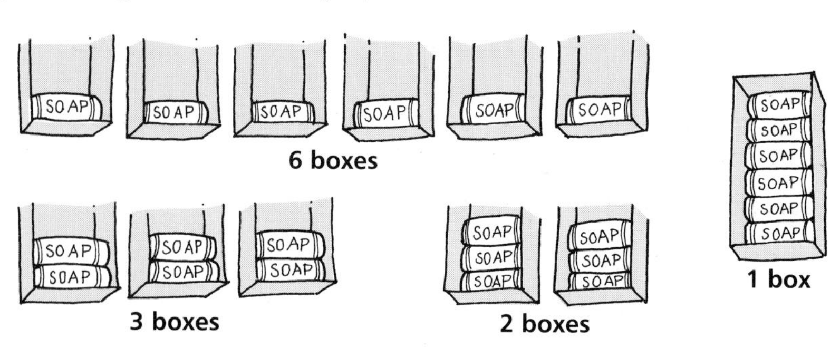


Point out that 6 can be divided exactly by **6**, by **3**, by **2** and by **1**. Introduce the term 'factor':

'6, 3, 2 and 1 are **factors** of 6.'

- Find factors of other numbers. For example,

  8 has factors 8, 4, 2 and 1.

  5 has factors 5 and 1.

In the discussion, emphasize

— 4, for example, is a factor of 8 because 8 divides exactly by 4

— 3, for example, is **not** a factor of 8 because 8 does not divide exactly by 3

— for a given number, the number itself and 1 are always factors.

- To find factors for larger numbers, such as 20, children should be encouraged to start at 1 then try 2, 3, 4 and so on, to obtain all the factors. Thus factors of 20 are 1, 2, 4, 5, 10 and 20.

## Textbook page 19 *Division: by 2, 3, 4, 5, factors*

UA3a/4 N3cde/4
PSE MD/C1 FPR/C1
P/4ab N/4gl A/4c

In question 3, some children, despite having carried out the introductory activity, may need help in listing factors.

In question 5, children should enter 24 in their calculators and divide by 2, 8, 1, etc. in turn to find which of these numbers are factors. Some children might be expected to work without a calculator.

**Problem solving**

In question 6, some children will systematically try each of the numbers 1, 2, 3, etc. in turn to look for factors of 100. They may, however, realize that when 100 is divided by 2 to give 50, then **both** 2 and 50 are factors, thus reducing the quantity of numbers to be tried.

Others may realize immediately from their knowledge of divisibility that 2, 5 and 10 are factors of 100, and then explore multiples of these such as 4 ($2 \times 2$), 25 ($5 \times 5$) and 20 ($2 \times 10$), and so on.

# DIVISION OF THOUSANDS, HUNDREDS, TENS AND UNITS BY 2, 3, 4 AND 5

UA3a/4 N3cde/4
MD/C1 MD/D3 FPR/D1
N/4gl N/5g A/4c

In Heinemann Mathematics 5, children were introduced to division of hundreds, tens and units by 2 to 9 through the use of structured base-ten materials. This led to the introduction of a pencil-and-paper algorithm which is now revised and extended to division of thousands, hundreds, tens and units by 2, 3, 4 and 5.

The Medic-Aid factory scenario continues with goods being packaged and dispatched.

## Introductory activities

### 1 Packing (*HTU divided by 2, 3, 4 and 5*)

- Discuss an example like the following to revise the language and recording for division:

  'Mog shares 443 plastic bottles equally among 3 crates. How many are in each crate and how many are left over?'

- For some children it may be helpful to represent the problem with base-ten materials.

  Put out materials for 443.

  Discuss the process of sharing the hundreds then the tens then the units, which results in 147 in each share with 2 left over.

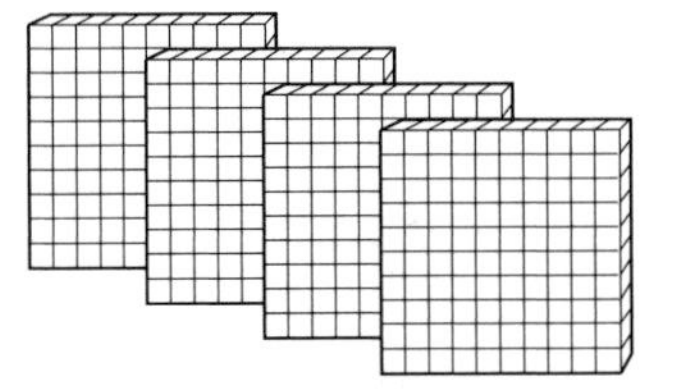

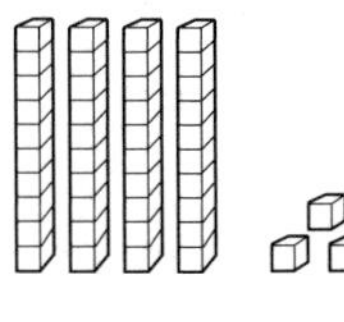

- The language and recording of the algorithm should match that developed earlier.

| **Recording** | **Language** |
|---|---|
| $\begin{array}{r} 1\phantom{4\,3} \\ 3\overline{)4^{1}4\;3} \end{array}$ | 'Share the hundreds<br>3 times what is 4?<br>3 times 1 is 3 and 1 left over.' |
| $\begin{array}{r} 1\;4\phantom{3} \\ 3\overline{)4^{1}4^{2}3} \end{array}$ | 'Share the tens.<br>3 times what is 14?<br>3 times 4 is 12 and 2 left over.' |
| $\begin{array}{l} \phantom{3)}1\;4\;7\text{ r }2 \\ 3\overline{)4^{1}4^{2}3} \end{array}$ | 'Share the units.<br>3 times what is 23?<br>3 times 7 is 21 and 2 left over.' |

There are **147** bottles in each crate and **2** left over.

- Repeat for other examples including some with zero difficulties. For example,

  $3\overline{)3\;0\;6}$ $\quad 4\overline{)3\;6\;0}$ $\quad 5\overline{)5\;1\;5}$ $\quad 4\overline{)5\;0\;0}$

### 2 Dispatching (*ThHTU divided by 2, 3, 4 and 5*)

- 'Superhero found 3624 spatulas. He sent $\frac{1}{2}$ of them to Edinburgh. How many spatulas were sent to Edinburgh?'

  The new teaching point in the division $2\overline{)3\;6\;2\;4}$ is the sharing of the thousands

Revise exchanging 1 thousand for 10 hundreds.

**Recording**

$$2\overline{)3^{1}624}$$ with 1 above

**Language**

'Share the thousands.
2 times what is 3?
2 times 1 is 2 and 1 left over.
Exchange 1 thousand for 10 hundreds.'

The recording and language continues as before, giving $2\overline{)3^{1}624}$ = 1812
**1812** spatulas were dispatched to Edinburgh.

- Repeat for other examples, such as:

$2\overline{)1057}$ $3\overline{)6098}$ $4\overline{)5470}$ $5\overline{)7000}$

**At this point the children could try Textbook page 20.**

## 3 Lazy Mog (*divisibility tests for 2, 3 and 5*)

- From earlier work the children should already know that

— even numbers divide exactly by 2, i.e. have 2 as a factor

— whole numbers with units digit 0 or 5 divide exactly by 5, i.e. have 5 as a factor.

Children who have covered the work on page E5 of the Extension Textbook in Heinemann Mathematics 5 should also know that

— a number divides exactly by 3 if the sum of its digits divides exactly by 3.

- Investigate divisibility by 3 by building up a table and looking for the pattern in the results. For example,

| Number | Divided by 3 | 3 as a factor? | Sum of digits | Divisble by 3? |
|---|---|---|---|---|
| 375 | 125 | YES | 15 | YES |
| 408 | 136 | YES | 12 | YES |
| 299 | 99r2 | NO | 20 | NO |
| 2709 | 903 | YES | 18 | YES |

- Introduce a problem that Mog has to solve. She has to find which of these pill bottles have numbers which are exactly divisible by 2 **or** 3 **or** 5.

Mog is lazy and uses quick ways to solve the problem.

Discuss using the divisibility tests to find that 1331 is the only number which is not divisible by 2 or 3 or 5, i.e. does not have 2 or 3 or 5 as a factor.

Number
Division
Superhero

UA2bc,3a/4 N3cde/4
PSE MD/C1 MD/D3 FPR/D1
P/4ab N/4gl N/5g A/4c

**Problem solving**

## Textbook pages 20 and 21

*Division: HTU and ThHTU by 2, 3, 4, 5*

On Textbook page 20, the two teaching panels are provided to **support** rather than replace the introductory activities, and should be discussed where appropriate.

The problem in question 6 can be solved in various ways. For example,

— the children could list multiples of 5 – 5, 10, 15, 20 . . . 50, 55, 60, 65, 70 – and then pick out the smallest of these, 60, which is also divisible by 2, 3 and 4.

— they could list in one row multiples of 2, in the next row multiples of 3 and so on, then look for the smallest number common to all four rows.

Others may think, wrongly, that the solution is found by multiplying $2 \times 3 \times 4 \times 5 = 120$, rather than $3 \times 4 \times 5 = 60$.

On Textbook page 21, all the questions involve division of thousands, hundreds, tens and units.

Question 3(b) may be difficult for some children. It would be worthwhile discussing different methods used by the children to find the answer. For example,

— since the answer to part (a) is 2738 r 1, the 1 pill left over and 2 **more** pills would give each shop 2739 and 3 more pills would give each shop 2740 pills

— since each shop receives 2738 pills and has to receive 2740 pills, then 2 pills need to be added for each shop, i.e. 6 pills altogether; but 1 pill is already available so $6 - 1 = 5$ pills are needed

— multiply 2740 by 3, giving 8220, then subtract 8215 to give 5 pills.

In question 4(a), the number of calculations required can be reduced by application of the divisibility tests outlined in the introductory activities:

8555 and 6850 have 5 as a factor.

7617, 7149 and 6237 have 3 as a factor.

The remaining numbers can be divided by 4 to find that 9904 and 3952 are exactly divisible by 4, i.e. have 4 as a factor.

**Problem solving**

In question 4(b), there is more than one possible answer for each new label number. For example,

7946 could become 7944 or 7948

8122 could become 8120 or 8124 or 8128

R6

# MENTAL DIVISION BY 6, 7, 8, 9 AND 10

N2a,3cde/4
MD/C1 MD/D2
N/4cgl

Mental division by 6, 7, 8 and 9 and a rule for dividing by 10 are now revised.

The Superhero context develops with a new scenario. Superhero and Mog chase away some robbers from the warehouse in which their loot has been hidden, and begin packing the loot for return to its rightful owners.

# Introductory activities



In addition to the activities suggested below, those given earlier for division by 2, 3, 4 and 5 can also be adapted for division by 6, 7, 8 and 9.

## 1 The loot *(6, 7, 8 and 9 times tables facts and remainders)*

- On the chalkboard, draw boxes and bags of loot which have been stolen by the robbers.

Ask a question such as

'The 6 robbers in the gang share each set of items equally. How many of each item does one robber receive and how many are left over?'

Answers should be found mentally using the 6 times table.

- Repeat the question for gangs of 7, 8 and 9 robbers.

## 2 Making necklaces *(a rule for mental division by 10)*

- This rule can be revised using a scenario of the robbers making necklaces with stolen beads. Each necklace is made using 10 beads.

Ask a question such as

'How many pearl necklaces can they make and how many pearls are left over?'

Although the children should be able to find answers mentally, it may help to highlight the pattern if the calculations are also recorded:

$$10\overline{)2634} = 263 \text{ r } 4 \qquad 10\overline{)5210} = 521 \text{ r } 0 \qquad 10\overline{)8342} = 834 \text{ r } 2$$

- Discussion of the answers should remind the children that
  - — whole numbers whose units digit is not zero do not divide exactly by 10

$$10\overline{)8342} = 834 \text{ r } 2$$

  - — the answer is obtained by moving the digits one place to the right
  - — the units digit becomes the remainder.

## Textbook page 22 *Division: by 6 to 10*

N2a,3cde/4
MD/C1 MD/D2 FPR/D1
N/4cgl A/4c

In questions 1 to 5 and in question 7, it is intended that only the **answers** are recorded. However, some teachers may prefer to request a fuller recording in question 3. For example,

$$6\overline{)21} = 3 \text{ r } 3 \quad \text{or} \quad \tfrac{1}{6} \text{ of } 24 = 4$$

# DIVISION OF THOUSANDS, HUNDREDS, TENS AND UNITS BY 6, 7, 8 AND 9, CONSOLIDATION OF DIVISION BY 2 TO 9

N3cde/4
MD/D3
N/4gl N/5g

The work begins by revising division of hundreds, tens and units by 6, 7, 8 and 9. The written algorithm is then extended to the division of thousands, hundreds, tens and units. Thereafter practice is provided in dividing a four-digit number by 2 to 9.

The Superhero context continues. Superhero and Mog complete the task of cleaning up the warehouse. They then chase and finally track down the robbers.

## Introductory activities

In addition to the activities suggested below, those given on pages 64 and 65 for division by 2, 3, 4 and 5 can be adapted for division by 6, 7, 8 and 9.

### 1 The car number *(division of HTU by 6, 7, 8 and 9)*

- Tell the children that one witness saw the number on the robbers' car. The number had three digits, 7, 8 and 3. Ask the children to write all the possible numbers the car could have had:

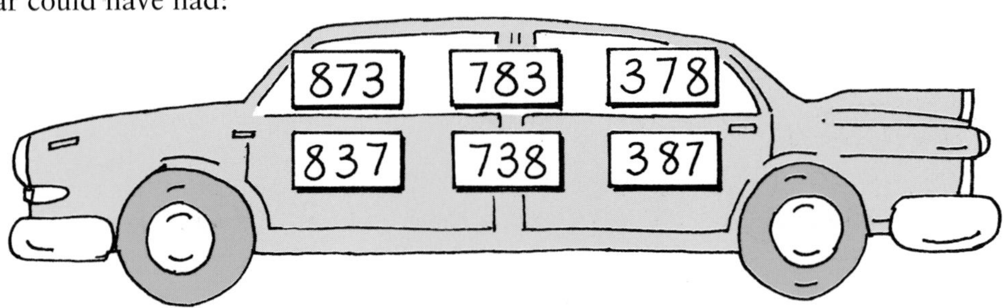


- 'The witness was sure that the number on the robbers' car was 387 because it had a remainder of 3 when divided by 6 or by 8. Was the witness correct?'

  Use these two examples to revise the language and recording for division of hundreds, tens and units, and conclude that the witness was correct.

  $$6\overline{)3^38^27}\quad = 64 \text{ r } 3 \qquad 8\overline{)3^38^67}\quad = 48 \text{ r } 3$$

- Details of suggested language and recording are provided on pages 64 and 65.

### 2 The stolen watch *(division of ThHTU by 6, 7, 8 and 9)*

- On the chalkboard, draw four identical stolen watches as shown.

  'One owner claims the watch numbered 3024 is his because he remembers that the number was exactly divisible by 6, 7, 8 and 9. Has he claimed the correct watch?'

  Discuss the language and recording for one or two of the examples with the children, who can then try the remaining examples on their own.

  $$9\overline{)3^30^32^54}\quad = 336$$

  If any further practice is necessary, the children could be asked to verify that 4368 ÷ 9, 1134 ÷ 8 and 2160 ÷ 7 all have remainders.

- Details of suggested language and recording are provided on page 65.

Number
Division
Superhero

## Textbook pages 23 and 24
*Division: HTU and ThHTU by 6, 7, 8, 9*

UA2bc/4 N3cde,4a/4
PSE MD/D3 FPR/D1
P/4abcd N/4gl N/5g

These two pages provide practice examples, word problems and some problem solving questions.

On Textbook page 23, in question 4, the children must realize that the 28 rings left over should be subtracted from 220 to give 192, the number of rings in 6 full trays. Hence each full tray holds 192 ÷ 6 = 32 rings.

**Problem solving**

In question 9, the children should reason that if one gold chain weighs the same as two silver chains, then three gold chains weigh the same as six silver chains and hence seven silver chains weigh 1162 grams. Thereafter direct calculation leads to weights of 166 grams for a silver chain and 332 grams for a gold chain.

**Problem solving**

On Textbook page 24, for some children, it may be necessary to discuss Colin Conman's note to clarify that the answers given are wrong and that they have to help Superhero to find the errors.

**Problem solving**

To answer question 3 the children must have deduced the robber's mistake, i.e. he has **added** the remainder to each share instead of discarding it. Repetition of this error leads to 55 in each share of the 405 pens.

R7

## Textbook page 25 *Division: ThHTU by 2 to 9, money*
## Workbook page 6

UA2bc,3a/4 N3cde,4a/4
PSE MD/D3 RN/D1
P/4a N/4gl A/4c

The examples provide practice in dividing four-digit whole numbers by a single-digit number with the emphasis on division by 6, 7, 8 and 9.

Discuss the scenario on Textbook page 25 in which Superhero and Mog transmit messages in code to one another as they track down the robbers and hunt for more of the stolen gems. Using their answers to the divisions, the children have to decode the messages to discover

— the robbers are hiding in a barn

— the gems are hidden in a cave.

The children must also use grid references, contained in Mog's message, to identify the hiding place for the gems on Superhero's map.

In question 1, Superhero's message is FIVE ROBBERS IN BARN, and in question 2, Mog's reply is GEMS AT THREE TWO.

On Workbook page 6, Superhero has caught the robbers and Mog is completing her files about them. Using their answers, the children find the robbers' file numbers and enter them on the robbers' photos. The children's working should be carried out in their jotters and only their answers recorded in the spaces provided on the Workbook page.

In question 2(d) the answer in every row to 999.

In question 2(e) the children also have to decode the robber's name – SNATCHER.

**Problem solving**

In question 3, a systematic approach might be:

— list the numbers which become 9950 when rounded to the nearest 10, i.e. 9945, 9946, 9947 . . . 9954. (9995 could also be included.)

— divide each of these by 9 to find which have 9 as a factor (9945, 9954)

— divide these two numbers by 7 to find that only 9954 also has 7 as a factor.

(Alternatively, first test for numbers which have 7 as a factor (9947, 9954) then test these two for 9 as a factor.)

The exact reward for catching Tee Leaf is £9554.

## DIVISION BY CALCULATOR: CHECKING EXACT ANSWERS, ROUNDING AND INTERPRETATION OF ANSWERS

N3cdh,4b/4 N4c/4→5
MD/C4 MD/D4 RN/D1
N/4jln A/4e

In Heinemann Mathematics 5, the children used calculators to divide hundreds, tens and units and sums of money greater than £10 by 2 to 10. All answers were exact.

This work is now extended to thousands, hundreds, tens and units divided by 2 to 9, and includes use of the inverse operation, multiplication, to check answers. Thereafter the children are introduced to divisions which do not have exact answers, and which require rounding of a calculator display to the nearest unit. Finally, the skill of rounding up or down to the nearest appropriate unit is applied in a number of word problems.

The Superhero context continues with Superhero and Mog going on holiday to Australia as a reward for catching the robbers.

## Introductory activities

### 1 Savings (*ThHTU divided by 2 to 9 and checking answers by multiplication*)

- 'Superhero and Mog have always tried hard to save up for a rainy day, for a holiday, and so on. In 3 years Superhero has saved £4164, and in 4 years Mog has saved £4784.'
  Ask the children to use a calculator to find the **average** amount each has saved in a year.
- Discuss how to check the calculations by using the inverse operation, multiplication. For example, to check 4164 ÷ 3

  Enter [4164.] Press [÷] [3] [=] to give [1388.]

  **Do not clear the calculator.**

  Multiply the answer by 3. Press [×] [3] [=] to give [4164.] again.

**At this point the children could try Textbook page 26.**

### 2 Number lines

- On the chalkboard, either draw a number line or attach a card number line and label it as shown.

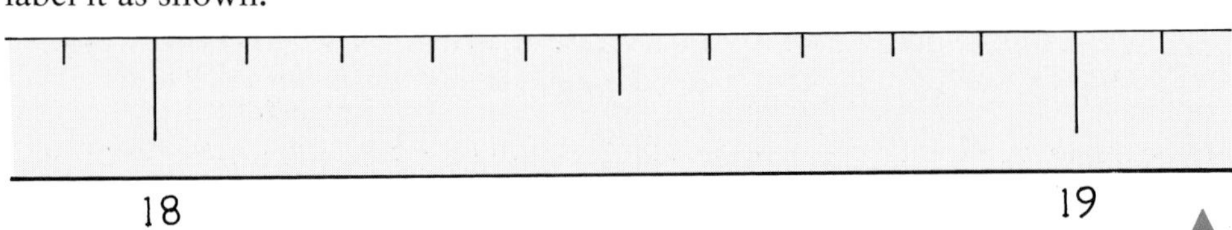


  Have available a card arrow for pointing to individual marks on the line.
- Through discussion and questioning, establish that each small division is one tenth or 0·1, and by counting on from 18 the sequence is 18·1, 18·2, 18·3 . . . 18·9, 19.

- Choose children to place the arrow on the number line to show the positions of specified numbers. For example,

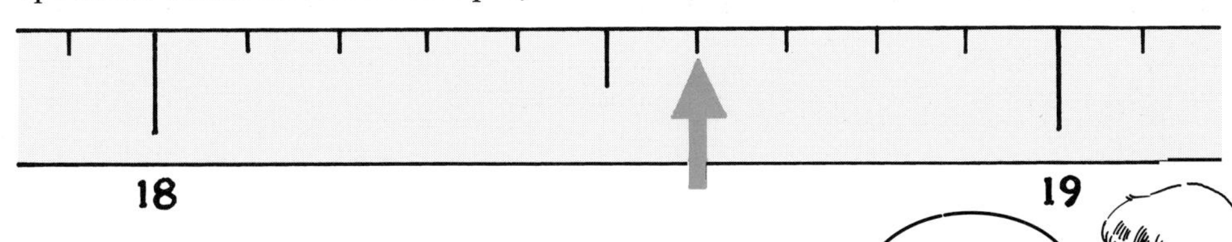


- Repeat until the children can place the arrow with confidence for a range of numbers.

## 3 Calculator displays

- Ask the children to divide 131 by 7 using a calculator.

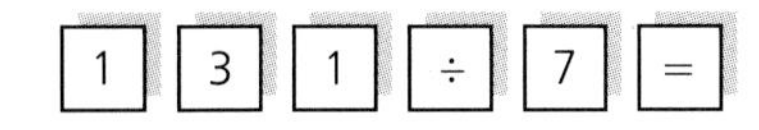


Discuss the display and establish that the answer

— is greater than 18 and less than 19

— is between 18 and 19.

Show the position of 18·7 on the number line.

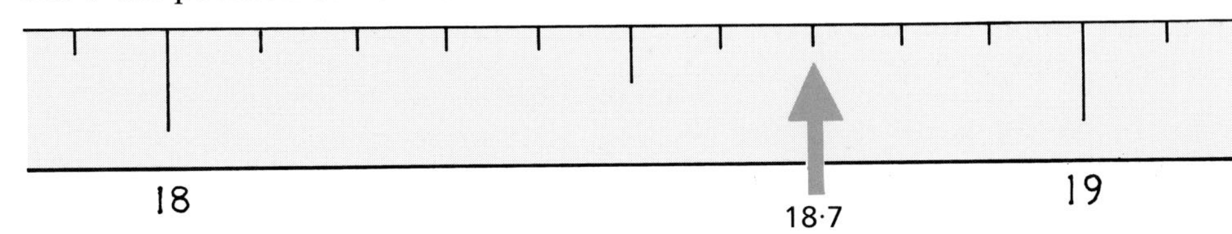


- Establish that the calculator display is 'a little bit more than 18·7 but not as much as 18·8'. Move the arrow a little to show the approximate position of 18·714285.

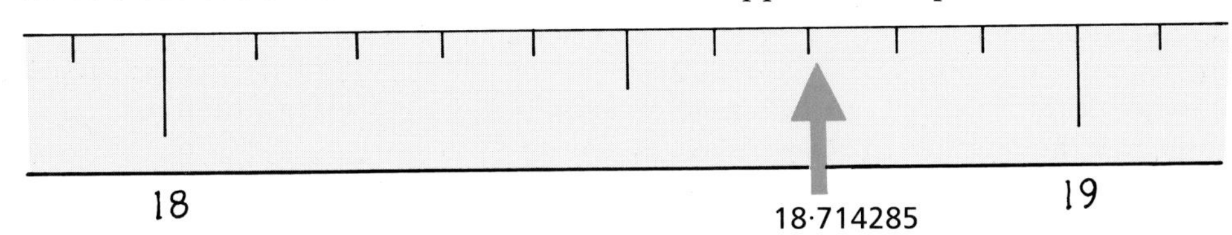


- Repeat this process for other divisions such as

$257 \div 8 = 32{\cdot}125$ $\quad$ $301 \div 6 = 50{\cdot}166666$ $\quad$ $121 \div 7 = 17{\cdot}285714$

## 4 Quotients *(rounding calculator displays to the nearest unit)*

- Ask the children to divide 143 by 9 using a calculator.

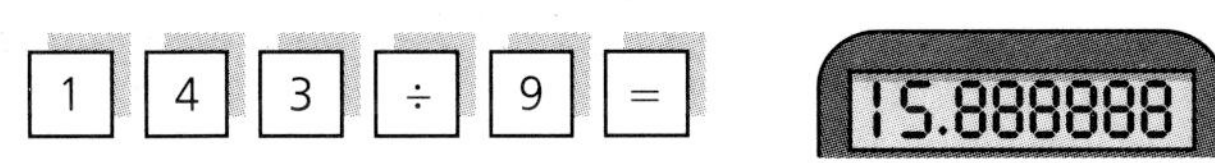


Show the approximate position of 15·888888 on the chalkboard number line.

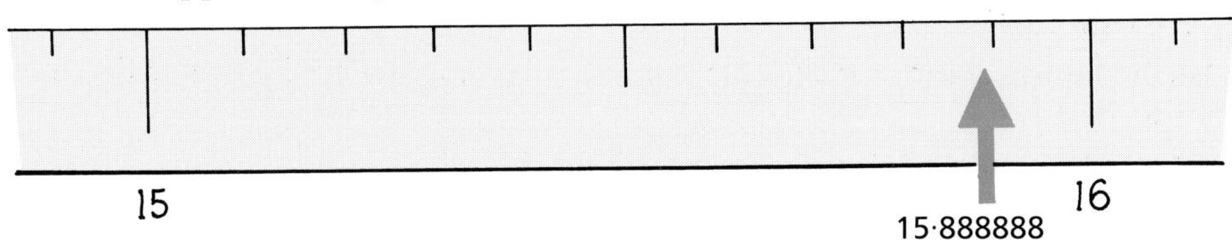


The language used in introductory activity 3 should be repeated and **extended** to describe 15·888888 as

— greater than 15 and less than 16

— between 15 and 16

— **nearer to 16**

— **16 to the nearest unit.**

- Repeat for 211 ÷ 7 = 30·142857

30 30·142857 31

Emphasize the new language, i.e. 30·142857 is

— **nearer to 30**

— **30 to the nearest unit.**

- Repeat for examples where the arrow's position is close to the halfway mark.

For example, 356 ÷ 9 = 39·555555 → 40 to the nearest unit

325 ÷ 7 = 46·428751 → 46 to the nearest unit.

**At this point the children could try Workbook page 7.**

## 5 Quotients (*rounding up or down in context*)

- Discuss several simple division word problems which lead to calculator displays containing a string of digits after the decimal point. In the context of the problem, the children must decide whether to round the number in the display up or down rather than simply round it to the nearest unit.

The examples which follow are all based on a beach scenario to link with the context on Textbook pages 27 and 28.

- '38 surfboards are shared equally among 3 trailers. How many boards are on each trailer?'

Using a pencil-and-paper technique, the children should conclude that there are 12 boards on each trailer and 2 boards left over.

12 r 2
3)38

Ask the children to carry out the same division using a calculator, then discuss how to interpret the number in the display:

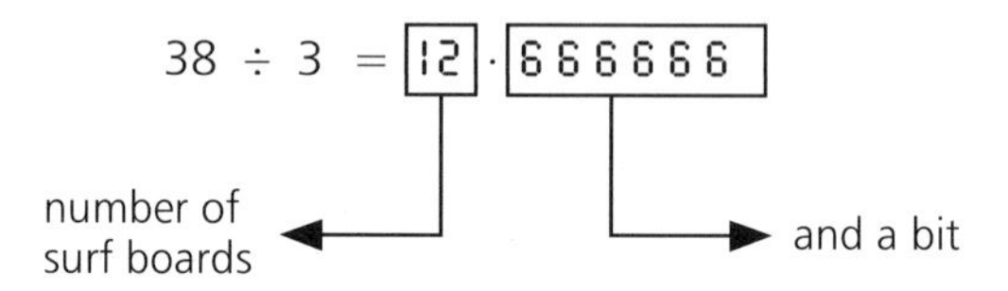


The display indicates a number between 12 and 13, i.e. **12 and a bit.**

In the context of the problem, it does not make sense to share 'bits' of a surfboard. Hence 12·666666 must be **rounded down** to give the required **whole** number, 12, of surfboards on each trailer.

- 'A minibus can carry 16 children. How many minibuses are needed to take 150 children to the beach?'

Discuss the interpretation of the calculator display as before.

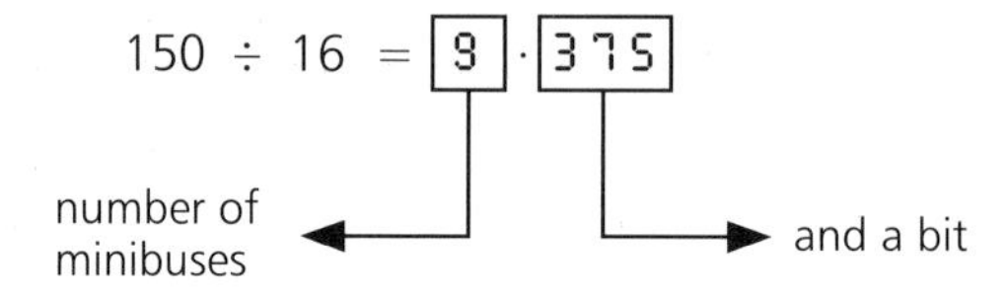


The display indicates a number between 9 and 10, i.e. **9 and a bit.**

In the context of the problem, it does not make sense to think about a 'bit' of a minibus. Hence 9·375 must be **rounded up** to give the required whole number, 10, of minibuses, so that **all** the children are taken to the beach.

- A selection of the following examples should be discussed in the same way. The children must realize that they should decide whether to round the number in the display **up or down by considering the context** in which the question is set.
  - 'How many burgers are placed on each barbecue when 68 burgers are shared equally among 7 barbecues?'
  - 'A trailer can carry 14 surfboards. How many trailers are needed to carry 156 boards?'
  - 'A raft can hold 36 people. How many rafts are needed for 250 people?'

## Textbook page 26

*Division: ThHTU by 2 to 9 using calculator and checking*

UA2bc/4 N3cdh/4 N4c/4→5
PSE MD/C4
P/4a N/4jl A/4e

The context in which the work is set should be discussed if this has not been done as part of the introductory activities. Superhero and Mog are flying to Australia for a holiday which is their reward for catching the robbers.

In question 2, despite the emphasis in the text, some children may not appreciate that the code refers to the **hundreds** digit of each answer. Application of the code reveals SOUTH EAST as the name of the airline.

Some children may need help with question 4. For example, in part (a), they should multiply 2316 by 3 to find whether or not this gives 6948.

In question 6 (a), the same idea is used, i.e. the number is found by **multiplying** 689 by 9.

**Problem solving**

The problem in question 6 (b) may cause difficulty.

$1317 \times 7$ gives 9219. To 9219 **add** 4 to give 9223.

This answer can be checked by doing the following calculation:

$(9223 - 4) \div 7 = 1317$

## Workbook page 7 *Division: rounding to the nearest unit*

N3cdh/4
MD/C4 RN/D1
N/4jn

Superhero and Mog are enjoying the flight to Australia. They help the captain to do 'important calculations' involving division of a three-digit number by a single-digit number.

In question 1, the children are expected to

- record the answer: for example, $101 \div 7 = 14{\cdot}428571$
- decide between which two consecutive whole numbers the answer lies
- draw an arrow on a number line to show the position of the answer.

In question 2, in addition to the above, the children round the answer to the nearest unit. Although this question is supported by two worked examples, it is important that teacher-led discussion takes place along the lines described in introductory activity 4.

N3cdh,4a/4
MD/C4 MD/D4 RN/D1
N/4jln

R9

## Textbook pages 27 and 28 *Division: interpretation of answers*

All the questions on these two pages are presented as word problems and lead to division. Some answers are exact, while other have to be rounded up or down according to the contexts in which they are set.

The contexts of the barbecue on a beach (Textbook page 27) and the preparation for the homeward flight (Textbook page 28) could be included as part of the discussion of the worked examples at the top of Textbook page 27.

On Textbook page 28, in question 2, division of the 167 hours of sunshine by 14 gives an answer just under 12 hours. Interpreting the graph suggests the city to be Sydney.

UA4bcd/4 N4a/4
PSE MD/E4
P/4cd N/4j A/4e

## Textbook page 29 *Other activity: division by calculator*

- This investigation can be attempted at any time after the division work on Textbook pages 19–28 has been completed.
- The children **must not** clear the calculator between steps when they are dividing by 13, then by 11, then by 7.
- In questions 1 and 2, the children should find that they return to the original number each time. They should try at least three 3-digit numbers before stating their conclusion. Encourage them to record their results.
  For example,

  423 423 ⟶ 423

  81 081 ⟶ 81
- In question 3, the children need to enter two zeros before repeating the digit: for example, 9009.

  Encourage the children to write the instructions in a similar form to those given in questions 1 and 2.
- Question 4 is designed to help the children to see why the instructions in questions 1, 2 and 3 work. They should deduce that

  in (a):

  $13 \times 11 \times 7$ gives 1001

  in (b):

  $1001 \times 215$ gives 215 215, the number used in question 1

  $1001 \times 67$ gives 67 067, the number used in question 2

  $1001 \times 9$ gives 9009, the number used in question 3.

The children are **not** expected to record their answer to part (c), but to explain orally, in their own words, why the instructions work. A possible explanation might be:

> 'For question 1 when you repeat the starting number you have 345 345. That is the same as $345 \times 1001$. When you divide by 1001 you will get back to the starting number. To make it a puzzle, instead of dividing by 1001, you divide by 13, then by 11 and then by 7 (because $13 \times 11 \times 7 = 1001$). For questions 2 and 3, you include zeros to make it the same as multiplying by 1001.'

# Merlin Castle

## A context for fractions and percentages

The work on

— fractions (Textbook pages 30–5 and Workbook pages 8–9)

— percentages (Textbook pages 52–3 and Workbook pages 12–14)

is set in the context of 'Merlin Castle'. The work on fractions is set in medieval times. The children consider the designs on the knights' shields, the Great Hall, the tower, the village near the castle, and the castle kitchen.

The work on percentages is set in the present time when the ruins of the castle have become a sightseeing location for tourists. Scenarios include floor and window designs, quilts in the royal bedroom, a restoration fund and visitors to the castle.

### Introducing the context

#### 1 Discussing the castles

The context could be introduced by asking the children what they know about castles. For example,

About the age of castle buildings: 'In what centuries were castles likely to be built?'

About the purposes of the castles: 'Why were castles built?'

About the location of castles: 'Where were castles built?'

About the exterior of castle buildings: 'What are features of the walls?'

'Are there other fortifications as well as walls?'

About the interior of castle buildings: 'What types of room do you find in a castle?'

'What might each of these rooms look like – size, types of window and door, decoration, furniture?'

About the people and their lives: 'Who lived in a castle?'

'Who might live near the castle but outside its walls?'

'What did people, who might have lived in the castle, wear?'

'How might people spend their time?'

About weapons: 'What weapons did these people have?'

#### 2 A model of a castle

The children could be asked to design or make a drawing or a model of a castle. They could discuss the shapes of containers and their suitability for use as turrets, towers, buildings and rooms in the model. Alternatively, the children could make turreted walls and tape them together to make the model.

#### 3 The use of ancient castles today

Ancient castles are used today for many different purposes which the children could research. Information can be gathered from books, videos and the children's own experiences. The children could talk about people visiting the ruins or sites of castles to learn more about how people lived in the past. Restoration could also be discussed, and how this might be paid for by fundraising or selling goods in a castle shop. A visit to a castle could be planned, and possibly implemented, with the class.

UA3b/4
N2c/4
N3g/4→5
RTN/D3
FPR/D1
N/3c
N/4g
N/5b

# Fractions

## Overview

This section

- revises and extends common fractions such as halves, quarters, fifths, sixths, sevenths, eighths, ninths and tenths
- introduces twentieths and hundredths and their notation
- introduces rules for finding equivalent fractions
- deals with finding a fraction of a whole number
- deals with changing a mixed number to an improper fraction and vice versa, involving halves and quarters only.



An extension activity is related to the above section of work as follows:



Teaching notes for the Extension Textbook are in a separate section at the end of the Teacher's Notes.

# Resources

## Useful materials

- coloured pencils or pens
- other materials suggested within the introductory activities

## Assessment and Resources Pack

### Assessment

*Number and Money Check-up 7*

Textbook pages 31–5
Workbook pages 8, 9
(equivalence, fraction of a whole number)

*Round-up 2*

Question 9

### Resources

*Problem Solving Activities*

7 Magic fractions
8 Get the point (equivalence, fractions and decimals)

*Resource Cards*

7 and 8 Equivalent fractions game

# Teaching notes

## FRACTIONS: REVISION, HUNDREDTHS

The concept of a fraction was introduced and developed in earlier stages of Heinemann Mathematics. This section begins by extending the children's understanding and introducing as new fractions the twentieth and the hundredth and their notation.

The children should be introduced to the context of Merlin Castle.

## Introductory activities

### 1 Shields

Show the children drawings or cut-out shapes which are designs for shields. Discuss the fractions coloured. For example,

| r | r |
|---|---|
| y | b |
| b | b |
| r | r |

yellow → one eighth → $\frac{1}{8}$

red → four eighths → $\frac{4}{8}$

blue → three eighths → $\frac{3}{8}$

The children should also be asked to find the fraction 'not coloured blue' (five eighths → $\frac{5}{8}$). Repeat for other shields of different shapes which involve fractions from halves to tenths.

### 2 Windows (*one twentieth*)

Ask the children to draw a coloured window design made up of twenty squares on grid paper. Through discussion identify that one square is one twentieth of each whole design. The children could establish that the notation for twentieths is written with 20 as the denominator: for example, $\frac{1}{20}$, $\frac{3}{20}$, $\frac{11}{20}$. Ask the children to find how many twentieths each colour is of their whole design and have a neighbour check the results.

### 3 Floor patterns (*one hundredth*)

Use a drawing or a cut-out shape to represent a floor pattern from the castle. Tell the children that the pattern has one hundred equal squares. Ask what fraction each small square is of the whole pattern. Discuss the notation for one hundredth as $\frac{1}{100}$, and one hundred hundredths as $\frac{100}{100}$.

Discuss the fraction of the pattern formed by each colour:

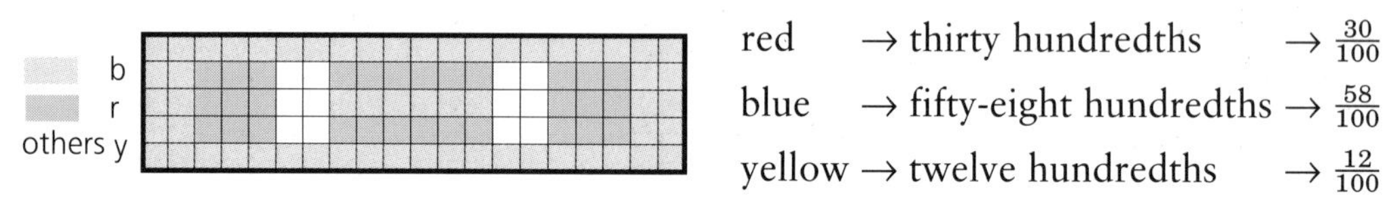


red → thirty hundredths → $\frac{30}{100}$

blue → fifty-eight hundredths → $\frac{58}{100}$

yellow → twelve hundredths → $\frac{12}{100}$

## Textbook page 30 *Fractions: revision, hundredths*

The context of Merlin Castle should be discussed before the children carry out the work on this page. The children could also discuss the drawings of the shields and other designs. They should realize that it is the **design** which forms the whole shape in each of the examples in question 2.

**Number**
Fractions
Merlin Castle

UA3b/4 N2c/4
RTN/D3
N/3c

# EQUIVALENT FRACTIONS: BY MULTIPLYING

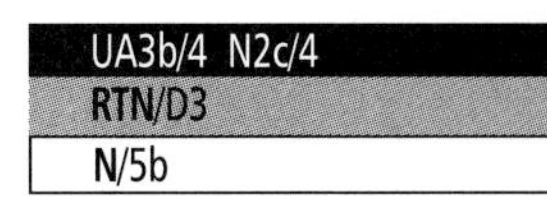


A limited range of equivalent fractions was introduced in an informal way in Heinemann Mathematics 5. This work is revised and extended to establish a rule for generating equivalent fractions by multiplying.
For example,

$$\frac{1}{2} \overset{\times 5}{=} \frac{5}{10} \quad (\text{numerator } \times 5, \text{ denominator } \times 5)$$

The context continues in the Great Hall of Merlin Castle.

## Introductory activity

### Wall hangings

Prepare a set of cards with coloured designs on the front and notation on the back.
For example,

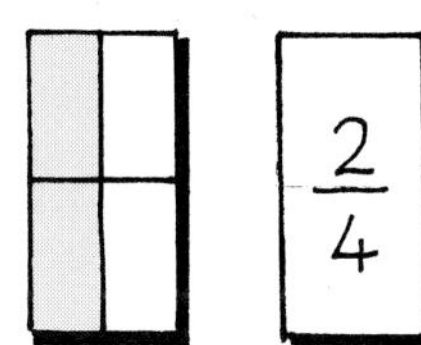


Display the set of coloured designs.

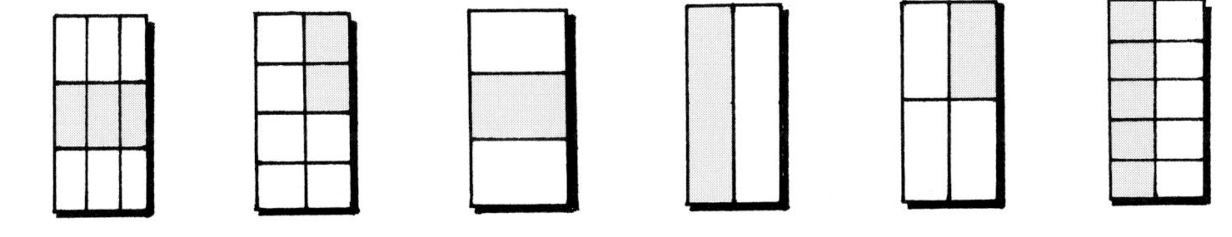

Ask the children to pick out pairs which have the same amount coloured.

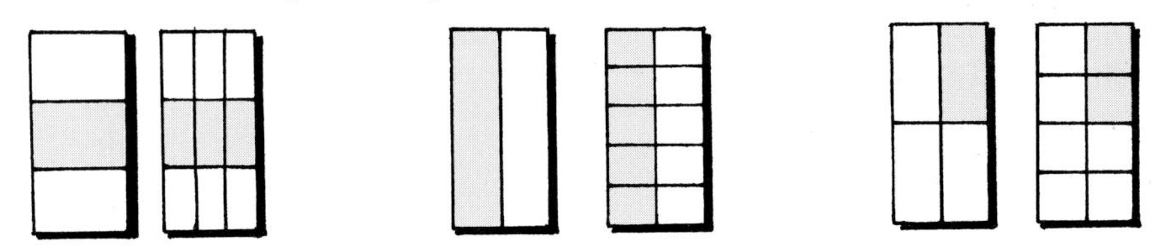

Establish that each pair shows equivalent fractions, and ask the children to identify each of the fractions. These could be confirmed by turning each pair over.

$\frac{1}{3}$ $\frac{3}{9}$ $\frac{1}{2}$ $\frac{5}{10}$ $\frac{1}{4}$ $\frac{2}{8}$

UA3b/4 N2c/4
RTN/D3
N/5b

## Workbook page 8 *Fractions: equivalence*
## Textbook page 31

The context for these pages is the Great Hall of Merlin Castle. On Workbook page 8, the illustration could be discussed with the children and the wall hangings, chairs and plates commented upon. The children should colour each pair of drawings to match, so that the equivalent fractions are obvious. The heavy lines in the diagrams are intended to help them to do this. For example in question 2(a),

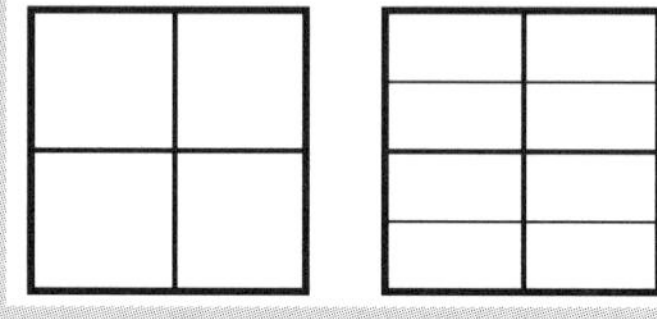

On Textbook page 31, after the children have completed question 1, record the pairs of equivalent fractions on the chalkboard:

$\frac{1}{2} = \frac{2}{4}$ $\frac{3}{4} = \frac{6}{8}$

$\frac{1}{2} = \frac{3}{6}$ $\frac{2}{3} = \frac{6}{9}$

$\frac{4}{5} = \frac{16}{20}$ $\frac{1}{4} = \frac{5}{20}$

Ask the children if they notice a pattern in the numbers for each pair of fractions. For example, how can they change the first fraction to become the second one in this pair?

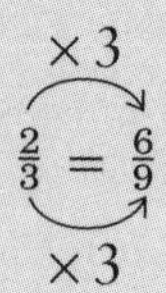


How can they change the 'top' number 2 to become 6?
By multiplying by 3.

How can they change the bottom number 3 to 9?
By multiplying by 3.

Discuss the other pairs of fractions to establish the general rule:

> 'To make an equal fraction, multiply the top and the bottom by the same number.'

Write a pair of equivalent fractions where the numerator of the second fraction is missing:

$\frac{1}{5} = \frac{\phantom{0}}{20}$

Ask the children about the rule to make an equal fraction.
What has the 5 been multiplied by to give 20? (4). The 'top' number, 1, must also be multiplied by 4, giving

×4
$\frac{1}{5} = \frac{4}{20}$
×4

Discuss several examples like these with the children:

$\frac{1}{3} = \frac{\phantom{0}}{9}$, $\frac{2}{5} = \frac{\phantom{0}}{10}$, $\frac{3}{4} = \frac{\phantom{0}}{12}$, $\frac{2}{3} = \frac{\phantom{0}}{6}$, $\frac{3}{5} = \frac{\phantom{0}}{20}$

The children should realize that they must look at the two denominators to identify what multiplier to use.

In question 5, the children are given the freedom to select their own equivalent fraction. They should realize that they can use any number as a multiplier to make an equivalent fraction.



# EQUIVALENT FRACTIONS: BY DIVIDING



A rule for generating equivalent fractions by dividing is introduced. For example,

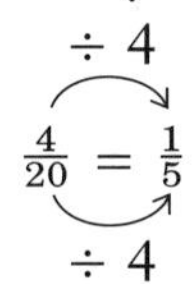

$$\frac{4}{20} = \frac{1}{5} \quad (\div 4)$$

The context continues in the Great Hall of Merlin Castle.

## Introductory activity

### Place mats

- Use drawings on card or paper to discuss pairs of equal fractions like these:

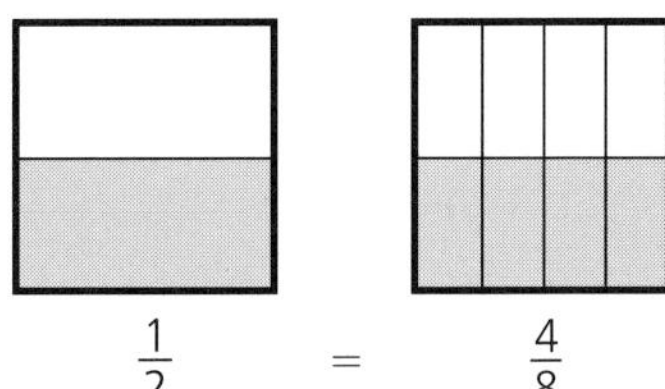

$\frac{1}{2} = \frac{4}{8}$

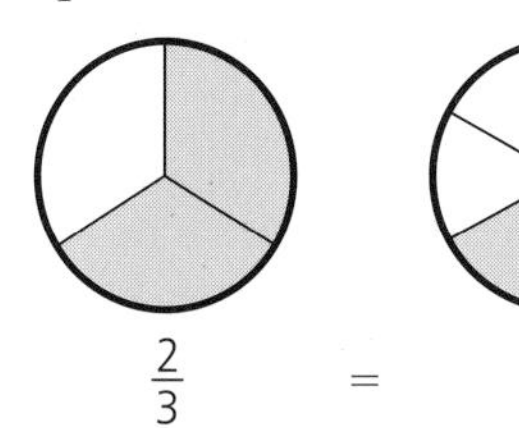

$\frac{2}{3} = \frac{4}{6}$

Interchange the drawings in each pair and establish that the two fractions are still equal. Show that the written statements can also be reversed. For example, $\frac{4}{6} = \frac{2}{3}$.

- Now use other pairs of place mats to give practice with equivalent fractions where the first fraction has the larger denominator. For example,

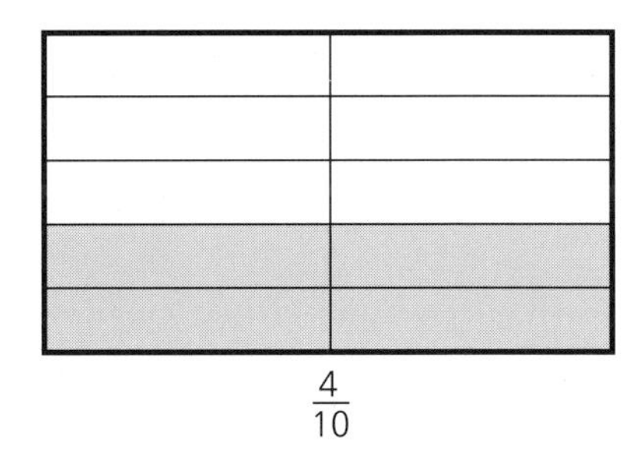

$\frac{4}{10} = \frac{2}{5}$

**Number**
**Fractions**
**Merlin Castle**

UA3b/4 N2c/4
RTN/D3
N/5b

## Workbook page 9 *Fractions: equivalence*
## Textbook page 32

The context for these pages continues within the Great Hall. On Textbook page 32, after the children have completed question 1, write the pairs of equivalent fractions on the chalkboard:

$\frac{2}{8} = \frac{1}{4}$ $\frac{2}{4} = \frac{1}{2}$

$\frac{4}{20} = \frac{1}{5}$ $\frac{5}{10} = \frac{1}{2}$

$\frac{8}{10} = \frac{4}{5}$ $\frac{6}{9} = \frac{2}{3}$

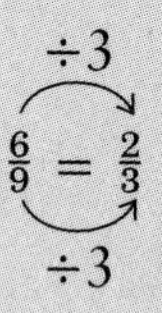

Ask the children if they notice a pattern in the numbers for each pair of fractions. For example, how can they change the first fraction to become the second one in this pair?

How can they change the 'top' number 6 to become 2?
By dividing by 3.

How can they change the bottom number 9 to 3?
By dividing by 3.

Discuss the other pairs of fractions to establish the general rule:

'To make an equal fraction, divide the top and the bottom by the same number.'

Write a pair of equivalent fractions where the numerator of the second fraction is missing:

$\frac{8}{12} = \frac{}{3}$

Ask the children about the rule to make an equal fraction.
What has the 12 been divided by to give 3? (4). The 'top' number, 8, must also be divided by 4, giving

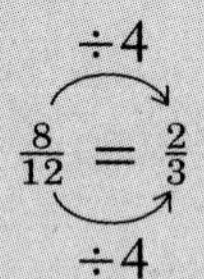

Discuss several examples like these with the children:

$\frac{5}{20} = \frac{}{4}$, $\frac{12}{16} = \frac{}{4}$, $\frac{9}{30} = \frac{}{10}$, $\frac{18}{20} = \frac{}{10}$, $\frac{6}{15} = \frac{}{5}$

In question 4, the children could be told that another name for making equivalent fractions by dividing is 'simplifying', because they are making the numbers smaller and the fraction simpler. Some children may need encouragement to simplify fractions fully. For example,

in question 4(i) $\frac{25}{100} \xrightarrow{\div 5} \frac{5}{20} \xrightarrow{\div 5} \frac{1}{4}$, i.e. $\frac{25}{100} = \frac{5}{20} = \frac{1}{4}$ (top and bottom ÷5, then ÷5)

(j) $\frac{20}{100} = \frac{2}{10} = \frac{1}{5}$ (top and bottom ÷10, then ÷2)

# FRACTIONS OF A SET

The division rule is used to simplify answers when dealing with a fraction of a set of objects.

The context continues in the tower of Merlin Castle.

## Introductory activities

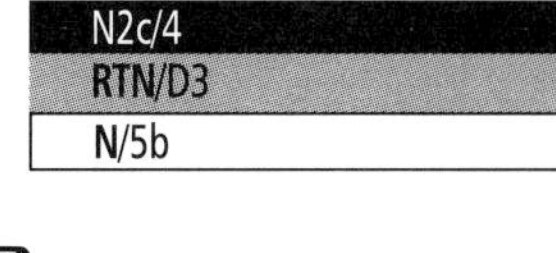


### 1 Soldiers' tunics (*fraction of a set **without** simplifying*)

- Use a set of drawings or cut-outs of soldiers' tunics, coloured red on one side and yellow on the other.

  Lay out two red and three yellow tunics.

  'What fraction of the tunics is red.?'

  2 tunics out of 5 are red. $\frac{2}{5}$ are red.

  Some children may need to be reminded that the fraction is found by counting the number of tunics to find the denominator, and then counting the number which are red to find the numerator.

- Repeat for other examples.

### 2 Soldiers' tunics (*fraction of a set **with** simplifying*)

Lay out four red and six yellow tunics.

'What fraction of the tunics is red?'

Discuss how this fraction could be simplified. For example,

$$\frac{4}{10} \overset{\div 2}{=} \frac{2}{5} \quad (\div 2)$$

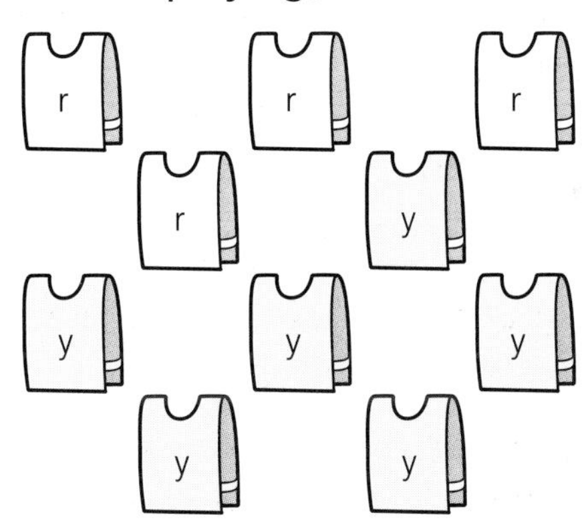


### 3 Towers

Tell the children that they should always look at any fraction answer to find out if it can be simplified. Discuss this example:

'The tower has 100 steps. 25 steps lead to the first floor, 35 steps lead to the second floor and 40 steps lead to the top floor. What fraction of the steps lead to
(a) the first floor
(b) the second floor
(c) the top floor?'

To obtain the answer to (a), the children might apply the division procedure twice to obtain the simplest answer: that is,

$$\frac{25}{100} \overset{\div 5}{=} \frac{5}{20} \overset{\div 5}{=} \frac{1}{4}$$

## Textbook page 33 *Fractions: equivalence*

For this page, the focus is on the tower of the castle. This scenario could be discussed with the children.

Discuss the worked example at the top of the page, and stress that the fractions are of the **total number** of flags. In questions 4 to 6, remind the children that they should look carefully at each fraction to see if it can be simplified. Some children may have to simplify in more than one step. For example, in question 6(b),

$$\frac{40}{100} = \frac{4}{10} = \frac{2}{5}$$

# FRACTIONS OF A WHOLE NUMBER

N3g/4→5
FPR/D1
N/4g

In Heinemann Mathematics 5, finding a fraction of a whole number was linked to the division process: for example, $\frac{1}{4}$ of 12 is calculated as 12 divided by 4.

This procedure is now revised using the scenario of the people who live in the village outside the castle walls.

## Introductory activities

### 1 Team game

Prepare cards displaying calculations such as

$\frac{1}{5}$ of 20 | $\frac{1}{9}$ of 36 | $\frac{1}{8}$ of 24 | $\frac{1}{7}$ of 14 | $\frac{1}{6}$ of 18 and so on.

Divide the children into teams. Each team, in turn, picks a card and calculates the answer by mental division. If the answer is correct, the team collects points equal to the number in the answer. The team with most points after all the cards have been picked wins.

### 2 Villagers (*written calculations*)

- 'There were 162 children in the village. $\frac{1}{6}$ of them were helping in the fields. How many were helping in the fields?'

  Suggest that, for such an example, the children should record like this:

  $\frac{1}{6}$ of 162
  =27

  *working:*
  $6\,\overline{)\,16^{4}2}$ = 27

  27 children were helping in the fields.

- Repeat for other examples, including a four-digit number: for example, $\frac{1}{5}$ of 1340.

N3g/4→5
FPR/D1
N/4g

## Textbook page 34 *Fractions of a whole number*

Discuss the village outside the castle walls and the jobs undertaken by the villagers. It is intended that questions 1 and 2 are calculated mentally, but written recording for division should be encouraged in questions 3, 4 and 5.

# MIXED NUMBERS, HALVES AND QUARTERS

In earlier stages of Heinemann Mathematics, the children have used mixed numbers in their measure work. For example, in Heinemann Mathematics 5, lengths were expressed as $1\frac{3}{10}$ m and areas as $22\frac{1}{2}$ cm$^2$.

The relationship between mixed numbers and improper fractions is now considered: for example, $2\frac{1}{2} = \frac{5}{2}$.The examples involve only halves and quarters. The context continues in the castle kitchen.

## Introductory activities

UA3b/4 N2c/4
RTN/D3
N/5b

### 1 The spit (*a length in half metres*)

Ask the children to mark out a length of $1\frac{1}{2}$ metres to show the length of a medieval spit.

The measurement should be made using

— a 1 metre stick and a $\frac{1}{2}$ metre stick, and

— three $\frac{1}{2}$ metre sticks.

The children's attention should be drawn to the fact that

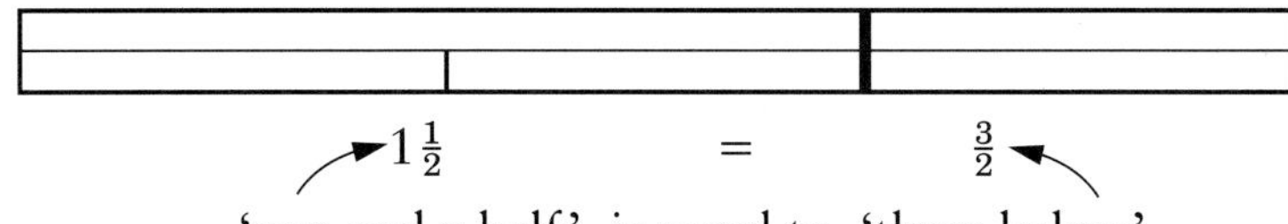


Other examples could be measured and/or discussed. For example,

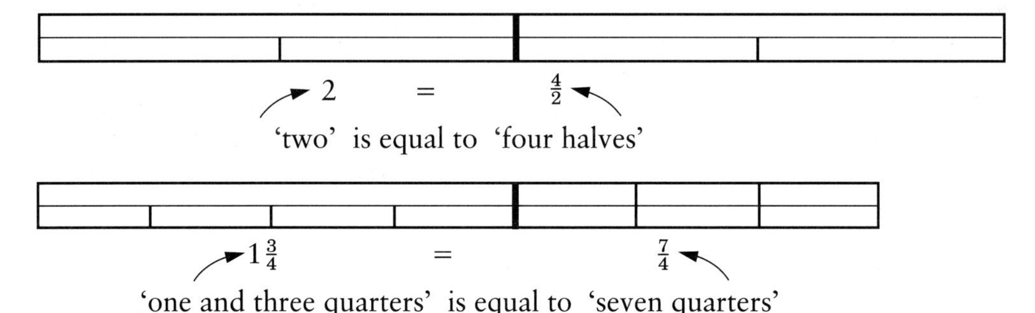


### 2 Pairs

This game is played by two players.

Prepare a set of eight cards, each displaying a whole number or a mixed number, and another set of eight cards, each displaying the equivalent improper fractions. Initially these should only involve halves. For example, they could include:

How to play:

- shuffle all sixteen cards
- deal four cards to each player and place the other cards in a face-down pile
- each player pairs any two cards which show equivalent fractions and puts them down for the other to see. For example,

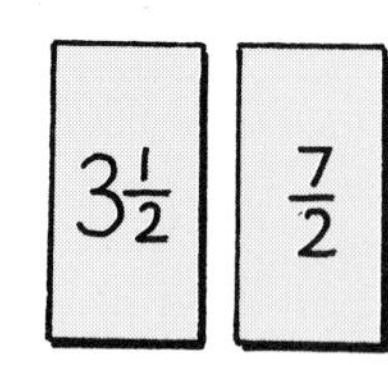


- the first player picks up the top card from the face-down pile and tries to make a pair with it; one card has to be discarded and placed face up beside the face-down pile
- each player, in turn, takes either the top face-down card or the top face-up card to try to form a pair, and then discards one card
- when all the face-down pile is used, the face-up pile should be placed face down
- the winner is the first player to have made pairs with all of their cards.

After several games, the sets of cards could be extended to include additional cards with fractions involving quarters.

N2c/4
RTN/D3
N/5b

## Textbook page 35

*Fractions: mixed numbers, halves and quarters*

The context of the castle kitchen should be discussed.

The children are expected to do the calculations using their knowledge of how many halves or quarters there are in each whole, and not by any formal algorithm.

In question 4, however, some children may realize that the number of wholes and halves left over can be found by division.For example,

$\frac{7}{2} = 3\frac{1}{2}$ since $7 \div 2 = 3$ r 1: that is, 3 wholes and 1 half $= 3\frac{1}{2}$

In question 7(h), some children might leave the answer as $3\frac{2}{4}$. They should be encouraged to simplify $\frac{2}{4}$ to $\frac{1}{2}$ to give $3\frac{1}{2}$.

# Alltmouth



## A context for decimals

The work on pages 36–51 of the Textbook and pages 10 and 11 of the Workbook is set in Alltmouth, the area around the estuary of the River Allt. This context involves the children in visiting different locations, including the fishing harbour at Allton, Point of Allt, the lighthouse and the suspension bridge.

### Introducing the context

The context could be introduced using an enlargement of the map on Textbook page 36. A wall map could be created by the children.

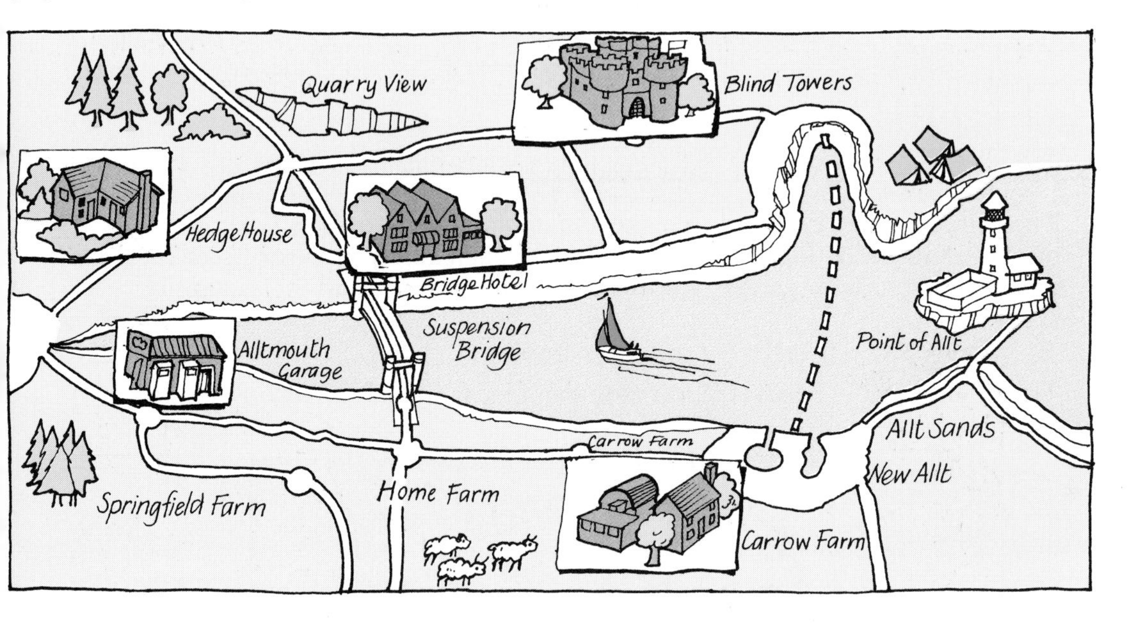


Discussion of activities related to an estuary could focus on

— different ways of crossing the estuary

— old and new ports. The original port is likely to have been Aberallt, while New Allt has developed as a ferry port; Allton is a fishing port

— the industries in the area: for example, fishing, farming, tourism.

Alternatively the context could be introduced by discussing and/or visiting a real estuary, preferably one known to the children.

After the introduction of the context, some of the following activities could be considered.

### 1 New Allt power station

An article in the *Allt News*, announcing plans to build a new power station on the outskirts of New Allt, could be used to provide an initial stimulus. The article could say that interested parties should respond to this proposal by a certain date. This would involve the children in exploring the effects of such a development on, for example, tourism, farming, fishing and conservation in the area.

## 2 Activity breaks (*tourism*)

The children could plan a holiday programme for a family of four camping at Alltbay campsite for one week. They could consider the activities they might do (sailing, walking, fishing, etc.), the places they might visit and the travel arrangements and costs. This information could be used to make an illustrated diary of the week's events, or a poster to encourage people to visit Alltmouth.

## 3 Aberallt (*farming*)

There are many types of farming undertaken in the estuary.

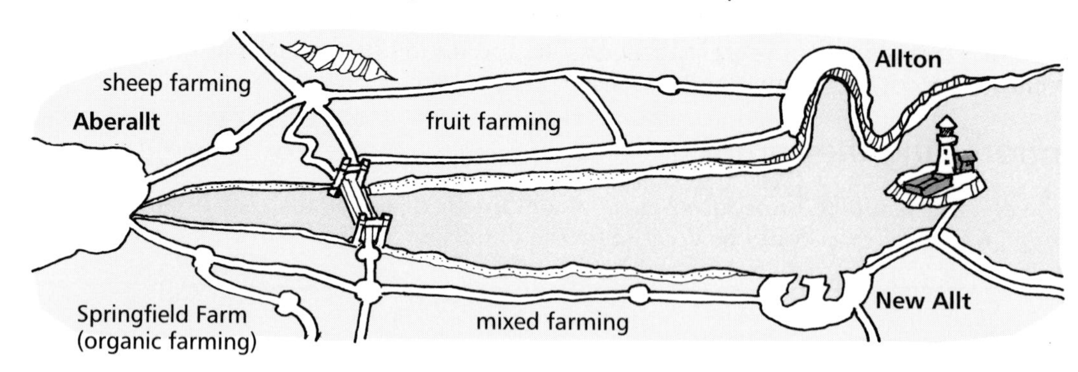


Aberallt is the main market town for the farmers.

The children could be asked to make a plan of Aberallt to show what ancillary businesses might be found there: for example, machinery repairers, seed merchants, cattle market, tractor garage, feedstores, livestock lorries and so on.

## 4 Save Alltmouth (*conservation*)

The children could collect information from local and national conservation organizations.

A display could be made showing the possible effects of the new power station on the local environment.

## 5 The response

The children could draft a response to the power station proposal, using some of the information from the above activities as the basis for their support or opposition.

## 6 Alltmouth Development Corporation

The class could be divided into groups and asked to devise a plan for the development of Alltmouth which takes account of

— the interests of the local people

— the needs of tourism, fishing and farming

— the needs of industry

— conservation.

Each group's plans could be displayed and discussed to compare and contrast the priorities selected. This is likely to highlight the conflicts that emerge when considering how to promote an area.

UA2abcd,3ab/4
N2b,3egh,4a/4
SSM4b/4
PSE
RTN/D3,4
AS/D2,3
MD/D2,3,4
RN/D1
ME/C6
P/4abd
N/3c
N/4dkln
M/3b
M/4a

# Decimals

## Overview

This section

- revises the decimal notation for tenths
- revises addition, subtraction, multiplication and division of one-place decimals
- deals with rules for multiplication and division by 10
- introduces decimal notation for hundredths
- introduces addition and subtraction of two-place decimals
- applies these two operations to money and length
- uses the calculator, involving addition, subtraction, multiplication and division of decimals, where appropriate, including applications to money.

| | Teacher's Notes | Textbook | Workbook | Reinforcement Sheets |
|---|---|---|---|---|
| **Alltmouth: a context for decimals** | 87 | | | |
| **First decimal place:** | | | | |
| **notation** | 92 | 36 | | |
| **addition and subtraction** | 93 | 37, 38 | | |
| **multiplication and division** | 95 | 39, 40 | | |
| **consolidation** | 97 | 41 | | |
| **Second decimal place:** | | | | |
| **notation, link to length and money** | 98 | 42, 43 | 10 | 12 |
| **tenths and hundredths** | 101 | 44 | 11 | 13 |
| **addition** | 104 | 45 | | |
| **subtraction** | 106 | 46, 47 | | 14 |
| **consolidation** | 108 | 48, 49 | | |
| **calculator** | 109 | 50, 51 | | |

Extension activities related to the above section of work are as follows:

| | Teacher's Notes | Extension Textbook |
|---|---|---|
| **Decimals: place value** | 283 | E10 |
| **Decimals: sequences, addition, subtraction** | 284 | E11 |

Teaching notes for the Extension Textbook are in a separate section at the end of the Teacher's Notes.

# Resources

## Useful materials

- decimal number line (tenths)
- squared paper ($\frac{1}{2}$ cm or 1 cm)
- metre stick calibrated in hundredths (centimetres)
- digit cards and place value cards
- calculators
- other materials suggested within the introductory activities

## Assessment and Resources Pack

### Assessment

*Number Check-up 8*

Textbook pages 36–41
(First decimal place: notation, addition, subtraction, multiplication, division)

*Number Check-up 9*

Textbook pages 42–44
Workbook pages 10, 11
(Second decimal place: length, money, tenths, hundredths)

*Number Check-up 10*

Textbook pages 45–49
(First decimal place: addition, subtraction multiplication, division
Second decimal place: addition and subtraction)

*Number Check-up 11*

Textbook pages 9, 10, 17, 18, 26, 50
(Calculator: addition, subtraction, multiplication, division, money)

*Round-up 3*

Questions 1(a), (b), 2(a), (b)

### Resources

*Problem Solving Activities*

8 Get the point (Fractions, decimals, percentages)

*Resource Cards*

9–11 Go-karting (Decimals: place value)
12 Five plus fifty (Second decimal place: addition, subtraction, multiplication)
13, 14 Tartan tours (Calculator: addition, subtraction, multiplication of money)

# Teaching notes

## FIRST DECIMAL PLACE: NOTATION

UA3a/4 N2b/4 SSM4b/4
RTN/D3 ME/C6
N/3c M/3b

In Heinemann Mathematics 5, the children were introduced to the decimal notation for tenths. This is now revised.

The 'Alltmouth' context is introduced with Stan and other drivers from the Moped Deliveries Company, delivering goods in the Alltmouth area.

## Introductory activities

In these activities the context is Alltmouth Moped Deliveries sponsoring the annual Alltbay Cycle Race.

### 1 Cycle race game (*revision of tenths*)

- Duplicate decimal race tracks for each player.

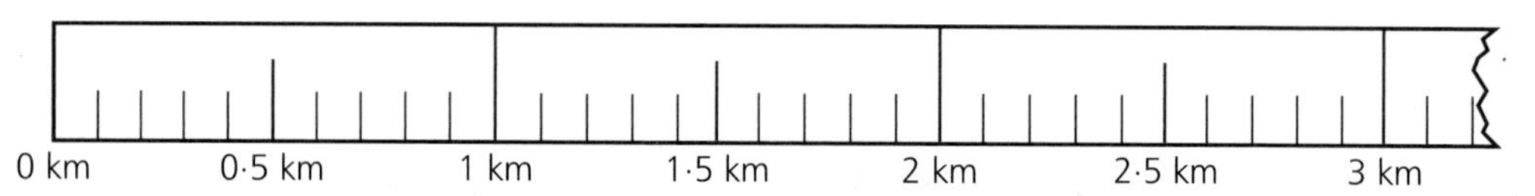


Prepare a 0 to 9 spinner, each number representing tenths of a kilometre.

Alternatively, prepare a set of cards, 0 to 9.

- In turns, the children
  — spin the spinner (or choose a card)
  — colour, on the race track, the number shown
  — record the total distance travelled so far.

For example,

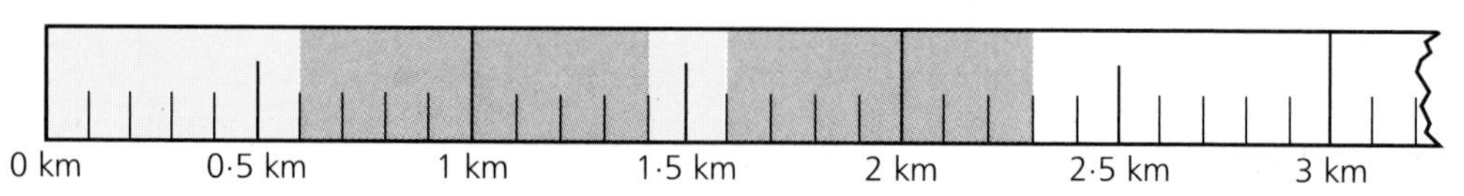


| **Spin** | **Colour** | **Record** |
|---|---|---|
| 6 | 6 tenths | 0·6 |
| 8 | 8 tenths | 1·4 |
| 2 | 2 tenths | 1·6 |
| 7 | 7 tenths | 2·3 |

- The winner is the first person to reach or pass an agreed target distance, such as 5 km.

### 2 Alltbay cycle track (*revision of first decimal place notation)*

- Draw a 'cyclometer' on the chalkboard. Discuss how it is used to record the distance, in kilometres, a cyclist has travelled.

8·9

- Prepare two differently coloured sets of cards displaying individually the numbers 0 to 9. One set of cards represents units and the other represents tenths.

The children take it in turns to select one card from each set of cards to represent the distance travelled along the track.

7 3

Record this distance on the cyclometer.

Discuss different ways of writing this number:

7·3

7 units and 3 tenths

70 tenths and 3 tenths

73 tenths

- Ask the children to show 31 tenths on the cylometer.

31 tenths = 3 units and 1 tenth = 3·1.

## 3 Winners

- Write a set of 'winning distances' on a poster or the chalkboard to show how far each cyclist travelled in a given time.

Ask the children to arrange the distances in order from the shortest to the longest, and discuss the results. The distances chosen should be such that the children have to consider carefully the units and tenths digits.

Winning distances

21·4 km 23·2 km

22·3 km 21·9 km

22·0 km 22·8 km

## 4 Games

- Snap and Pelmanism-type games could be used for further practice.

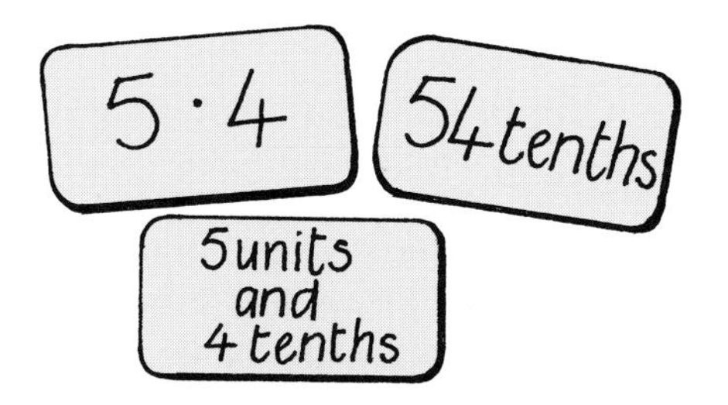


Number

Decimals

Alltmouth

UA3a/4 N2b/4 SSM4b/4

RTN/D3 ME/C6

N/3c M/3b

### Textbook page 36 *Decimals: first decimal place, notation*

The map shows various features of Alltmouth, the area around the estuary of the River Allt. The context begins with Stan delivering parcels around the area.

In question 1, make sure the children realize that each small interval on the fuel gauges represents 1 tenth.

In question 2(c), some children may need to be reminded that 20 tenths can be recorded as 2·0 or 2.

In question 3(f), some children may experience difficulty when recording thirty and eight tenths, and may incorrectly record 3·8.

In question 4, some children may need to be reminded that '*l*' represents 'litres'.

# FIRST DECIMAL PLACE: ADDITION AND SUBTRACTION

UA3a/4 N2b,3g/4

RTN/D3 AS/D2

N/3c N/4k

Addition and subtraction of first-place decimals, introduced in Heinemann Mathematics 5, is revised.

The context is based on the distances travelled and the weights of parcels carried by Alltmouth Moped Deliveries.

# Introductory activities

## 1 Aberallt bus routes (*addition and subtraction of one-place decimals*)

- Use a large map or diagram of Aberallt showing
  - the main bus routes in the town
  - the distance between selected bus stops.

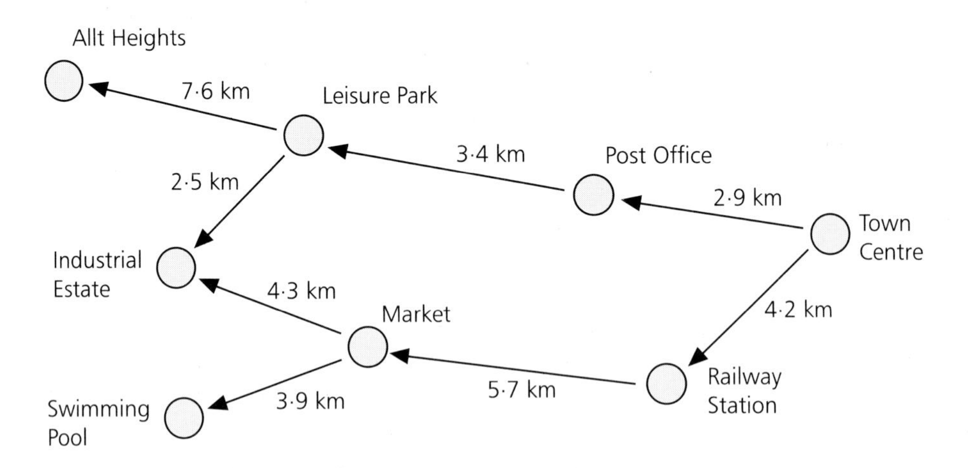


This will generate addition and subtraction examples for discussion. For example,

— 'How far is it from the Town Centre to the Leisure Park?'

| Recording | Language |
|---|---|
| 2·9 km<br>+ 3·4 km<br>6·3 km<br>1 | '**Add the tenths.** 9 and 4 is 13 tenths.<br>Exchange for 1 unit and 3 tenths.<br>**Add the units.** 1 and 3 is 4 and 2 is 6.'<br>The distance is **6·3 km.** |

Some children may need to be reminded of the importance of lining up the decimal points and the digits in columns.

— 'How far is it from the Town Centre to the Swimming Pool via the Railway Station?'

4·2 km
5·7 km
+ 3·9 km
13·8 km
1

The distance **13·8 km.**

**At this point the children could try Textbook page 37.**

- 'Which is further from the Post Office – the Leisure Park or the Town Centre? By how much?'

| Recording | Language |
|---|---|
| 2 1<br>~~3~~·4 km<br>– 2·9 km<br>0·5 km | '**Subtract the tenths.** 4 take away 9, I cannot.<br>Exchange 1 unit for 10 tenths.<br>14 take away 9 leaves 5 tenths.<br>**Subtract the units.** 2 take away 2 leaves 0 units.'<br>The Leisure Park is **0·5 km** further from the Post Office. |

Remind the children to put a zero in the empty units column.

The language and recordings shown are guidelines. Schools should use methods consistent with their own policy.

Number
Decimals
Alltmouth

UA2d,3ab/4 N2b,3g,4a/4
PSE AS/D2
P/4abd N/4k

## Textbook pages 37 and 38

*Decimals: first decimal place, addition and subtraction*

Discuss the map on Textbook pages 37 and 38 with the children.

On Textbook page 37, in question 1, the children must realize that, to find the *total* distance for each journey, they need to add together all the distances shown along the route. For example,

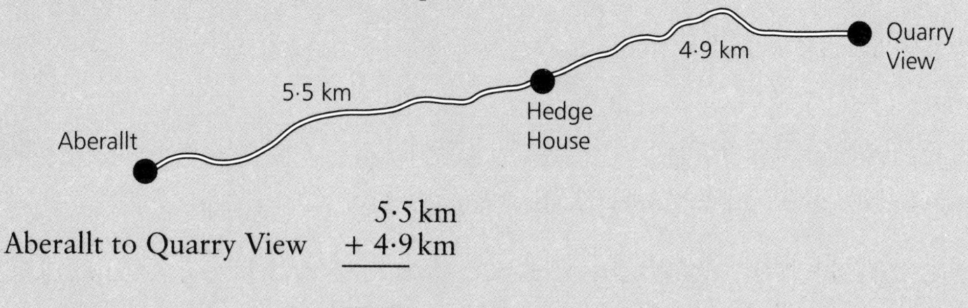


Aberallt to Quarry View $\begin{array}{r} 5{\cdot}5\,\text{km} \\ +\ 4{\cdot}9\,\text{km} \\ \hline \\ \hline \end{array}$

For Fred's journey, two answers are possible depending on the route used from Bridge Hotel to Quarry View. In the final example in question 1, three distances have to be added together. Two answers are again possible, depending on the route used from Quarry View to Bridge Hotel.

In question 2, some children may not appreciate that they have to add the two distances travelled by Stan in question 1.

In questions 4(h) and (k), some children may need to be reminded of the importance of lining up the decimal points and the digits in columns, and writing numbers such as 12 as 12·0.

On Textbook page 38, in question 4(d), the need to record a zero in the units column may have to be emphasized.

**Problem solving**

In question 5, some children may not realize that *any* total weight of parcels up to 20 kg can be carried. Some children may find the total weight of all the parcels (60 kg) and, knowing that 3 times 20 is 60, might think that Stan needs to take 3 trips. However, when combing the weights of the parcels, the children should find that four trips are required. One possible solution is

— reels (9·7 kg) and hooks (10 kg)

— cord (6·4 kg) and tape (11·8 kg)

— nets (12·3 kg)

— tins (9·8 kg).

The children could be encouraged to find more than one solution for four trips.

# FIRST DECIMAL PLACE: MULTIPLICATION AND DIVISION

UA3a/4 N3eg/4
MD/D2,3
N/4l

Multiplication and division of one-place decimals by 2 to 9 is revised. Rules for multiplication of a one-place decimal by 10 and for division of a whole number by 10 with an answer in tenths (e.g. 19 ÷ 10 = 1·9) are also revised.

The context continues with preparations for a fishing trip from Allton harbour.

# Introductory activities

## 1 Loading fish boxes (*multiplication of a one-place decimal*)

Each empty box weighs 2·7 kg. What is the weight of 6 boxes?

**Recording**

$$\begin{array}{r} 2{\cdot}7\text{ kg} \\ \times 6\phantom{\text{ kg}} \\ \hline 16{\cdot}2\text{ kg} \\ {\scriptstyle 4}\phantom{\text{ kg}} \end{array}$$

**Language**

'Put the decimal point in the answer space.
**Multiply the tenths.** 6 times 7 is 42 tenths.
Exchange for 4 units and 2 tenths.
**Multiply the units.** 6 times 2 is 12 and 4 is 16.'

6 boxes weigh **16·2 kg.**

Repeat for other examples.

**At this point the children could try Textbook page 39.**

## 2 Weight of ice (*division of a one-place decimal*)

Sam shares 20·7 kg of ice equally among 3 boxes.
What weight of ice is put in each box?

**Recording**

$$3\overline{)20\cdot{}^{2}7}\quad\text{answer: }6\cdot9$$

**Language**

'**Share the units.** 3 times what is 20?
3 times 6 is 18 and 2 units left over.
Exchange 2 units for 20 tenths.
**Share the tenths.** 3 times what is 27?
3 times 9 is 27.'

There are **6·9 kg** of ice in each box.

Examples of type 32÷5, which can be set down as $5\overline{)32\cdot0}$, should be discussed.

UA3a/4 N3eg/4
MD/D2,3
N/4l

## Textbook pages 39 and 40

*Decimals: first decimal place, multiplication and division*

On Textbook page 39, the multiplications by single-digit numbers are graded as follows:

Question 1 – tenths only
Questions 2 and 3 – units and tenths, with no exchange
Questions 4 and 5 – units and tenths, with exchange
Question 6 – tens, units and tenths, with exchange.

Questions 1, 2 and 3 could be done mentally.

Question 7 reminds children about the rule for multiplication by 10 before applying it in question 8. Further discussion of the rule should take place to ensure that the children fully understand the **digit** shift which occurs.

On Textbook page 40, the divisions by single-digit numbers are graded as follows:

Question 1 – tenths only, with no exchange (this question could be done mentally)
Questions 2, 3 and 4 – tens, units and tenths, mostly requiring exchange
Question 5 – whole numbers with quotients involving tenths
Question 6 – miscellaneous examples.

In question 6, some children may need to be reminded to record like this: $4\overline{)38\cdot0}$, inserting a zero in the tenths column.

Question 7 reminds children about the rule for division by 10 before applying it in questions 8 and 9. Further discussion of the rule should take place to ensure that the children fully understand the **digit** shift which occurs.

# FIRST DECIMAL PLACE: CONSOLIDATION

Mixed examples are provided covering addition, subtraction, multiplication and division involving one-place decimals.

The mathematical activities are set within the context of the fishermen returning from a fishing trip.

UA2abcd,3a/4 N3eg,4a/4
PSE AS/D2 MD/D3
P/4ab N/4kl

## Textbook page 41 *Decimals:* +, −, ×, ÷

Some children may have difficulty in deciding which of the four operations to use in each question. Encourage the children to identify key words, such as 'total'.

In questions 1, 2(a) and 2(c), they may need to be reminded that 24 m can be written as 24·0 m, in order to carry out the operation required.

Note that the answers to question 5(a) are used to find the answers to question 5(b).

**Problem solving**

The problem in question 6 is challenging. The information given and the fact that → is multiplied by 7 leads to

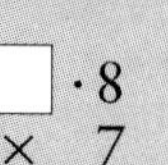


```
  □ □ · 8
    ×   7
? 6   0 · 6
↗     5
even number
```

There are now **two** possible approaches.

- Continue **multiplying.** Using a guess and check strategy, the units digit must be 5, giving →

```
  □  5 · 8
      × 7
? 6  0 · 6
  4  5
```

  and the tens digit must be 6, giving →

```
 65·8
  ×7
460·6
 45
```

  The length of the fish is **65·8 cm.**

- **Use division.** Divide possible **products** 260·6, 460·6, 660·6, 860·6 each by 7 to find which gives an answer with a **tenths digit of 8.**

  460·6 ÷ 7 gives 65·8

  The length of the fish is **65·8 cm.**

# SECOND DECIMAL PLACE: NOTATION, LINK TO LENGTH AND MONEY

UA3ab/4 N2b/4 SSM4b/4
RTN/D3, 4
N/4d

Hundredths and fractional notation – for example, $\frac{4}{100}$, $\frac{20}{100}$, $\frac{28}{100}$ – first appear in the Fractions section of Heinemann Mathematics 6.

The children are now introduced to the corresponding decimal notation: for example, $\frac{4}{100} \rightarrow 0{\cdot}04$, $\frac{20}{100} \rightarrow 0{\cdot}20$, $\frac{28}{100} \rightarrow 0{\cdot}28$. This is then extended to include decimals involving units, such as 3·14.

The relationship of centimetres to metres and pence to pounds in terms of hundredths is illustrated: for example, 3 cm = 0·03 m and 47p = £0·47.

The work on notation for hundredths is set in the context of road traffic signs on the suspension bridge which crosses the Alltmouth estuary. The scenario for length and money is that of preventing beach erosion at Allt Sands.

## Introductory activities

### 1 Signs *(introducing the second decimal place)*

- Introduce the idea of a sign consisting of 100 light bulbs or light cells in a variety of colours. It could display a logo outside a shop or an attractive pattern. Examples could be drawn on grids on the chalkboard, on squared paper or on an overhead projector transparency. For example,

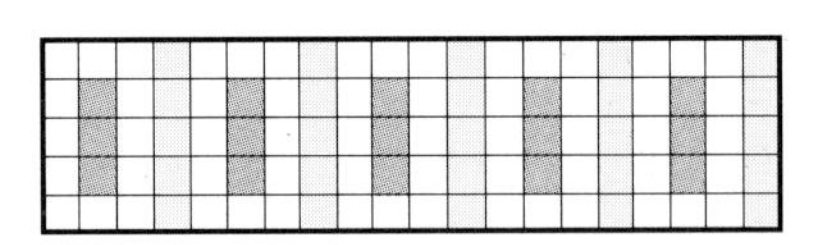

- Discuss each sign and ask the children to state
  - — how many lights or cells make up the sign
  - — that each is **one hundredth** of the sign
  - — how many of them are coloured red, blue, etc. For example

    '15 hundredths are coloured red.'
  - — that 15 hundredths can be written as $\frac{15}{100}$.
- Show how $\frac{15}{100}$ can be written in decimal form as 0·15, emphasizing that 0·15 means 0 wholes (units) and 15 hundredths. Include in the discussion examples such as

  $\frac{1}{100} \rightarrow 0{\cdot}01$ and $\frac{30}{100} \rightarrow 0{\cdot}30$.
- This activity can be extended to include signs consisting of more than one whole, as shown.

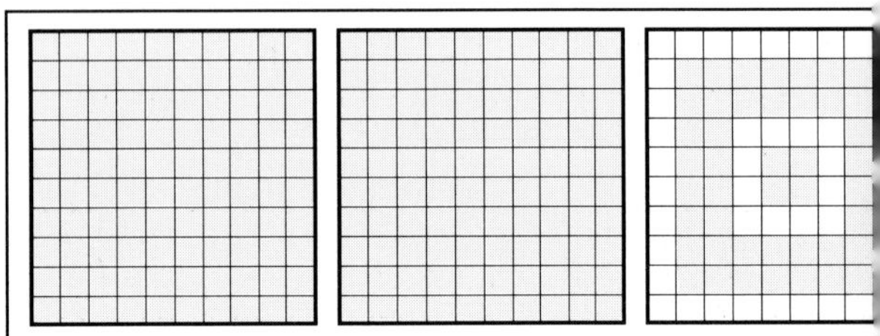


- The Allt Bridge traffic signs which appear on Textbook page 42 could be introduced at this point and examples discussed in a similar way. For example,

  'There are 100 light cells altogether.
  Each cell is one hundredth or 0·01 of the sign.
  48 cells are lit. The fraction of the sign lit is
  48 hundredths or $\frac{48}{100}$ or 0·48.'

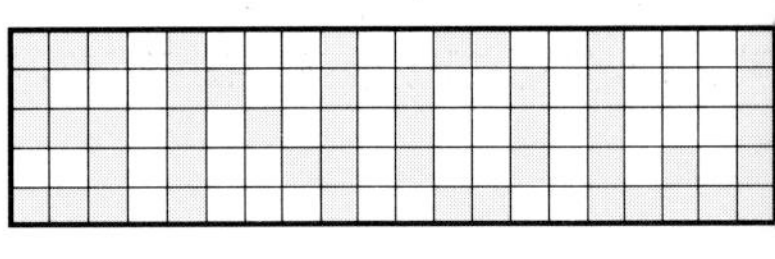

## Colouring signs

Give the children $\frac{1}{2}$ cm or 1 cm squared paper and ask them to draw squares and rectangles consisting of 100 small squares. They could then make coloured patterns or traffic signs of their own, labelling them as shown.

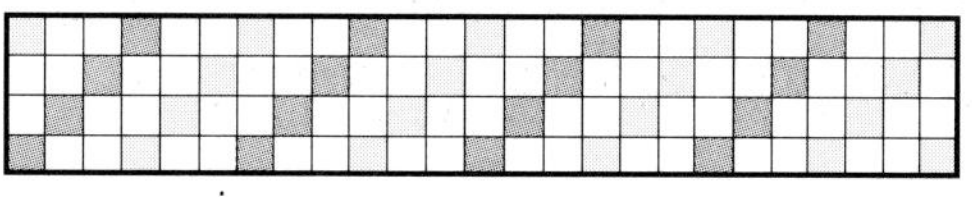

The fraction coloured red is $\frac{22}{100}$ or 0·22.

Provide ready-drawn 100-squares and 100-rectangles and specify in decimal form the fraction of each which must be coloured. Include examples involving decimals greater than one. For example,

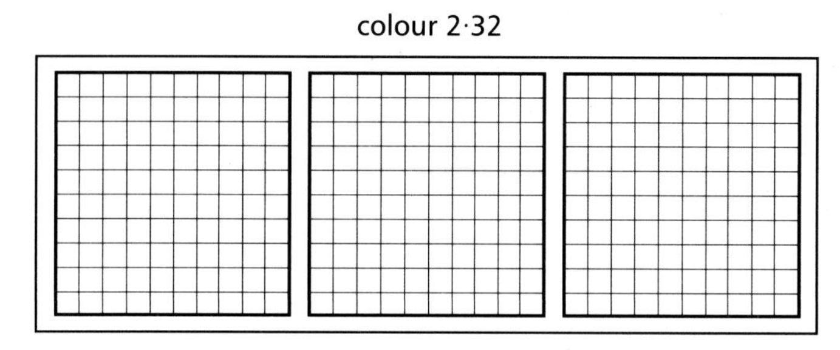


**t this point, the children could try Textbook page 42 and Workbook page 10.**

## Hundredths of a metre (centimetres)

Remind the children that there are 100 cm in 1 metre. Establish that 1 cm is therefore 1 **hundredth** of 1 metre.

$$1\text{ cm} = \tfrac{1}{100}\text{ m} = 0{\cdot}01\text{ m}$$

Use a metre stick. Ask the children to say or write numbers of centimetres as decimal fractions of a metre.

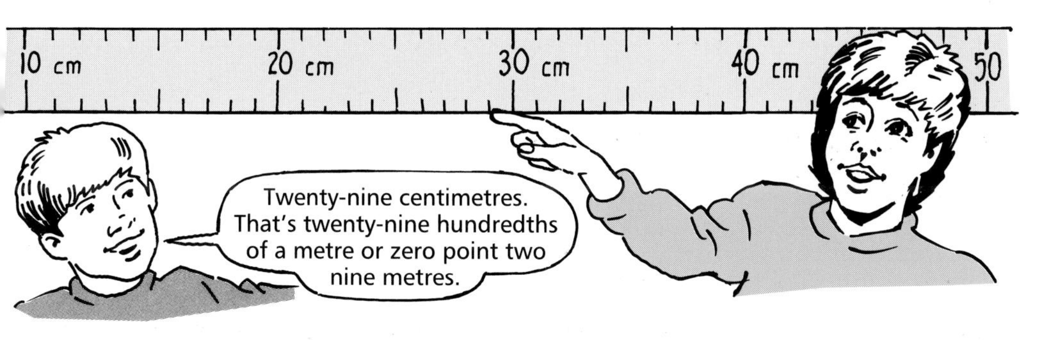


Include examples such as 8 cm = 0·08 m
and 30 cm = 0·30 m

Ask the children to measure some lengths in the classroom using metre sticks or tapes. These measurements should then be converted to a decimal form of metres. For example,

| height of window | 95 cm | 0·95 m |
|---|---|---|
| length of table | 1m 42cm | 1·42 m |
| width of door | 80 cm | 0·80m |

## 4 Hundredths of a pound (pence)

- Remind the children that there are 100 pence in £1. Establish that 1p is therefore 1 hundredth of £1.

$$1\text{p} = £\tfrac{1}{100} = £0{\cdot}01$$

- Ask the children to say or write amounts of pence as decimal fractions of £1.

Include examples like 30p = £0·30 and 3p = £0·03.

- Establish that a sum of money such as £3·85, which was previously regarded as £3 and 85 pence, is also £3 and 85 hundredths of a pound.
- Give the children various prices in pence or in pounds and pence. Ask them to make up the till receipt where all the prices are in £s.

UA3ab/4 N2b/4
RTN/D3,4
N/4d

**Textbook page 42** *Decimals: second decimal place, notation*
**Workbook page 10**

**Textbook page 43** *Decimals: length, money*

The scenario of traffic signs used in question 1 on Textbook page 42 should be discussed if it has not already been met in an introductory activity. Emphasize that each sign consists of 100 equal parts.

In question 4, some children may find it helpful to read the given decimals as fractions. For example, in part (a),

'38 hundredths, 31 hundredths . . .'

On Workbook page 10, in questions 1 and 3, a single colour should be used within each sign, although the selected colour may vary from sign to sign. The small squares to be coloured could be chosen randomly. However, it is much easier to keep track of the number coloured if they are coloured in rows or columns.

On Textbook page 43, the Alltmouth context continues. The scenario of preventing beach erosion at Allt Sands should be discussed. This can be done by referring to the illustration at the top of the page.

The mathematical information contained in the two 'fence' panels should also be discussed.

Question 4 deals with pence being expressed as hundredths of a pound, and again discussion of the examples in the appropriate panels may be required.

R12

# SECOND DECIMAL PLACE: TENTHS AND HUNDREDTHS

UA3ab/4 N2b/4 SSM4b/4
RTN/D3,4 ME/C6
N/4d

The children know that a number such as 2·48, which has two decimal places, means 2 units and 48 hundredths. The work is now concerned with the **place value** of the two digits after the decimal point. The children have to learn that 2·48 is 2 units, 4 tenths and 8 hundredths.

The conservation theme continues.

## Introductory activities

### 1 The 100 square (*introducing tenths and hundredths*)

- Draw a square enclosing 100 small squares on grids on the chalkboard, on squared paper or on an overhead projector transparency.

  Colour 34 hundredths (0·34) as shown, to give 3 strips (i.e. 3 tenths) and 4 small squares.

  Write: 34 hundredths (0·34) = 3 tenths and 4 hundredths.

  Repeat for other examples.

- On another 100 square, colour 6 strips (i.e. 6 tenths) and 5 small squares, to give 65 small coloured squares.

  Write: 6 tenths and 5 hundredths = 0·65 (65 hundredths).

  Repeat for other examples.

- It would be worthwhile duplicating '100 squares' for children to colour in **strips** and small squares. For example,

  'Colour 0·47.

  Write 0·47 = ____ tenths and ____ hundredths

  = ____ hundredths.'

### 2 The metre stick (*tenths and hundredths*)

Use a metre stick.

- Ask the children

  'How many centimetres are in 1 tenth of a metre, 2 tenths . . .?'

  'How many tenths of a metre is 30 cm, 60 cm, 80 cm?'

- Ask a child to point to, say, the 24 cm mark.

  Show that 24 cm = 20 cm and 4 cm
  = **2 tenths and 4 hundredths** (of a metre).

  Establish through discussion that, since 24 cm is also 0·24 m (24 hundredths),
  0·24 = **2 tenths and 4 hundredths.**

  Discuss other examples including, for example,
  0·40 = 4 tenths and 0 hundredths (= 0·4)
  1·34 = 1 unit, 3 tenths and 4 hundredths.

## 3 Comparisons

Use a set of digit cards numbered 0, 1, 2 . . . 9 and a place-value card.

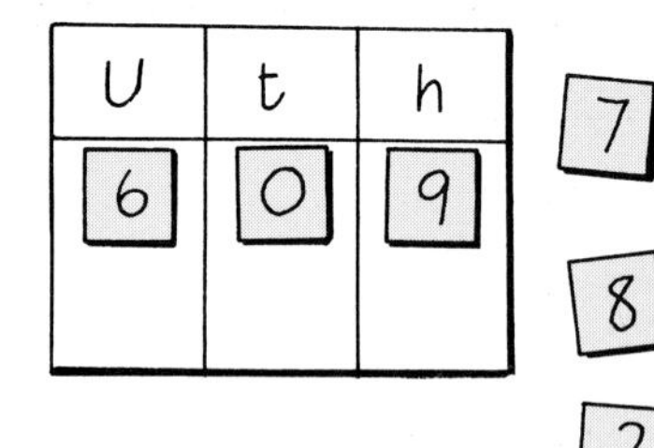


Discuss the column headings and ask the children to
- pick three digit cards and position them on the place-value card
- write the number shown as a decimal (6·09)
- arrange the three digits to make other numbers and record each number
- order the numbers from largest to smallest.

Repeat for other sets of three digit cards.

## 4 Game

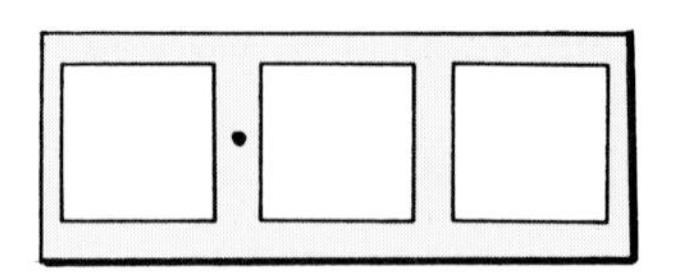

Use a set of digit cards comprising three of each of the digits 0, 1, 2 . . . 9. Each player requires a recording base made from card or drawn on paper.

The digit cards are shuffled and spread out, face down. Each player chooses a card and places it in one of the boxes on their recording base, where it must remain. The winner of the game is the player who has the largest number after all three cards have been placed. The game can also be played to find who has the smallest number.

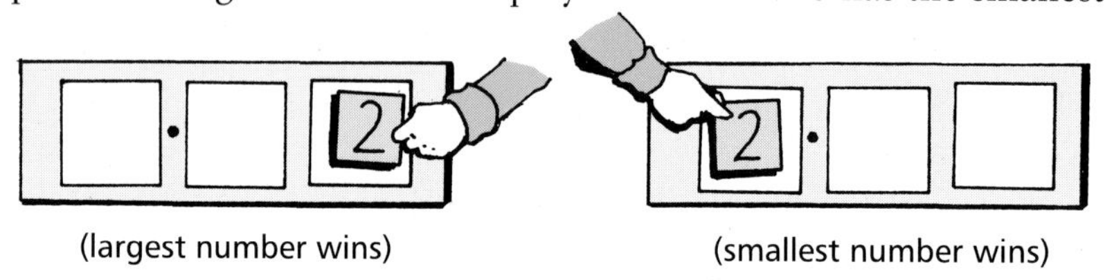


(largest number wins) (smallest number wins)

Repeat for other sets of three digit cards.

## 5 Matching

Make up a set of cards where various decimal numbers are shown in different forms. For example,

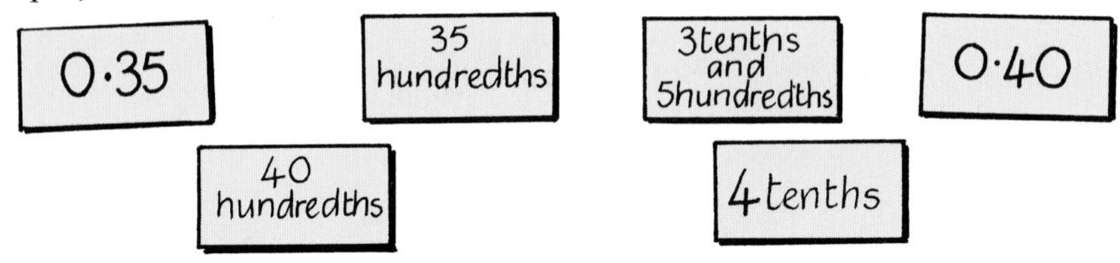


These can be used for Snap and other matching activities.

**At this point, the children could try Textbook page 44.**

## 6 Calculator targets

Each child requires a calculator.

- Discuss examples where the number in the display can be changed to a target number by adding or subtracting **one number**. For example,

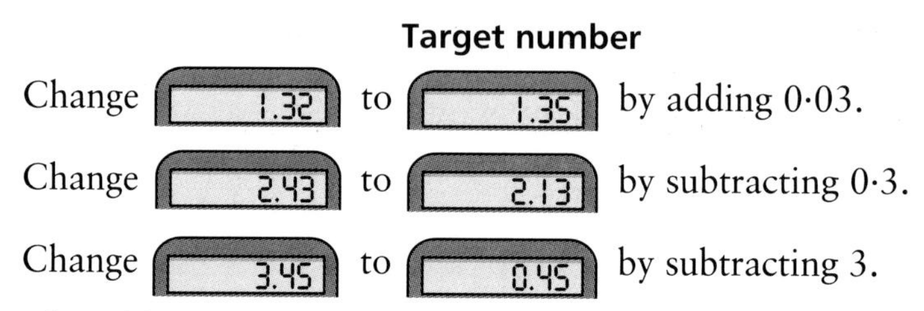


The addition or subtraction should affect only **one** place value in the number.

- Each member of an environmental protection group was asked to try to raise £5 for the funds. Joe raised £3·73.

  Ask the children to make this up to the target of £5:

  **Target £5**

  Starting number → 3.73

  Add 7 hundredths (0·07) → 3.80

  Add 2 tenths (0·2) → 4.00

  Add 1 unit (1) → 5.00

**The target number is reached by adding the hundredths, then the tenths, then the units.**

Do several examples with the children. Each step could be recorded like this:

'Add 0·07, 0·2, 1.'

The target number could be varied as well as the starting amount raised.

UA3a/4 N2b/4 SSM4b/4
RTN/D3,4 ME/C6
N/4d

## Textbook page 44 *Decimals: tenths and hundredths*

If appropriate, point out that the metre stick illustrated is marked differently from those normally used in school. Each tenth, or 10 cm mark, is labelled as a decimal. For example, 40 cm appears as 0·4 m. This should help the children to see that the log, which measures 68 cm (0·68 m) is, in fact, the same as 6 tenths (0·6) and 8 hundredths (0·08).

The difficulties which can arise with place value in relation to decimals should not be underestimated. Some children may think that the **size** of the digit rather than its **position** is what determines the value of the number. For example, '1·02 is smaller than 0·99.' Other children may think that a three-digit number must be greater than a two-digit one. For example, '5·56 is greater than 5·6.'

Questions 6 and 7 attempt to address these difficulties.

R13

UA2d/4 N2b/4
RTN/D4
N/4d

## Workbook page 11 *Decimals: place value using a calculator*

The Alltmouth context continues with Alex measuring post-lengths sticking out of the sand to compare yearly erosion changes.

In question 1, make sure the children understand what Alex is doing. In the table, some children may omit to enter the addition (+) or subtraction (–) sign. They also have to realize that only **one** digit needs to be changed in each example.

In question 2, it may be necessary to emphasize that the hundredths **then** the tenths, **then** the units have to be added. It may be worthwhile discussing the worked example to show how the numbers have to be recorded. Some children may find this aspect difficult.

# SECOND DECIMAL PLACE: ADDITION

Initially the examples involve the mental addition of hundredths only. Thereafter, a written technique is developed, examples being graded as follows:
— addition of tenths and hundredths with exchange of hundredths
— addition of units, tenths and hundredths, then tens, units, tenths and hundredths, with exchange of hundredths and tenths.
Addition of money in decimal form is included.

The work is set in the context of queuing to cross the Alltmouth estuary on the small car ferry, a 'landing craft' type of vessel which can carry a maximum of about 12 vehicles.

## Introductory activities

### 1 Mental addition of hundredths

- Draw a 100-square on a chalkboard grid, squared paper, or an overhead projector transparency to illustrate tenths and hundredths. For example,
colour 7 hundredths of the square red
then a further 8 hundredths green
to complete 1 strip (1 tenth) and 5 small squares (5 hundredths).

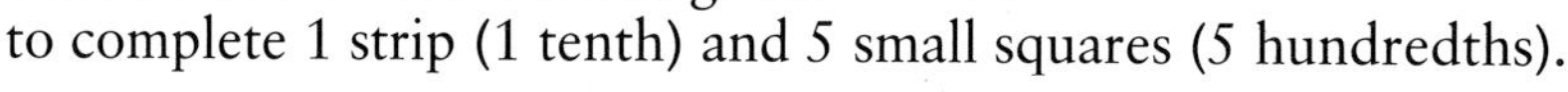

Discuss with the children the addition represented, namely

7 hundredths + 8 hundredths is **15 hundredths**
or **1 tenth and 5 hundredths.**

- Repeat for several similar examples.

### 2 Signs (*a written algorithm*)

- The car ferry traffic signs which appear on Textbook page 45 could be used as the basis for a discussion in which a pencil-and-paper technique for the addition of two-place decimals is introduced.

'This sign has 100 light cells.

29 hundredths of the sign is red.

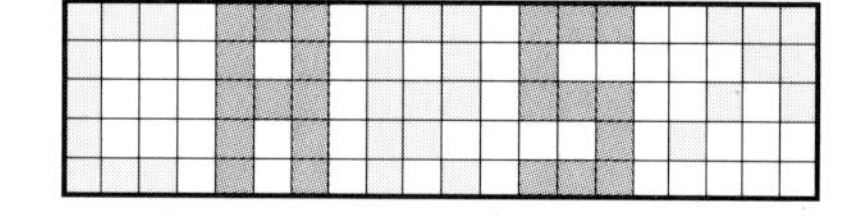

23 hundredths of the sign is green.

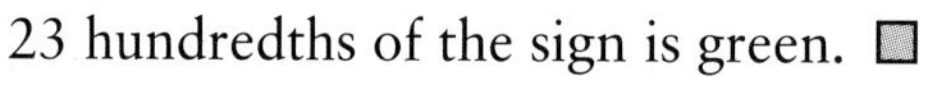

What decimal fraction of the sign is lit?'

Emphasize both the recording and the accompanying language which should develop from the method used for the addition of one-place decimals.

| **Recording** | **Language** |
|---|---|
| 0·29<br>+ 0·23<br>· 2<br>1 | '**Add the hundredths.** 9 and 3 is 12 hundredths.<br>Exchange for 1 tenth and 2 hundredths.' |
| 0·29<br>+ 0·23<br>0·52<br>1 | '**Add the tenths.** 1 and 2 is 3, and another 2 is 5 tenths.<br>**Add the units.** 0 and 0 is 0.'<br>The answer is **0·52**. |

The discussion should cover

— the importance of lining up the decimal points and the digits in columns

— the need to record the zero in the units column

— the use of the words 'hundredths', 'tenths' and 'units'.

- Examples such as

1·83 + 3·48 involving the addition of units, tenths and hundredths
and 10·27 + 29·85 involving the addition of tens, units, tenths and hundredths should also be discussed. Examples like these do not require the use of any new technique other than the addition of hundredths.

## 3 Tickets *(addition of money)*

- 'What is the cost of a single journey on the Alltmouth ferry for a car and its driver?'

**Alltmouth Ferry**

| | Single | Return |
|---|---|---|
| Car | £6·75 | £12·85 |
| Driver | £3·15 | £ 5·40 |
| Dog | 27p | 50p |

In earlier stages of Heinemann Mathematics, the children have added money amounts involving pounds and pence **either** by using a calculator **or** by converting the pounds and pence to pence. For example,

```
 £6·75 →   675p
+£3·15 → + 315p
          ----
           990p → £9·90
```

They should now be introduced to the following shorter technique:

```
 £6·75
+£3·15
------
 £9·90
------
   1
```

- Other examples can be devised from the above table of charges, which includes costs for dogs, priced in pence only. This will ensure that the children experience writing amounts such as 27p and 50p in decimal form, i.e. £0·27 and £0·50.

UA2bd,3a/4 N3g,4a/4
PSE AS/D2
P/4ab N/4k

## Textbook page 45 *Decimals: second decimal place, addition*

The scenario of queuing for the car ferry to cross the Alltmouth estuary should be discussed if it has not already been met in an introductory activity.

In question 4, some children may need to be reminded that amounts given in pence such as 48p should be written in the form £0·48 for written calculation.

**Problem solving**

Although the problem in question 5 can be approached in different ways, most children are likely to employ a guess and check strategy, finding the total cost for different combinations of three sandwiches. Some children might realize that only one of the sandwiches can cost more than £1 – the two cheapest in this category total £2·21, leaving 29p, which is not enough for a third sandwich. If they find the total cost of two of the cheapest sandwiches then add one of the dearest, the solution, 48p + 74p + £1·28, should be found, i.e. Helen buys a lettuce, a cheese and a beef sandwich.

# SECOND DECIMAL PLACE: SUBTRACTION

The written technique suggested involves the decomposition method of subtraction, although in the initial examples no exchange is required.

Thereafter, the examples are graded as follows:

— with exchange: from tenths to hundredths
from units and tenths to tenths and hundredths
involving zero difficulties.

The Alltmouth context continues with vehicles and passengers boarding the ferry.

## Introductory activities

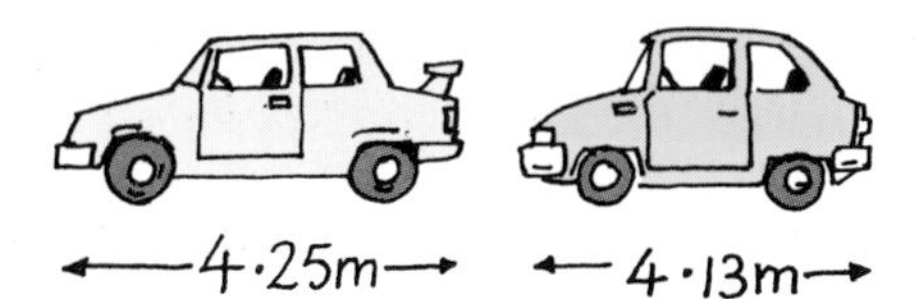

### 1 Car lengths (*no exchange*)

Sketch or show pictures of two cars.

Ask the children to find the difference between the cars' lengths. Link the recording of the technique and the language used.

| Recording | Language |
|---|---|
| 4·25 m<br>− 4·13 m<br>0·12 m | '**Subtract the hundredths.** 5 take away 3 leaves 2.<br>**Subtract the tenths.** 2 take away 1 leaves 1.<br>**Subtract the units.** 4 take away 4 leaves 0.'<br><br>The difference in their lengths is **0·12 m.** |

### 2 Vehicle lengths (*exchanging tenths for hundredths*)

- Sketch or show pictures of a lorry and a car.

  'How much longer is the lorry than the car?'

  Develop a language and recording for the calculation as outlined below.

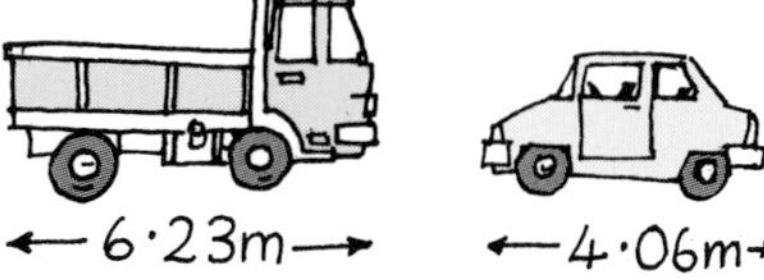


| Recording | Language |
|---|---|
| 6·~~2~~[1]3[1] m<br>− 4·06 m<br>2·17 m | '**Subtract the hundredths.** 3 take away 6 I cannot.<br>Exchange 1 tenth for 10 hundredths.<br>13 take away 6 leaves 7.<br>**Subtract the tenths.** 1 take away 0 leaves 1.<br>**Subtract the units.** 6 take away 4 leaves 2.'<br><br>The lorry is **2·17 m** longer than the car. |

Emphasize the importance of lining up the decimal points and the digits in columns.

- Do several examples like this with the children, including examples of type

$$\begin{array}{r} 5{\cdot}3 \\ -\,2{\cdot}17 \\ \hline \\ \hline \end{array} \qquad \begin{array}{r} 5{\cdot}2 \\ -\,3{\cdot}07 \\ \hline \\ \hline \end{array} \qquad \begin{array}{r} 3{\cdot}14 \\ -\,2{\cdot}7 \\ \hline \\ \hline \end{array}$$

where zeros have to be inserted in the hundredths column.

**At this point the children could try Textbook page 46.**

## 3 Ferry fares *(exchanging units and tenths for tenths and hundredths)*

'An adult pays £7·20 and a child £3·75 to sail on the ferry. How much more does the adult pay?'

Show how this problem can be solved using the following recording of the subtraction.

$$\begin{array}{r} \pounds\,{}^{6}\not{7}\cdot{}^{11}\not{2}\,{}^{1}0 \\ -\,\pounds\,3\cdot7\,5 \\ \hline \pounds\,3\cdot4\,5 \\ \hline \end{array}$$

An adult pays **£3·45** more than a child.

Further examples like this may need to be discussed.

## 4 Change from ferry fares *(subtraction involving zeros)*

'A van driver pays £21·65 to sail on the ferry. What is his change from £25?'

Through discussion, establish that the subtraction required is
Emphasize the need for zeros to be inserted.

$$\begin{array}{r} \pounds25{\cdot}00 \\ -\,\pounds21{\cdot}65 \\ \hline \\ \hline \end{array}$$

Link the recording of the technique and the language used.

| Recording | Language |
|---|---|
| $\begin{array}{r} \pounds2\,{}^{4}\not{5}\cdot{}^{1}0\,0 \\ -\,\pounds2\,1\cdot6\,5 \\ \hline \\ \hline \end{array}$ | '**Subtract the hundredths.** 0 take away 5 I cannot. There are no tenths to exchange. Exchange 1 unit for 10 tenths, giving 4 units and 10 tenths.' |
| $\begin{array}{r} \pounds2\,{}^{4}\not{5}\cdot{}^{1\,9}\not{0}\,{}^{1}0 \\ -\,\pounds2\,1\cdot6\,5 \\ \hline \pounds\;\;3\cdot3\,5 \\ \hline \end{array}$ | 'Exchange 1 tenth for 10 hundredths, giving 9 tenths and 10 hundredths. **Subtract the hundredths.** 10 take away 5 leaves 5,' and so on. The van driver's change is **£3·35**. |

Several examples like this should be discussed. Some children may be able to combine the two steps by directly exchanging 5 units for 49 tenths and 10 hundredths.

The method, language and recording shown in the above are guidelines. Schools should use methods consistent with their own policy.

UA2bd,3a/4 N3g,4a/4
PSE AS/D2
P/4ab N/4k

## Textbook pages 46 and 47

*Decimals: second decimal place, subtraction*

On Textbook page 46, in questions 1 and 2, no exchange of units or tenths is required. Some children may need help in putting the subtractions in a vertical setting.

Questions 3 to 6 require exchange from the tenths to the hundredths, as shown in the worked example. Some children may need to be reminded to insert a zero in the empty hundredths column in the examples in question 6.

On Textbook page 47, exchanging units for tenths, and tenths for hundredths, is required in questions 1 to 3.

Questions 4 and 5 have the added difficulty of zeros in both the tenths and hundredths columns.

In question 7, the children have to add together the fares for an adult and 3 children, giving £13·95, and then subtract this amount from £33·45, giving an answer of £19·50. From the ferry fares table, they should realize that the vehicle used was a minibus.

**Problem solving**

In question 8, some children may forget that the **car** has to be paid for on the ferry. £23·95 – £12·85 leaves £11·10 for the fares for the people travelling in the car. Using a simple guess and check strategy, the children should find that 1 adult and 2 children are the passengers in the car.

R14

# SECOND DECIMAL PLACE: CONSOLIDATION

A variety of word problems which involve addition, subtraction, multiplication and division of one-place decimals, and addition and subtraction of two-place decimals, is provided. Most of the data required for solving the problems has to be interpreted from the tables and illustrations provided. The context is Alltbay Marina, a leisure centre on the estuary, with sailing, fishing and camping facilities.

UA2abcd,3a/4 N3eg,4a/4
PSE AS/D2 MD/D3
P/4ab N/4kl

## Textbook pages 48 and 49

*Decimals: consolidation, 1 and 2 decimal places*

On Textbook page 48, in question 2, 'metre length' refers to 4·7 m, 7·6 m, etc. The calculations, such as 9 × 4·7, produce a number of pounds with 1 decimal place, such as £42·3. Discussion may be necessary to remind some children that this should be written as £42·30.

**Problem solving**

In question 5(b), the cheapest way to buy six hours of lessons is £27·64 (£17·64 + £10). The children are likely to find other ways which are more expensive. These should be discussed.

**Problem solving**

On Textbook page 49, in question 3, the children should realize, after reading all three clues, that the length of the trout is a three-digit number with 2 decimal places. Interpreting the first two clues will establish that 4 is the hundredths digit. The third clue then determines the answer to be 0·34 metres.

**Problem solving**

In question 5(c), a guess and check strategy, in which different groups of 3 items are tried, should result in the answer of barbecue, heater and TV set.

# SECOND DECIMAL PLACE: CALCULATOR

N3h/4
AS/D3 MD/D4 RN/D1
N/4dkln M/4a

In earlier stages of Heinemann Mathematics, with the exception of applications in money, the use of a calculator was for whole number and first decimal place work. Children are now required to use a calculator for second decimal place work. This involves using the operations of addition, subtraction, multiplication and division in applications

— in number: for example, averages, rounding a calculator display to the nearest whole number

— in measure: for example, using the metre/centimetre relationship

as well as in money.

The setting for the work is Point of Allt, a rocky island at the mouth of the estuary and the site of a lighthouse.

## Introductory activities

### 1 Revision

As no new mathematical concepts are introduced, many children should be able to attempt the work with minimal guidance. For some children it may be appropriate to revise some or all of the following:

— entering money amounts such as 47p as 0.47

and 7p as 0.07

and interpreting 12.6 as £12·60.

— rounding a display such as 60.714285 to the nearest whole number, i.e. 61.

— the meaning of the method of calculating an 'average' or 'mean'

— the metre/centimetre relationship and associated notation such as 2·46 m = 2 m and 46 cm.

### 2 Using a calculator's memory

An opportunity exists, particularly in relation to the work on Textbook page 50, to introduce the use of the memory facility, at least to some children.

- Tell the children there is a small tearoom for visitors to the lighthouse. Ask them to find the total cost of this order:

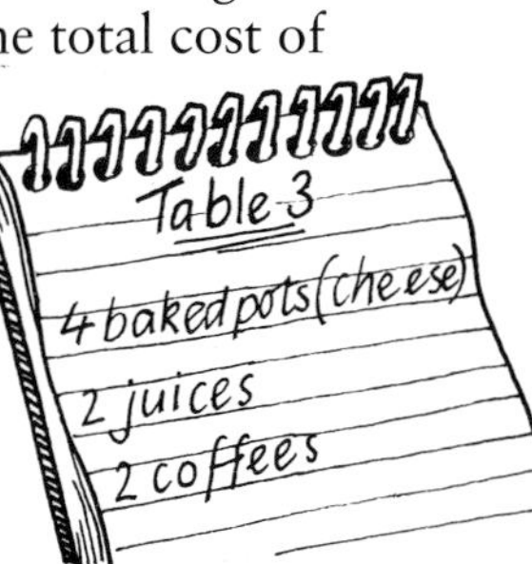


LIGHTHOUSE TEAROOM.
Baked Potatoes
Plain £1·20
Cheese £1·84
Tuna £2·32
Drinks
Juice 65p
Tea 60p
Coffee 75p

They are likely

**either** to add, entering each of the eight items individually

**or** to calculate, by multiplying, the total cost for each **type** of item, record these totals **on paper**, then add to find the grand total.

**Number**
**Decimals**
**Alltmouth**

- Discuss how the memory of the calculator can be used instead. The steps are
  — clear the display and memory
  — calculate the total for the baked potatoes, i.e. 4 × £1·84
  — use [M+] to add this to the memory
  — calculate the total for the juices, i.e. 2 × 65p (2 × £0·65)
  — use [M+] to add this to the memory
  — calculate the total for the coffees, i.e. 2 × 75p (2 × £0·75)
  — use [M+] to add this to the memory
  — use [MR] to recall the grand total from the memory.
- Other ways of using the memory for this type of calculation could be discussed. For example, the last two steps can be replaced by [+] [MR] [=].
- The method of clearing the memory varies from one make of calculator to another. It may be necessary to press [MC] or [AC] or [C] [CM] or [M$^{R}_{C}$] [M$^{R}_{C}$] or . . .
- The children must realize that the memory is empty only when no small 'M' appears in the display.

There is something in the memory

2.6

There is nothing in the memory

Switching off does not always clear the memory for some makes of calculator.

- Further examples, in which the total cost of several items at one price and a different number of items at another price has to be found, should be discussed until the children are confident in using the memory facility.

UA2abcd/4 N3h/4
PSE AS/D3 MD/D4 RN/D1
P/4ab N/4dkln M/4a

## Textbook pages 50 and 51

*Decimals: second decimal place, calculator*

The context should be discussed. Point of Allt is a rocky island at the end of a sand spit at the extreme mouth of the estuary. The island is reached by crossing a causeway when the tide is low enough. Visitors to the island must pay to park their cars and to view the lighthouse, which stands on the edge of the cliffs at the end of the island.

On Textbook page 50, in question 1, some children may need help to interpret the table. They may not know what a 'causeway' is.

In question 3, although the depth of the water is less than 1 m 50 cm a little before 9 am and a little after 3 pm, the answer expected is 'between 9 am and 3 pm'.

In question 5, after multiplying, intermediate totals should be written on paper to facilitate the final addition to obtain a grand total – unless the children use the memory facility as described in the introductory activities.

Problem solving

To solve the problem in question 6, the children should first deduct the minibus parking fee of £1·36, leaving a total of £11·64 as the cost of entrance. Thereafter they can systematically check the cost for 1 adult and 12 children, 2 adults and 11 children, 3 adults and 10 children, and so on, until the solution, 5 adults and 8 children, is found. Some children are likely to use a simple guess and check strategy.

Problem solving

On Textbook page 51, in question 4, ensure that the children can interpret 'a 4-digit **starting number** with 2 decimal places'. They should notice that whatever starting number is chosen the answer, the date the lighthouse was built, is always the same, i.e. 1902.

UA3ab/4
N2c/4
N3g/4→5
RTN/D4
FPR/D1
N/4f
N/5ch

# Percentages

## Overview

This section introduces

- the concept of a percentage and the notation %
- the idea that 100% means one whole
- the relationship between some common fractions and percentages
- calculation of 50%, 25% and 10% of a quantity.

| | Teacher's Notes | Textbook | Workbook | Reinforcement Sheets |
|---|---|---|---|---|
| **Merlin Castle: a context for fractions and percentages** | 75 | | | |
| **Percentages: concept** | 114 | 52 | 12 | 15 |
| **Percentages: 100% is 1 whole** | 115 | | 13 | |
| **Linking percentages and fractions** | 117 | | 14 | |
| **Calculating 50%, 25% and 10% of a quantity** | 118 | 53 | | 16 |
| *Other activity* | *119* | *54* | | |

An extension activity is related to the above section of work as follows:

| | Teacher's Notes | Extension Textbook |
|---|---|---|
| **Percentages: fractions as percentages** | 286 | E13 |

Teaching notes for the Extension Textbook are in a separate section at the end of the Teacher's Notes.

# Resources

## Useful materials

- $\frac{1}{2}$ cm or 1 cm squared paper
- other materials suggested within the introductory activities

## Assessment and Resources Pack

### Assessment

*Number and Money Check-up 12*
Textbook pages 52, 53
Workbook pages 12–14
(100% as one whole, % of a quantity)

*Round-up 3*
Question 2(a), (b)

### Resources

*Problem Solving Activities*
8 Get the point (fractions and percentages)

*Resource Cards*
15–17 Castle attack
(100% as one whole, % of a quantity)

# Teaching notes

This is the first time the children meet percentages. The fractional notation for hundredths is introduced in the Fractions section and used in the Decimals section of Heinemann Mathematics 6. This notation is now linked to the notation for percentages.

The context is a modern-day version of Merlin Castle, first introduced in its ancient form in the Fractions section.

For a detailed introduction to this context, see page 75 of these notes.

## PERCENTAGES: CONCEPT

UA3ab/4 N2c/4
RTN/D4
N/4f

The context introduces the modern version of Merlin Castle with the focus on floor and window designs.

## Introductory activities

### 1 Window design

Draw a large 100 square 'window' on a chalkboard or on an overhead projector transparency.

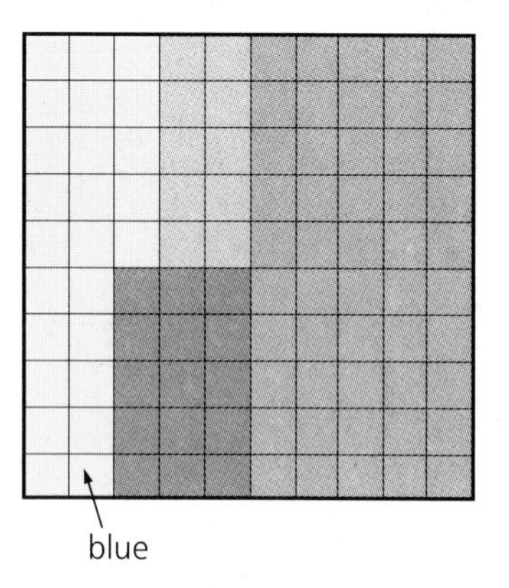


Using an example like this, establish that

— the large square has 100 small squares, each equal in size

— 25 out of the 100 squares are coloured blue

— $\frac{25}{100}$ of the window is blue.

Introduce the language

'25 per cent is blue'

and write this as

'25% is blue.'

Repeat for other colours in the window.

Establish that the **whole** window or **100%** of the window is coloured.

### 2 Colouring percentages

Issue each of the children with a 100 square. Ask them to

— make a window design using different colours

— record the percentage of each colour used: for example,
green – 20%
yellow – 40%
blue – 30%
red – 10%

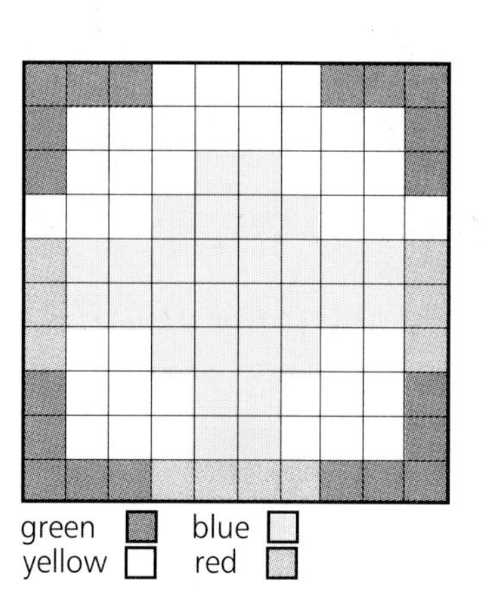


A wall display could be made using the children's designs.

### 3 Percentages poster

Ask the children to make a collection of advertisements in which percentages appear, using newspapers, magazines or other sources. For example,

The advertisements can be used to reinforce the language and notation of percentages. The children could be asked to read out some of the advertisements and to give the meaning of the percentages used. For example,

Sale
30%
off

— 30% is read as 'thirty per cent'

— 30% means 30 out of 100 or $\frac{30}{100}$ (thirty hundredths)

— 30% is taken off the normal price in the sale.

UA3ab/4 N2c/4
RTN/D4
N/4f

## Textbook page 52 *Percentages: concept*
## Workbook page 12

Introduce the present-day Merlin Castle context and ensure that the children are familiar with percentage notation before they attempt the work on these pages.

On Textbook page 52, in question 1, it is expected that the children will find the percentage not coloured by counting the number of squares not coloured. Some children may find the answer by subtracting the percentage coloured from 100%.

In question 2, the children could be encouraged to check their answers by totalling the percentages for each colour. They should find that the total for each window is 100%.

On Workbook page 12, in question 3, the children need not produce a symmetrical design and may choose to leave some squares **not** coloured.

R15

# PERCENTAGES: 100% IS 1 WHOLE

UA3b/4 N2c/4
RTN/D4
N/4f

The work now emphasizes 100% as meaning one whole.

The context continues with fundraising for the restoration of Merlin Castle.

## Introductory activities

### 1 Percentage cards

■ Make a number of cards where a given percentage of each is coloured.

For example, these cards could be made by sticking coloured gummed paper on white card.

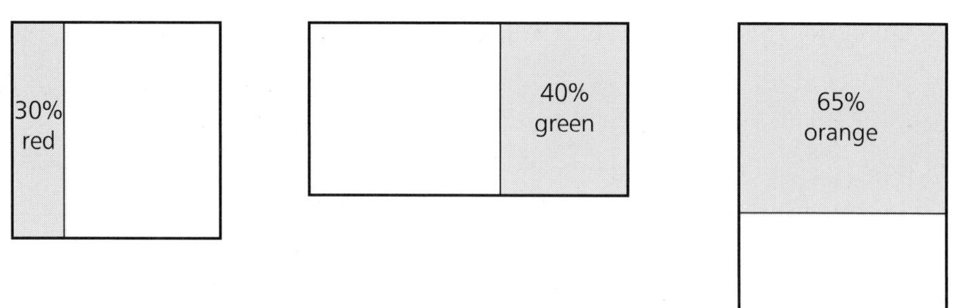


Ask the children how they can find the percentage of each card which is **not** coloured. For example,

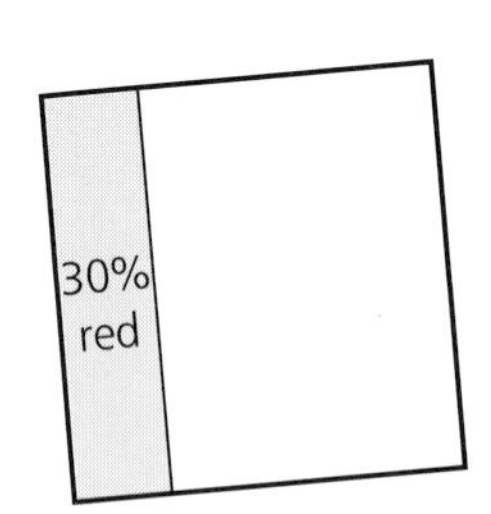


The whole card is 100%
30% is coloured red.
100% – 30% = 70%.
So 70% is **NOT** coloured.

Through discussion, establish that the percentage coloured and the percentage **not** coloured gives a total of 100%.

Oral examples could be given to consolidate this concept. For example,

'25% of a card is coloured. What percentage is not coloured?'

- Discuss this type of work in other familiar situations. For example,

  '30% of a class take a bus to school. What percentage do not take a bus?'

  '55% of a jumper is wool. What percentage is not wool?'

  '47% of children have school dinners. What percentage do not have school dinners?'

## 2 Childhelp

- Draw a scale like this on the chalkboard or on an overhead projector transparency.

  Tell the children that the charity Childhelp has been collecting bottle tops and wants to sell them to pay for children to visit the castle. The scale shows the percentage of bottle tops which has already been collected.

  Ask the children to

  — give the percentage already collected

  — give the percentage still to be collected.

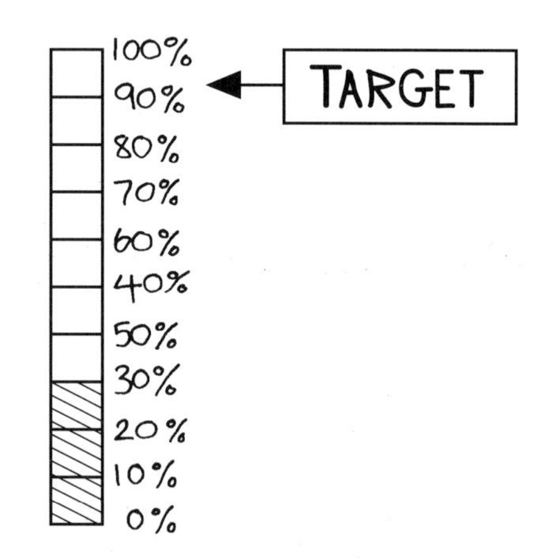


- Change the percentage of bottle tops collected by colouring more of the scale and asking these questions again.

  Establish that

  — the percentage of bottle tops collected plus the percentage of bottle tops still to be collected give a total of 100%

  — the percentage still to be collected can be found by subtracting the percentage collected from 100%.

## Workbook page 13 *Percentages: 100% is 1 whole*

Introduce the idea of a 'Castle Fund' which raises money to restore the castle.

Discuss the chart in the worked example to ensure the children understand that the amount raised plus the amount still to be raised gives a total of 100%.

In question 1(a), in the final chart, the children should estimate where to colour 83% on the scale, since the scale is marked in intervals of 5%.

In question 1(b), some children may find the answers by reading the amount not coloured from the charts in question 1(a), but it is intended that they subtract the 'percentage raised' from 100% to find the 'percentage still to be raised'.

In question 2, some children may have to be reminded that the total for each item is 100%.



UA3b/4 N2c/4
RTN/D4
N/4f

# LINKING PERCENTAGES AND FRACTIONS

The link is now made between percentages and some simple common fractions.

The Merlin Castle context continues with quilts in the Royal Bedroom.

UA3b/4 N2c/4
RTN/D4
N/4f N/5c

## Introductory activities

### 1 The Royal Quilt

- Draw a 100 square 'quilt' on the chalkboard or on an overhead projector transparency, and colour $\frac{1}{4}$ of it as shown.

  Establish that $\frac{1}{4} = \frac{25}{100} = 25\%$ by

  — observing that $\frac{1}{4}$ of the quilt is coloured and that 25 out of 100 squares are coloured. So

  $$\frac{1}{4} = \frac{25}{100} = 25\%$$

  — changing the common fraction into an equivalent fraction with denominator 100.

  $$\frac{1 \times 25}{4 \times 25} = \frac{25}{100} = 25\%$$

- A similar approach should be taken for changing percentages into common fractions. For example,

  $$25\% = \frac{25}{100} = \frac{1}{4}$$

UA3b/4 N2c/4
RTN/D4
N/4f N/5c

## Workbook page 14 *Percentages: link with fractions*

In each diagram, the children should colour blocks of squares as indicated by the bold lines in the diagrams. For example, in question 4, $\frac{3}{4}$ might be coloured like this.

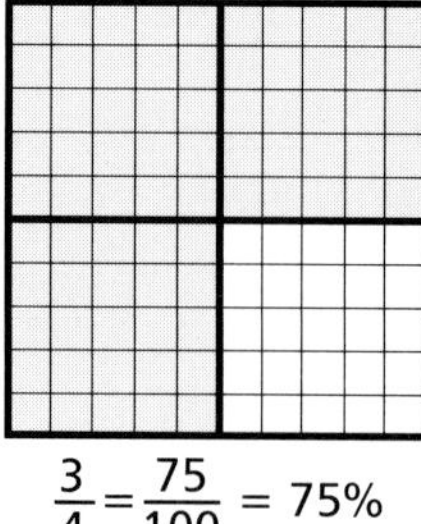


Colouring in other ways will make the corresponding common fraction less obvious.
In questions 1 to 4, some children may use equivalent fractions to find corresponding percentages. For example,

$\frac{1}{2} = \frac{50}{100} = 50\%$

Others may count the number of small coloured squares to obtain '50 out of 100' and record as

$\frac{50}{100} = 50\%$

In questions 5 and 6, some children may simplify to find the equivalent fraction. For example,

$25\% = \frac{25}{100} = \frac{5}{20} = \frac{1}{4}$

It is more likely, however, that the children will find the fraction by counting the number of blocks of a particular colour: for example, 1 red block out of 4.

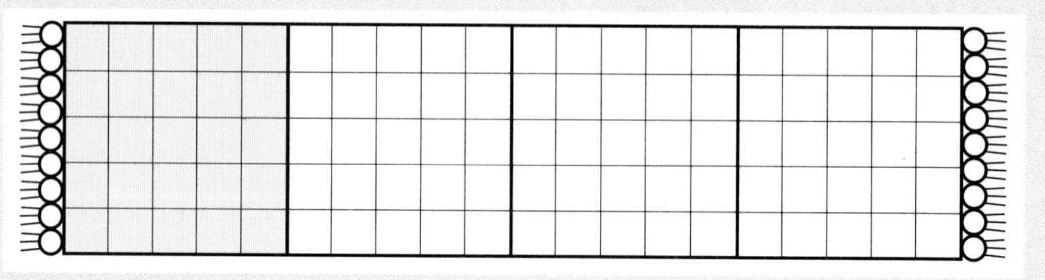

# CALCULATING 50%, 25% AND 10% OF A QUANTITY

N2c/4,3g/4→5
RTN/D4 FPR/D1
N/4f N/5ch

The work of this section finishes by making use of the fractional equivalences $50\% = \frac{1}{2}$, $25\% = \frac{1}{4}$, $10\% = \frac{1}{10}$ to find percentages of a quantity.

The Merlin Castle context continues with busloads of tourists visiting the castle to celebrate its 500th anniversary.

## Introductory activities

### 1 Castle visitors

Tell the children that there is a gift shop in the castle and ask what types of gift could be on sale. Responses may include pens, pencils, rulers, rubbers, badges and posters.

Explain that a group of 40 visitors bought some gifts from the shop.

50% bought a Merlin Castle pencil.

25% bought a mug.

10% bought a painted shield.

Discuss how to calculate the number of visitors buying each item. For example, 50% bought a pencil:

50% of 40

$= \frac{1}{2}$ of 40

= 20

**20 visitors** bought a pencil.

N2c/4,3g/4→5
RTN/D4 FPR/D1
N/4f N/5ch

## Textbook page 53 *Percentages: of a quantity*

Before attempting the work of this page, the children should be familiar with finding '$\frac{1}{2}$ of', '$\frac{1}{4}$ of', and '$\frac{1}{10}$ of' a quantity.

Although questions 1 to 6 can be done mentally, the children should be encouraged to record their answers in a similar way to the worked example at the top of the page.

In question 5, the children should realize that the number of adults is found by subtracting the number of children from the total number of passengers on each bus.

In question 7, a pencil-and-paper method is likely to be needed. For example,

Sunday 25% of 260

$= \frac{1}{4}$ of 260 $\quad 4\overline{)26^{2}0}$ = 6 5

= 65

**65 visitors** were given a free badge.

R16

UA2a/4,4cd/5
PSE RTN/D3, 4
P/4ab N/5bc

## Textbook page 54 *Other activity: fractions*

- The children should attempt this activity only after the work on common fractions, decimals and percentages in Heinemann Mathematics 6 has been completed.
- In question 1, the children could make small paper 'yellow cards', or use counters with a fraction marked on each, so that they can take each 'card' along the track and make a decision at each box. The answers to question 1 are 75%, 0·65 and 0·2. Part (c) is likely to be the most challenging, as the numbers on the cards are expressed in common fraction, decimal or percentage form.
- In question 2, only two cards remain, 30% and 15%, after the cards have passed through the question boxes. To eliminate one of them, some children might answer using a decimal clue – for example, 'greater than 0·25' – as it is the decimal form which has not already been used. Of course, other answers are also possible.

UA3ab,4b/4
N3ac/4
PSE
PS/C1
PS/D1
PS/E1
FE/C1
FE/D1
FE/E1
P/4ac
A/3d
A/4bd
A/5abc

# Pattern

## Overview

This section

- revises function machines
- introduces the strategy and a layout for the tabulation of number patterns
- includes recognition and continuation of simple shape and number patterns
- introduces simple word formulae in context
- introduces square and triangular numbers.

| | Teacher's Notes | Textbook | Workbook | Reinforcement Sheets |
|---|---|---|---|---|
| **Function machines, tabulation** | 121 | 55 | 15 | |
| **Word formulae** | 124 | 56 | 16 | 17 |
| **Square and triangular numbers** | 127 | 57, 58 | | |
| *Other activity* | *128* | *59* | | |

An extension activity is related to the above section of work as follows:

| | Teacher's Notes | Extension Textbook |
|---|---|---|
| **Pattern: problem solving** | 286 | E14, 15 |

Teaching notes for the Extension Textbook are in a separate section at the end of the Teacher's Notes.

## Resources

### Useful materials

- counters
- other materials suggested within the introductory activities

### Assessment and Resources Pack

**Assessment**

*Round-up 3*
Question 7(a), (b)

**Resources**

*Problem Solving Activities*
6 Number cells
9 Let's face it
17 It's a grey area

# Teaching notes

**Number**
**Pattern**

This section consolidates earlier work on function machines using the four operations with whole numbers, and includes recognition and extension of simple shape and number patterns. Tabulation, simple word formulae to describe number patterns, and square and triangular numbers are introduced.

## FUNCTION MACHINES, TABULATION

N3a/4
PS/C1 FE/C1 FE/D1 FE/E1
A/3d A/4bd

Initially the layout of the function machines is developed to lead towards tabulation, and to help children recognize the relationship between number pairs within a table. Some application of inverse operations (working backwards) is also included. The work is extended to finding the rule for a machine, and then to finding the rule which connects number pairs in a table.

## Introductory activities

### 1 The function machine

- Prepare a set of **rule cards** like these:

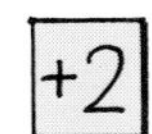
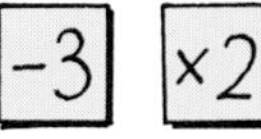

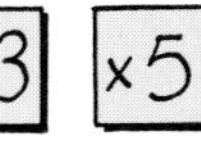

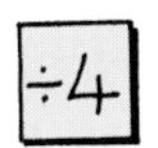


Draw a function machine on the chalkboard:

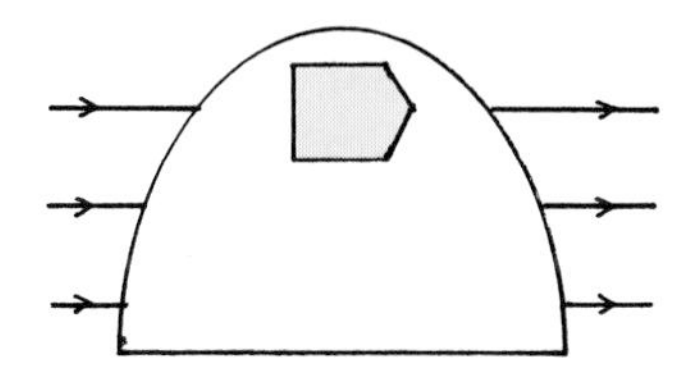

Write numbers and attach a rule card to the machine.

Ask the children to supply the missing input/output numbers.

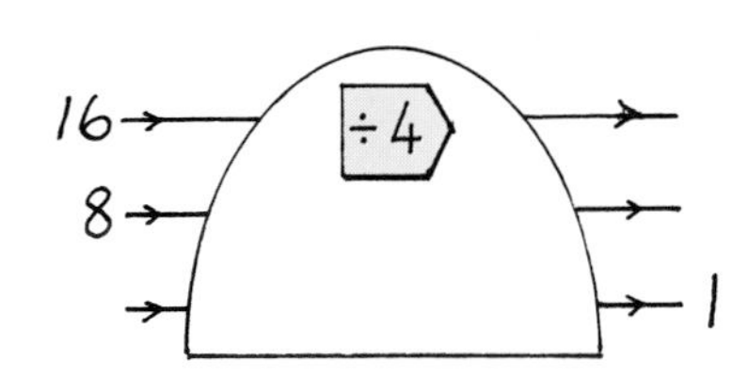


For the last row, to help the children to 'work backwards', discuss the first two answers by asking

This should help them to see that 'divide by 4' is the inverse of 'multiply by 4'.

Repeat for other examples.

■ Discuss examples where both numbers are given. Lead the children to find the rule for the machine. For example, ask for a rule which connects this number pair:

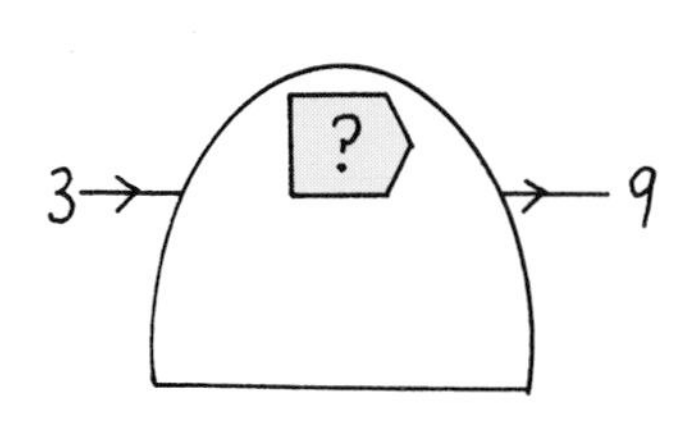


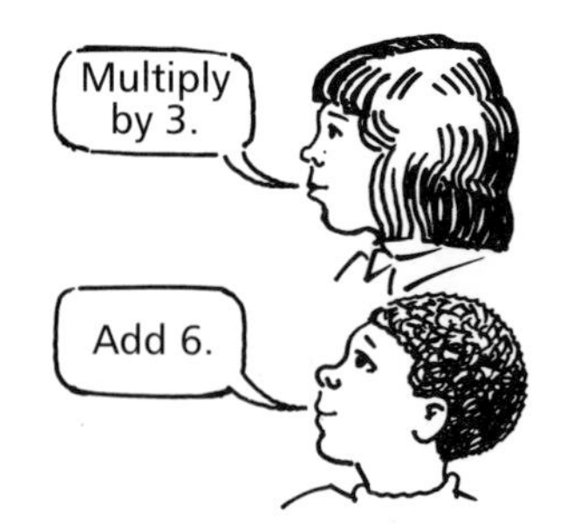


Insert another number pair and test both rules.

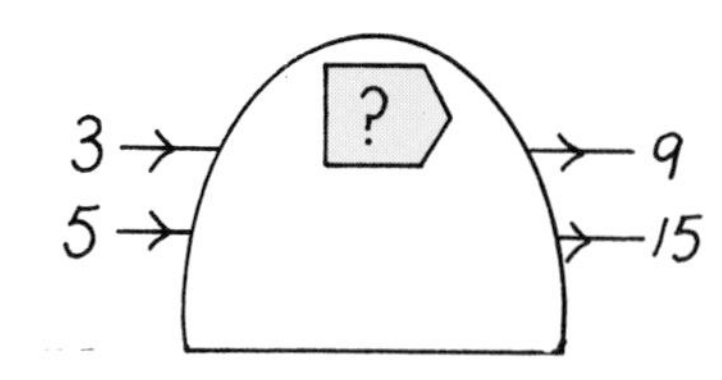

Emphasize that, to be reasonably sure of a rule, they should test that it works for at least 3 **pairs** of numbers. Insert a third pair:

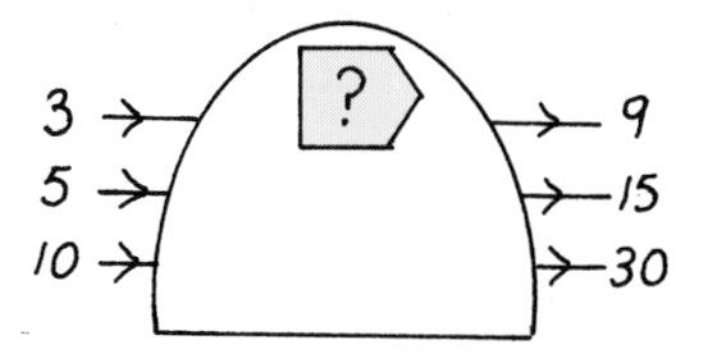


The rule x 3 can now be attached to the function machine.

Ask the children to find the rule for each of these machines.

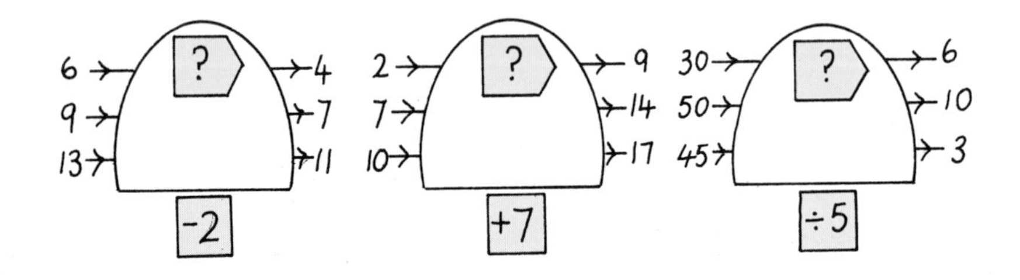


## 2 Two-rule machines

Discuss a function machine with two rules.

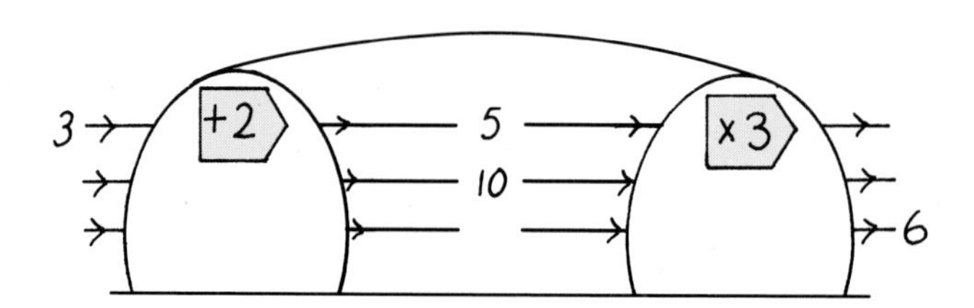

In the last line emphasize that

— 'multiply by 3' is reversed by 'divide by 3', giving → 2 → 6

— 'add 2' is reversed by 'subtract 2', giving 0 → 2 → 6

Repeat for other examples.

## 3 Tables

■ Introduce the idea of not having a function machine but just pairs of numbers instead. For example,

56 ⟶ 7
32 ⟶ 4
24 ⟶ 3

Show the children how to set this out in tabular form.

| in | out |
|---|---|
| 56 | 7 |
| 32 | 4 |
| 24 | 3 |

Discuss the format:

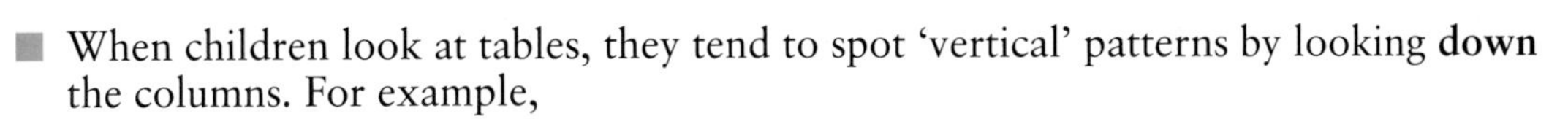


■ When children look at tables, they tend to spot 'vertical' patterns by looking **down** the columns. For example,

| | in | out | |
|---|---|---|---|
| 'add 1' | 3 | 18 | 'add 6' |
| 'add 1' | 4 | 24 | 'add 6' |
| | 5 | 30 | |

Emphasize the need to read **across** the table when trying to find a rule.

| in | out |
|---|---|
| 3 | 18 |
| 4 | 24 |
| 5 | 30 |

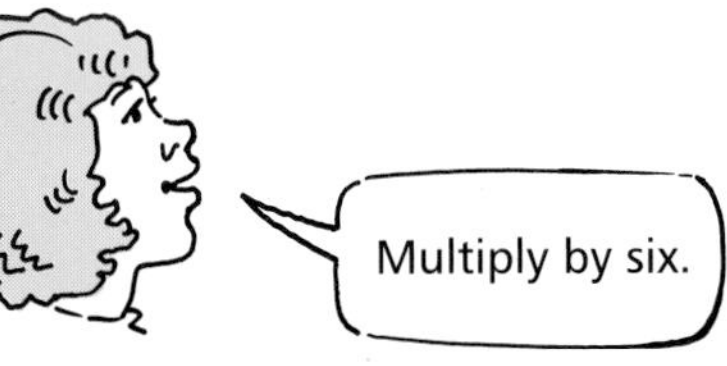


The tendency to look at a 'vertical' number pattern can be reduced by avoiding examples with consecutive 'in' numbers. For example,

| in | out |
|---|---|
| 5 | 30 |
| 3 | 18 |
| 4 | 24 |

rather than

| in | out |
|---|---|
| 3 | 18 |
| 4 | 24 |
| 5 | 30 |

Number
Pattern

N3a/4
PS/C1 FE/C1 FE/D1 FE/E1
A/3d A/4bd

**Textbook page 55** ***Pattern: function machines***
**Workbook page 15**

The children are immediately directed from Textbook page 55 to Workbook page 15.

Workbook page 15 includes examples requiring the application of inverse operations. Some function machines with two rules are included at the bottom of the page.

On Textbook page 55, the children are not expected to copy the diagrams or tables.

In question 3, where the use of the arrows is discontinued in parts (e) to (h), it may be necessary to emphasize that the children continue to read **across** the table and not vertically.

# WORD FORMULAE

UA3b,4b/4 N3a/4
PS/C1 FE/D1
A/4bd

In previous work, the children have recognized and continued shape and number patterns. They are now introduced to writing a word formula to describe a simple shape or number pattern.

## Introductory activities

### 1 Matchstick fun

- Show the children patterns made with spent matchsticks, straws or rods. Ask them to continue the pattern and describe the formulae in their own words.

For example, make this pattern of 'peaks'.

1 peak

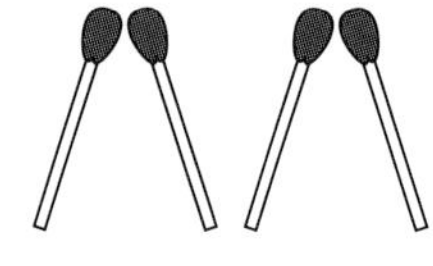

2 peaks

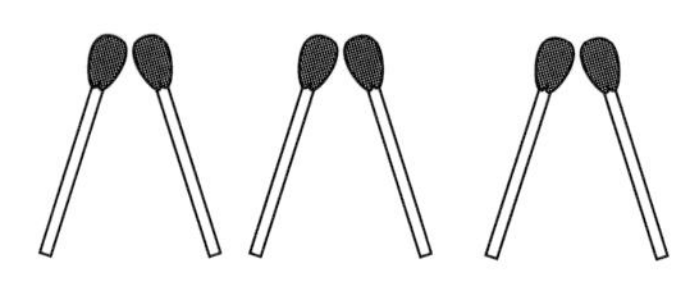

3 peaks

Draw a table like this on the chalkboard.

| Number of peaks | Number of matches |
|---|---|
| 1 | 2 |
| 2 | 4 |
| 3 | 6 |
| 4 | |
| 5 | |

Ask the children how many matches are needed for 4 peaks and 5 peaks, then ask them to describe any patterns they see. They are likely to refer to the 'peak' pattern and the '2 more matches' pattern each time – shown in the vertical columns. Accept these answers, but encourage them to look across the columns for a **rule** connecting the number of peaks and the number of matches.

| Number of peaks | Number of matches |
|---|---|
| 1 | 2 |
| 2 | 4 |
| 3 | 6 |
| 4 | 8 |
| 5 | 10 |

A rule such as 'times 2' or 'multiply by 2' should be suggested. Show the children how to record the 'rule' in the same format as on Workbook page 16:

'The number of matches is 2 times the number of peaks.'

Ask the children to use the rule to find the number of matches needed for 10 peaks. Allow them to check by continuing the pattern to 10 peaks.

- Repeat for different patterns. For example,

1 kite

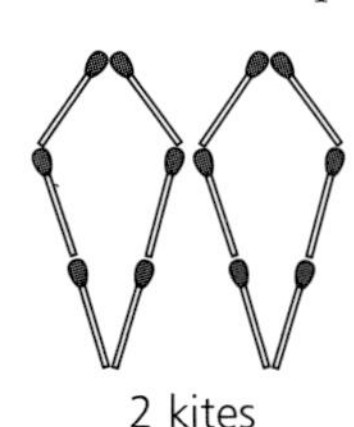
2 kites

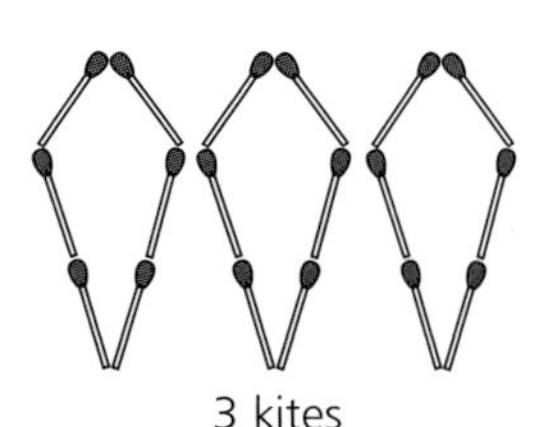
3 kites

'The number of matches is six times the number of kites.'

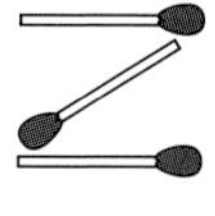
1 zed

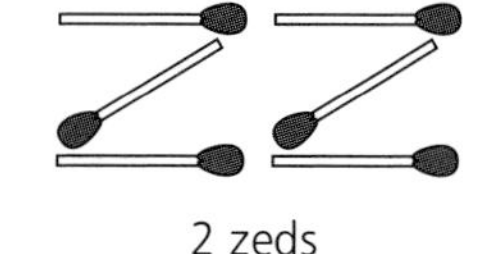
2 zeds

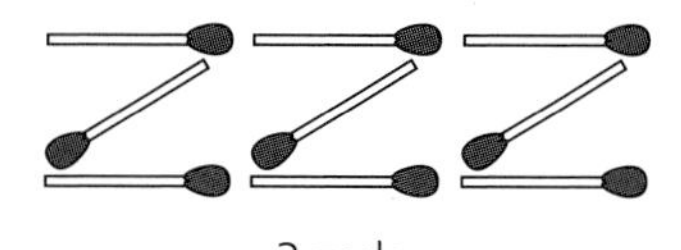
3 zeds

'The number of matches is 3 times the number of zeds.'

**At this point the children could try Workbook page 16.**

## 2 Arranging desks

- Tell the children that they are to investigate different ways of arranging desks and chairs.

On the chalkboard, draw these arrangements, avoiding the numbers of desks being consecutive.

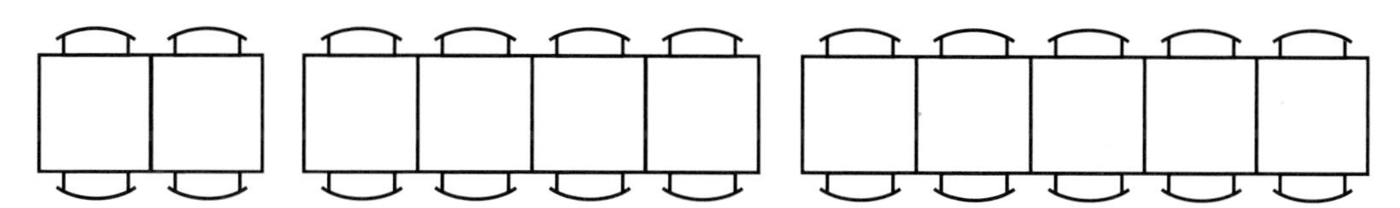

Draw a table like this:

| Number of desks | Number of chairs |
|---|---|
| 2 | |
| 4 | |
| 5 | |
| 10 | |

Discuss with the children how to complete the table and describe the rule.

'The number of seats is 2 times the number of desks.'

- Repeat for other examples, such as:

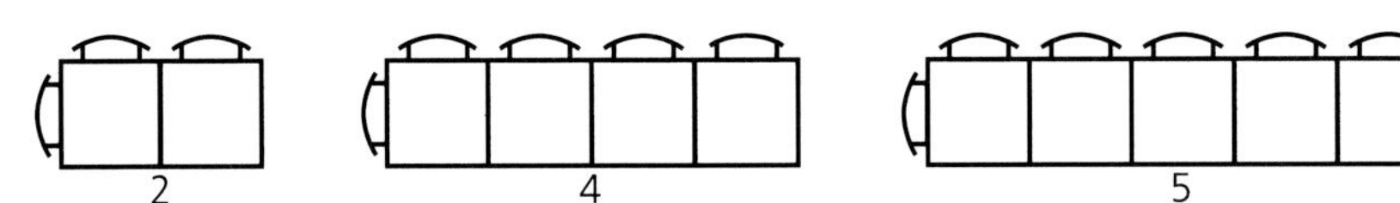


'The number of seats is the number of desks, add 1.'

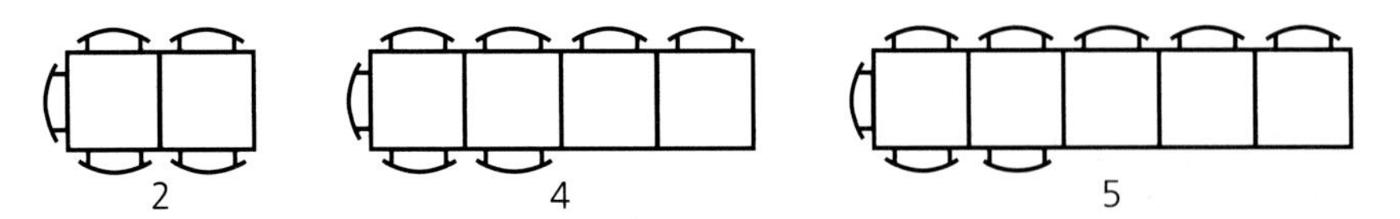


'The number of seats is the number of desks, add 3.'

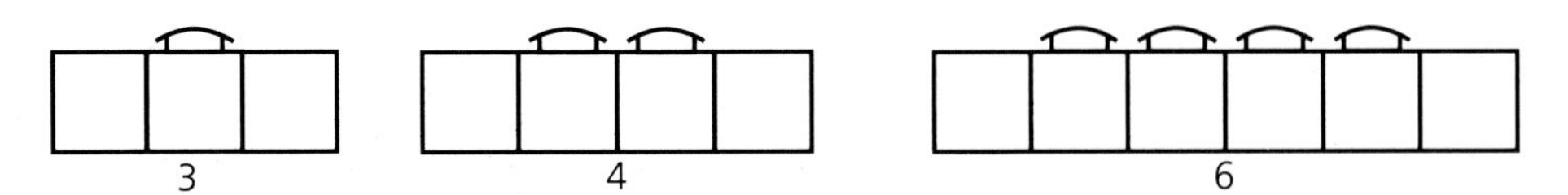


'The number of seats is the number of desks, subtract 2.'

## Workbook page 16 *Pattern: word formulae*
## Textbook page 56

On Workbook page 16, in questions 1(c) and 2(c), encourage the children to check their answers by drawing the patterns.

On Textbook page 56, it may be necessary to explain to the children that they have to find formulae to describe the different patterns in the dining hall.

The word 'sections' is used to avoid confusion of 'tables' in the dining hall with 'tables' of results. The children should record the number pattern in words in the usual way. For example, for question 1(b), 'the number of seats is six times the number of sections.'

# SQUARE AND TRIANGULAR NUMBERS

The children's experience of number patterns is extended to square and triangular numbers.

UA3ab,4b/4 N3ac/4
PSE PS/D1 PS/E1
P/4ac A/5abc

## Textbook pages 57 and 58

***Pattern: square and triangular numbers***

On Textbook page 57, in question 2(a), the children should list the numbers as a sequence:

1, 4, 9, 16, 25, 36, 49, 64, 81, 100

In question 2(b), the children are expected to use a guess-and-check approach. The pairs are:

9 and 16 $\longrightarrow$ 25, 36 and 64 $\longrightarrow$ 100.

Make sure the children realize that, in the table in question 3, the 'total' column lists the square numbers.

**Problem solving**

If help is needed with question 4, it may be worth suggesting that the children add more detail to the table in question 3 as shown:

| Odd numbers | Total | |
|---|---|---|
| 1 | 1 | $\longleftarrow (1 \times 1)$ |
| 1 + 3 | 4 | $\longleftarrow (2 \times 2)$ |
| 1 + 3 + 5 | 9 | $\longleftarrow (3 \times 3)$ |

It may be necessary to draw their attention to the fact that

— the total of the first two odd numbers is 2 multiplied by itself

— the total of the first three odd numbers is 3 multiplied by itself, and so on.

So the total for the first twenty odd numbers is

$20 \times 20 = 400$

On Textbook page 58, in question 3(a), the children should list the sequence

1, 3, 6, 10, 15, 21, 28, 36, 45, 55

In question 3(c), encourage the children to work systematically by adding

— the first and second triangular numbers ($1 + 3 = 4$)

— the second and third triangular numbers ($3 + 6 = 9$)

— the third and fourth triangular numbers ($6 + 10 = 16$)

and so on.

The square numbers should soon become obvious.

UA4c/4 N3a/4
PS/D1 PS/E1 FE/D1
A/4bc A/5b

## Textbook page 59 *Other activity: sequences and relationships*

- This activity involves number sequences and can be attempted at any time after the pattern work on Workbook pages 15 and 16 and Textbook pages 55–8 has been completed.
- The children must realize that each sequence has four numbers arranged vertically or horizontally. The 'holes' in the grid ensure that there are only 20 possible sets of four numbers to check.
- There is only one possible sequence for each part of question 1. The odd numbers 9, 11, 13, 15, 19 in the top row may cause difficulty as only 9, 11, 13, 15 are **consecutive.**

The sets of numbers are indicated on this diagram.

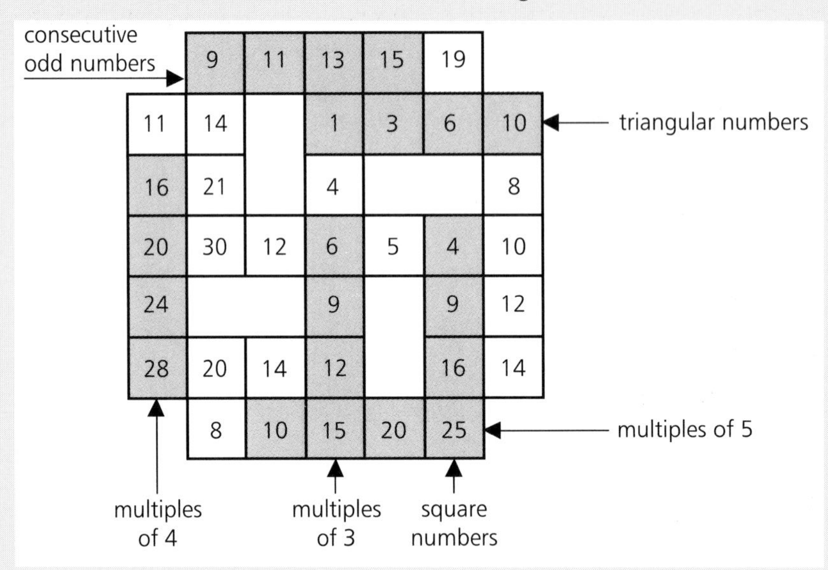


The children should list the sets **without** copying the grid.

- In question 2, some children may need to copy the sets of numbers from the large grid in list form to help them find the rules:

| **divide by 2** | **subtract 5** |
|---|---|
| 16 ⟶ 8 | 9 ⟶ 4 |
| 20 ⟶ 10 | 14 ⟶ 9 |
| 24 ⟶ 12 | 21 ⟶ 16 |
| 28 ⟶ 14 | 30 ⟶ 25 |

# ynchester Airport

## A context for multiplication and division by a two-digit number

he work on Textbook pages 60–7 is set in the context of Lynchester Airport, a large nternational airport.

s the children progress through the Textbook and Workbook pages, they are ntroduced to the following features and facilities associated with the airport:

- car parking
- vans carrying airmail sacks
- freight trolleys
- passenger numbers on different small aircraft
- customs
- caterers loading in-flight meals
- refreshment items and free gifts
- parties visiting the airport
- helicopter trips.

## ntroducing the context

### Discussing airports

sk the children if they have been to an airport. Discuss aspects such as

- **airport facilities:** check-in desks, facilities in the lounge areas, catering, baggage handling, passports, customs, arrival and departure screens
- **airport workers:** at information desks, in the airport shops, in the cafeterias, baggage handlers, cleaners, pilots, stewards
- **the aircraft:** seating, refreshments, entertainment, safety.

### 2 Visit to an airport

 there is an airport near the school, it may be possible to arrange a visit to ensure hat all the children are familiar with the context.

ossible activities might include:

- interpreting data from arrival and departure boards
- listing the airlines which use the airport, and the countries they fly to
- buying refreshments in the cafeteria and comparing prices in the various food outlets in the airport
- collecting publicity material about the airport
- finding out about the aircraft: types, seats, number of passengers, cabin crew, catering
- investigating procedures followed by arriving and departing passengers, including security.

UA2b/4
N3d/4→5
PSE
AS/C3
MD/C2, 3
P/4b
P/4il

# Multiplication

## Overview

This section

- revises multiplication of two-digit numbers by multiples of 10
- revises multiplication by a two-digit number in two stages using the distributive law
- consolidates a written technique for TU multiplied by TU
- illustrates other methods for TU multiplied by TU.

| | Teacher's Notes | Textbook | Workbook | Reinforcement Sheets |
|---|---|---|---|---|
| **Lynchester Airport, a context for multiplication and division by a two-digit number** | 129 | | | |
| **Multiplication:** | | | | |
| **by multiples of 10** | 132 | 60 | | |
| **by two-digit numbers, the distributive law** | 133 | 61 | | |
| **written technique for tens and units by tens and units** | 134 | 62, 63 | | 18, 19 |
| **tens and units by tens and units, other methods** | 135 | 64 | | |

An extension activity is related to the above section of work as follows:

| | Teacher's Notes | Extension Textbook |
|---|---|---|
| **Other methods of multiplication** | 287 | E16 |

Teaching notes for the Extension Textbook are in a separate section at the end of the Teacher's Notes.

# Resources

## Useful materials

- calculator
- other materials suggested within the introductory activities

## Assessment and Resources Pack

### Assessment

*Number and Money Check-up 13*
Textbook pages 60–67
(Multiplication and division by TU)

### Resources

*Problem Solving Activities*
1 Corner Conquest (Multiplication of TU by TU)

# Teaching notes

Multiplication of tens and units by a two-digit number was introduced in Heinemann Mathematics 5 and included multiplication by multiples of 10. This work is now revised.

The activities are set in the context of 'Lynchester Airport'.

## MULTIPLICATION BY MULTIPLES OF 10

N3d/4→5
AS/C3 MD/C2, 3 MD/E
N/4il

The context is introduced by looking at transportation within Lynchester Airport.

### Introductory activity

#### Passengers on buses

- Discuss the following example:

  'Each airport bus can carry 31 passengers.
  How many passengers can 20 buses carry?'

  Card could be used to represent the buses, with the number of passengers written on each card.

  Illustrate 20 busloads of 31 passengers, or 20 × 31, by laying out twenty cards and partitioning them into 2 sets of 10 like this.

31 31 31 31 31 31 31 31 31 31    31 31 31 31 31 31 31 31 31 31

**10 × 31**    **10 × 31**

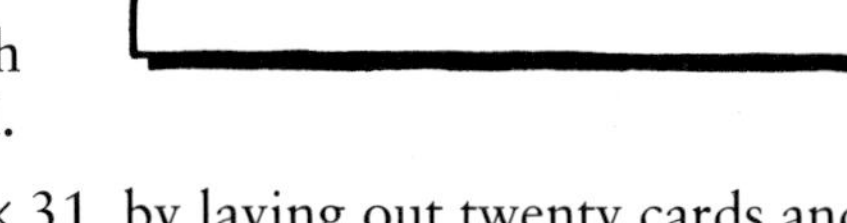

  Explain that, since the number of passengers in each set of 10 buses is the same, the answer can be found by

  **multiplying by 10 and then multiplying by 2.**

  There are **620 passengers.**

- Show the children how to record as a single calculation. First multiply by 10. This can be indicated by placing a zero in the units column of the answer, so that subsequent digits move one place to the left.

$$\begin{array}{r} 31 \\ \times\ 20 \\ \hline 0 \\ \hline \end{array}$$

  Now multiply by 2 and record in the appropriate column.

$$\begin{array}{r} 31 \\ \times\ 20 \\ \hline 620 \\ \hline \end{array}$$

  There are **620 passengers.**

  If necessary, discuss other examples.

- Extend the rule to include other multiples of 10 – by considering, for example, the number of passengers on 40, 60 and 90 buses.

Number
Multiplication by TU
Lynchester Airport

## Textbook page 60 *Multiplication: TU by multiples of 10*

N3d/4→5
MD/C2, 3
N/4il

In question 2, whether the example is given as 64 × 90 or 90 × 64, the children should record their working like this:

$$\begin{array}{r} 64 \\ \times\, 90 \\ \hline \end{array} \quad \text{and not} \quad \begin{array}{r} 90 \\ \times\, 64 \\ \hline \end{array}$$

In question 4, one of the numbers being multiplied represents the cost in pence of parking for one hour. The children should be encouraged to do the calculation without including the 'p' for pence. This should be inserted at the end, and the final cost changed to pounds and pence.

# MULTIPLICATION BY TWO-DIGIT NUMBERS, THE DISTRIBUTIVE LAW

N3d/4→5
AS/C3 MD/C2, 3
N/4il

n Heinemann Mathematics 5, multiplication by other two-digit numbers was ecorded as a two-step process using the distributive law. This is now revised.

The context of Lynchester Airport continues with freight being loaded on to aircraft.

## ntroductory activity

### Washing machines

- Consider the following example:

  'Each washing machine weighs 41 kg.

  There are 25 on each trailer.

  Find their total weight.'

  Discuss a two-step approach with the children, where they find the weight of 5 machines and 20 machines separately and then add.

  For example,

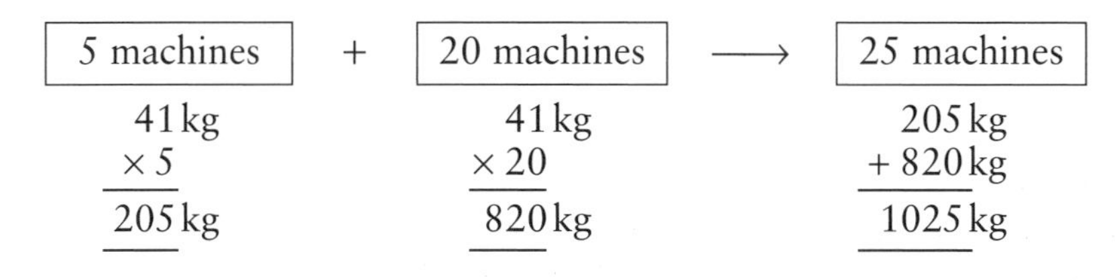


The total weight is **1025 kg**.

- Repeat for other examples.

## Textbook page 61 *Multiplication: TU by TU, distributive law*

N3d/4→5
AS/C3 MD/C2, 3
N/4il

Discuss the context of loading trailers at an airport, or loading fork-lift trucks. Some of the items are perishable goods, which have to reach their destination quickly in order to be in good condition on arrival. Hence the reason for sending them by air freight.

The final recording of answers to questions 1 and 3 should show the unit of weight: for example, 525 kg.

# MULTIPLICATION: WRITTEN TECHNIQUE FOR TENS AND UNITS BY TENS AND UNITS

N3d/4→5
AS/C3 MD/C2, 3 MD/E
N/4il

In Heinemann Mathematics 5, a limited amount of work was done involving a standard written technique for multiplying by 11 to 19. This is now revised and extended to multipliers up to 99.

The context continues with the children being introduced to some of the aircraft which use Lynchester Airport.

## Introductory activities

The earlier introductory activities involved passengers on buses or weights of washing machines. Either of these ideas could be used to develop the written technique, as **one calculation**, for tens and units multiplied by tens and units.

### 1 Passengers on buses *(TU multiplied by 11 to 19)*

- Pose the question,

  'How many passengers can 18 buses carry if they can each carry 37 people?'

  Record the calculation as on Textbook page 61.

| 8 buses | + | 10 buses | ⟶ | 18 buses |
|---|---|---|---|---|
| $\begin{array}{r} 37 \\ \times 8 \\ \hline 296 \end{array}$ | | $\begin{array}{r} 37 \\ \times 10 \\ \hline 370 \end{array}$ | | $\begin{array}{r} 296 \\ +370 \\ \hline 666 \end{array}$ |

There are **666 passengers.**

- Now discuss how the recording can be set down as **one calculation.**

$$\begin{array}{rr} & 37 \\ & \times 18 \\ \hline 8 \times 37 \longrightarrow & 296 \\ 10 \times 37 \longrightarrow & +370 \\ \hline \text{Add to find } 18 \times 37 \longrightarrow & 666 \end{array}$$

There are **666 passengers.**

The children should set down only the boxed part of the recording.

Where schools prefer the children to multiply by the tens first, the recording would be as shown.

$$\begin{array}{r} 37 \\ \times 18 \\ \hline 370 \\ +296 \\ \hline 666 \end{array}$$

There are **666 passengers.**

- Repeat for other examples, such as

  'How many passengers can be carried by 13 buses, 17 buses. . .?'

## 2 Crates *(TU multiplied by 21 to 99)*



Discuss the example:

'There are 27 crates of food to be loaded on each of 35 jumbo jets. How many crates is this altogether?'

The new feature is that multiplication by a multiple of 10 is recorded in one line. For example,

$$\begin{array}{rr} & 27 \\ & \times\,35 \\ \hline & 135 \\ (30 \times 27) \longrightarrow & +\,810 \\ \hline & 945 \\ \hline \end{array}$$

There are **945 crates** altogether.

N3d/4→5
AS/C3 MD/C2, 3 MD/E
N/4il

### Textbook pages 62 and 63
### *Multiplication: TU by 11 to 19 and by 21 to 99*

The work on Textbook page 62 deals with multipliers from 11 to 19. The context used is that of numbers of passengers carried by small aircraft. The last question brings in cost, but involves only whole numbers of pounds. All products are less than 1000.

On Textbook page 63, the multipliers range from 21 to 99. The context used is that of customs. Discuss the idea of passengers paying duty or tax on items brought back from abroad.

In question 4(a), the children should multiply 65 × 28 to find that Alex's pay for 28 days is £1820. In question 4(b), they should add on £65 to give £1885 and then another £65 to give £1950 and then another £65 to give £2015. So he must work another 3 days.

The children should **not** use a calculator.

**R18, 19**

# MULTIPLICATION: TENS AND UNITS BY TENS AND UNITS, OTHER METHODS

N3d/4→5
AS/C3 MD/C2, 3
N/4il

In the Heinemann Mathematics 5 Extension Textbook, on pages E12 and E13, two other methods of multiplying were introduced. One of these involved forming a doubles table, selecting the appropriate products and adding. For example,

37 lots of. . . = 32 lots of . . . + 4 lots of. . . + 1 lot of. . .

The other method made use of factors in the multiplication. For example,

$18 \times 23 = 3 \times 6 \times 23$

Two other methods, appropriate to particular multipliers, are now introduced. The context used is that of in-flight meals.

# Introductory activities

## 1 Multiples of 10 times multiples of 10

- The children know how to multiply by 10 and by multiples of 10. This 'other method' involves multiplying two multiples of 10 in the following way:

$$70 \times 30$$
$$= (7 \times 3)00$$
$$= 2100$$

This can be introduced as described below.

- The first step is to emphasize that **tens times tens give hundreds**, by starting with

$10 \times 10 = 100$ so 1 ten $\times$ 1 ten = 1 hundred

$10 \times 20 = 200$ so 1 ten $\times$ 2 ten = 2 hundreds

$10 \times 30 = 300$ so 1 ten $\times$ 3 ten = 3 hundreds, and so on.

Now develop

$20 \times 10 = 200$ so 2 tens $\times$ 1 ten = 2 hundreds

$20 \times 20 = 400$ so 2 tens $\times$ 2 tens = 4 hundreds, and so on.

After a few of these examples, the children should realize that to find an answer it is only necessary to multiply the tens digits and call the answer hundreds. Each answer ends with zeros in the tens and units places.

- The above products can be readily demonstrated using a calculator. The children could multiply multiples of 10 together and try to spot the rule.

## 2 Multipliers with 9 as units digit

- The technique which the children are expected to use is similar to the two-step process involving the distributive law, used earlier, but this time subtraction is involved. For example,

'How many passengers can be carried on 29 planes, each carrying 34 people?'

29 lots of 34 = 30 lots of 34 – 1 lot of 34

$$\begin{array}{r} 34 \\ \times 30 \\ \hline 1020 \\ \hline \end{array} \quad \longrightarrow \quad \begin{array}{r} 1020 \\ -\ 34 \\ \hline 986 \\ \hline \end{array}$$

29 planes can carry **986 passengers.**

- Repeat for other examples.

| UA2b/4 N3d/4→5 |
|---|
| PSE AS/C3 MD/C2, 3 |
| P/4b N/4il |

## Textbook page 64 *Multiplication: TU by TU, other methods*

The questions involve catering staff calculating the numbers of lunches and bottles of mineral water required for the in-flight meals.

In question 5, the children should use both methods practised in questions 1 to 4, and so be able to find the answer with a minimum of working. For example, $39 \times 20$ involves finding $40 \times 20$ and then subtracting $1 \times 20$. Some children may be able to find these answers mentally.

**Problem solving**

In question 6, the children have to use the information given earlier on the page about the number of lunches on a tray and the number of bottles in a crate. They should find that Yasmin has enough. In fact there are 4 extra bottles.

N3e/4 N3d/4→5
AS/C3
MD/D2
MD/E
N/4cil

# Division

## Overview

This section

- introduces division of a two-digit number as repeated subtraction
- deals with both the sharing and grouping aspects of division by a two-digit number.

| | Teacher's Notes | Textbook | Workbook | Reinforcement Sheets |
|---|---|---|---|---|
| **Lynchester Airport, a context for multiplication and division by a two-digit number** | 129 | | | |
| **Division by a two-digit number:** | | | | |
| **repeated subtraction** | 139 | 65 | | |
| **shares of 10** | 140 | 66 | | 20 |
| **grouping** | 142 | 67 | | 21 |
| *Other activity* | *143* | *68* | | |

## Resources

### Useful materials

- calculator
- other materials suggested within the introductory activities

### Assessment and Resources Pack

Assessment

*Number and Money*

*Check-up 13*

Textbook pages 60–67

(Multiplication and division by TU)

Resources

# Teaching notes



This section introduces written techniques for division by two-digit numbers based on repeated subtraction. The examples are graded to ensure that not more than four subtractions are necessary to find any answer.

The Lynchester Airport context continues.

## DIVISION BY A TWO-DIGIT NUMBER: REPEATED SUBTRACTION

N3e/4 N3d/4→5
AS/C3 MD/E
N/4il

The Lynchester Airport context continues with sharing food and gifts among passengers whose flights are delayed.

## Introductory activity

### 1 Sharing the sandwiches *(shares of one)*

- Discuss a problem such as

  '31 sandwiches have to be shared equally among 13 children.

  How many will each receive and how many will be left over?'

  Establish that the answer can be found by dividing 31 by 13. Record as shown and discuss the sharing process, using language similar to that indicated below.

| | | | |
|---|---|---|---|
| 13 ⟌ 31 | | ⟵ | 'Share the 31 sandwiches. |
| − 13 | 1 | ⟵ | Give 1 to each. $13 \times 1 = 13$. 13 sandwiches used. |
| 18 | | ⟵ | Subtract. 18 sandwiches are left. |
| − 13 | 1 | ⟵ | Give another 1 to each. |
| 5 | 2 | ⟵ | Subtract. 5 sandwiches are left. |
| 5 left over ↗ | ↖ 2 sandwiches each | | There are not enough left to share.' |

  Each child receives **2 sandwiches** and there are **5 left over.**

- Repeat for other examples, such as $65 \div 17$, $43 \div 14$ and $67 \div 15$, to ensure that the children understand the procedure.

## 2 Sharing the sandwiches *(simplified language)*

As the children gain confidence in the technique, the language can be reduced. A common language that will work for all examples should be developed.

'Share 50 sandwiches equally among 16 children.

How many will each receive and how many are left over?'

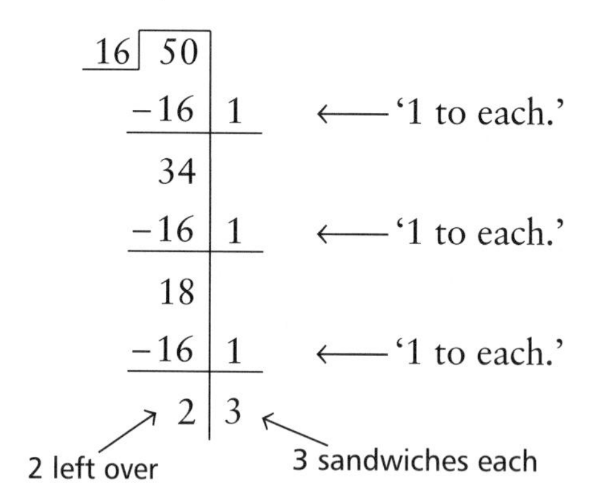


Each child receives 3 **sandwiches** and there are 2 **left over.**

N3e/4 N3d/4→5
AS/C3 MD/E
N/4il

### Textbook page 65 *Division: by TU*

Discuss the table in question 1 to make sure that the children can translate each example into a division.

For example,

| beef |
|---|
| 54 |
| 27 |

leads to $27 \overline{)\,54}$

In question 3, money has to be shared, but the answer is an exact number of pounds.

In question 4, the children have to work out that there are 49 adults and 15 children before they can attempt parts (b) and (c).

# DIVISION BY A TWO-DIGIT NUMBER: SHARES OF 10

N3d/4→5
AS/C3 MD/C2 MD/E
N/4cil

The process of repeated subtraction can become impractical if only shares of 1 are used. Examples where it is sensible to subtract one or more shares of 10 are now introduced.

The context continues with the distribution of travel gifts.

## Introductory activities

### 1 Sharing the gifts *(shares of 10)*

- Extended language should be used to ensure that the children fully understand the process. For example, discuss the problem:

  'The airline shares 150 duty-free gifts among 13 stewards.

How many does each receive and how many are left over?'

```
13⟌150             ← 'Share the 150 gifts.
  −130 | 10        ← Give 10 to each.  13 × 10 = 130.  130 gifts used.
    20 |           ← Subtract. 20 gifts are left.
   −13 |  1        ← Give another 1 to each.
     7 | 11        ← Subtract.  7 gifts are left.
                     There are not enough left to share.'
7 left over   11 gifts each
```

Each steward receives **11 gifts** and there are **7 left over.**

- Repeat for other examples, such as 149 ÷ 12, 320 ÷ 26, and 418 ÷ 19.

## 2 Sharing the gifts *(simplified language)*

- This language should then be shortened as shown.

'The airline shares 340 gifts equally among 16 stewards.

How many does each receive and how many are left over?'

```
16⟌340
  −160 | 10        ← '10 to each.'
   180 |
  −160 | 10        ← '10 to each.'
    20 |
   −16 |  1        ← '1 to each.'
     4 | 21
4 left over   21 gifts each
```

Each steward receives **21 gifts** and there are **4 left over.**

- If necessary, repeat for other examples.

N3e/4 N3d/4→5
AS/C3 MD/C2 MD/E
N/4cil

## Textbook page 66 *Division: by TU*

Throughout the page, the examples are chosen so that no more than four subtractions are necessary to find the answers.

The technique which the children have been using can be applied to any division situation. So far all the examples have been based on sharing, but the examples on Textbook page 67 are about the grouping aspect of division.

The context continues with a tour of the airport.

## Introductory activity

### Visitors to the airport

- Discuss a few examples where the grouping language of division is used. The recording looks the same as before, but the language is different. Each answer should be given in the context of the question. For example,

  'Visitors are taken round the airport in groups of 12.

  How many groups are needed for 132 people?'

| | | |
|---|---|---|
| 12 ⟌ 132 | | ⟵ '132 people. |
| −120 | 10 | ⟵ 10 groups would take 120 people. |
| 12 | | ⟵ 12 people are left. |
| −12 | 1 | ⟵ 1 other group of 12 people.' |
| 0 | 11 | |
| ↑ 0 left over | ↑ 11 groups | |

  **11 groups** are needed.

- This language should be shortened once the children understand the process. For example,

  'How many groups of 36 are in 432?'

| | | |
|---|---|---|
| 36 ⟌ 432 | | |
| − 360 | 10 | ⟵ '10 groups.' |
| 72 | | |
| − 36 | 1 | ⟵ '1 group.' |
| 36 | | |
| − 36 | 1 | ⟵ '1 group.' |
| | 12 groups | |

  There are **12 groups.**

  The children should appreciate that the recording is the same for both sharing and grouping.

**They should not write the language beside their working.**

## Textbook page 67 *Division: by tens and units*

All the examples on this page involve the grouping language. Some of the examples have remainders.

In question 2, the division produces the following working.

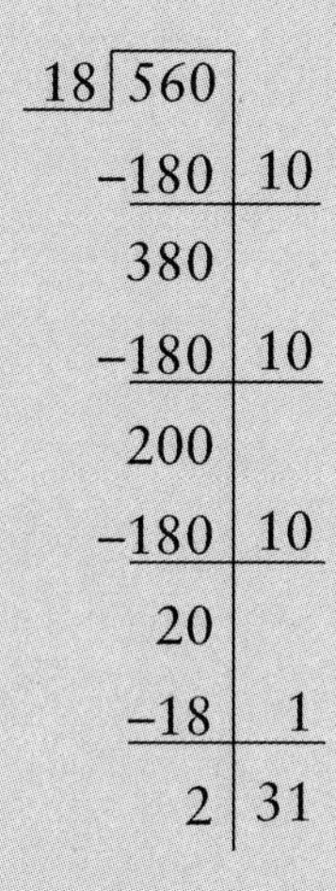

The children have to interpret the answer as **31 rows** and **2 cars** in the next row. In question 6, £5 has to changed to 500p before dividing. The answer will be a **number of tickets** and not money, as some children might think. The amount of money left is, of course, expressed in pence.

**Number**
**Division by TU**
Lynchester Airport

N3e/4 N3d/4→5
AS/C3 MD/C2 MD/E
N/4cil

R21

## Textbook page 68 *Other activity: pattern*

- This activity is probably best attempted after the children have completed the multiplication work on Textbook pages 60–4. However, as the use of a calculator is recommended, some children could attempt it earlier.
- Make sure the children understand the instructions. They can choose any two-digit number and then add 1 to it and subtract 1 from it to give the other two numbers. The three consecutive numbers are then listed in order. The reason for this way of forming three consecutive numbers is to establish the pattern for forming the other number sequences in questions 3, 4 and 5.
- The product differences are:

  — when 1 is added and 1 subtracted the difference is 1,

  — when 2 is added and 2 subtracted the difference is 4,

  — when 3 is added and 3 subtracted the difference is 9, and so on.

  Hopefully children will recognize 1, 4 and 9 as the start of the sequence of square numbers.
- In question 5, the difference is the next square number, 16.

N3a,4a/4
AS/C3 MD/C2
N/4cil

The Measure part of Heinemann Mathematics 6 has five sections, each with an Overview and accompanying notes.

# Animal Protection and Education Centre (APEC)

## A context for length, weight and volume

The work on

— length (Textbook pages 69–74, Workbook page 17)

— weight (Textbook pages 75–8, Workbook page 24)

— volume (Textbook pages 81–4, Workbook page 28)

is set within the context of the Animal Protection and Education Centre (APEC), an organization concerned with animal welfare and conservation.

### Introducing the context

The context could be introduced by discussing animal protection and the need to educate people about the reasons for caring for animals and the reasons for the possible extinction of some species. Other activities are suggested below.

### 1 Animals in danger

- The children could be asked to find out about animals in danger of extinction: for example, panda, Indian tiger, rhinoceros and osprey. Information could be obtained from books, videos or information factsheets published by organizations such as WWF (World Wide Fund for Nature), RSPB (Royal Society for the Protection of Birds), Friends of the Earth or Greenpeace.
- It would be useful to discuss with the children the type of information to be collected. This might include

  — a description of the animal

  — where the animal lives

  — how many of these animals are left

  — the reasons why they are endangered

  — possible solutions to the problem.

Animal data files could be made using the information, and these could be displayed, together with the pictures of the animals and other factsheets, on a wall map.

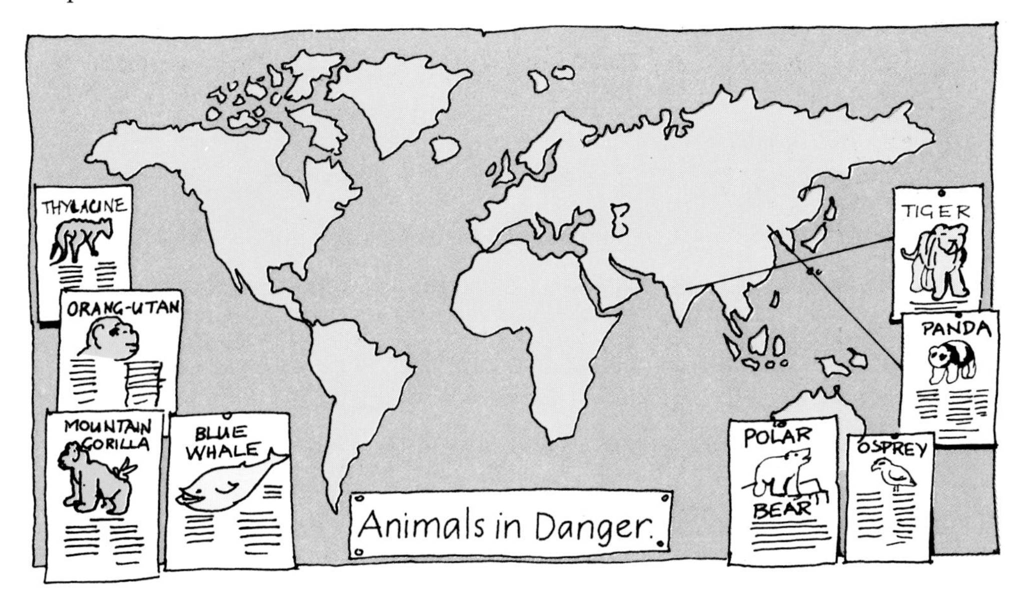


Alternatively, this information could be stored using an appropriate database.

## 2 Visit to an animal sanctuary

A bird sanctuary, wildlife centre, veterinary hospital or zoo may be located nearby. A visit could be arranged to allow the children to find out about the animals, how they are cared for, and why a sanctuary is necessary.

During the visit, the children could

- — make sketches of some of the animals they observe
- — make a note of the animals' habitat, the food that they eat, and so on
- — find out about the size and weight of the animals
- — observe the way the animals move and the tracks they leave
- — identify the animals using appropriate reference books
- — interview the people who work at the sanctuary
- — find out how to become more involved with the work of the sanctuary.

## 3 A plan or model of APEC

The children could be asked to design and make a plan or model of what they think APEC would be like.

They should think about

- — the location and types of buildings and enclosures needed
- — the location of the visitors centre
- — the inclusion of 'hides' so that the animals can be observed, and so on.

Measure
Length
APEC

UA2ad,3b/4
N4a/4
SSM4c/4
SSM4a/4→5
PSE
AS/C3
AS/D2
MD/D1
FE/D1
ME/C5
ME/D1, 7
PFS/D1
PFS/E3
P/4abd
N/4ik
A/4d
M/4abc
M/5a

# Length

## Overview

This section

- consolidates practical measurement in metres and centimetres
- revises conversion of lengths expressed in metres and centimetres to centimetres, and vice versa
- provides practice in addition and subtraction of lengths expressed in metres and centimetres
- consolidates calculation of perimeter and introduces a simple word formula for the perimeter of a square
- introduces the concept of scale and calculation of 'true lengths'
- introduces the kilometre and conversions of kilometres and metres to metres, and vice versa
- provides practice in addition and subtraction of lengths expressed in kilometres and metres.



An extension activity is related to the above section of work as follows:



Teaching notes for the Extension Textbook are in a separate section at the end of the Teacher's Notes.

# Resources

## Useful materials

- ruler, metre sticks and tapes (100 cm or 150 cm) calibrated in centimetres
- long metric tape (10 m, 20 m or 25 m) or trundle (click) wheel
- two marker cones or bean bags
- a watch
- centimetre square dotty paper
- reference book or atlas containing distances in kilometres between towns
- other materials suggested within the introductory activities

## Assessment and Resources Pack

### Assessment

*Measure Check-up 1*
Textbook pages 69–74
Workbook page 17
(Length: metres and centimetres, addition and subtraction, scale, kilometres)

*Round-up 1*
Question 3

### Resources

*Problem Solving Activities*
13 Peter's pipes (Lengths in metres)
14 Taking sides (Adding lengths to find perimeters)
15 Scales (Measuring in centimetres and calculating true lengths)

*Resource Cards*
18 APEC plan (Length: scale drawing, calculating true lengths and perimeters)
23,24 True or false cards (Practical work)

# Teaching notes

## METRES AND CENTIMETRES, PERIMETER

N4a/4 SSM4c/4 SSM4a/4→5
AS/C3 ME/C5 PFS/D1
N/4i M/4ab

In Heinemann Mathematics 5 the children were given the opportunity to practise their measuring skills using metre sticks, tapes and rulers. They added and subtracted lengths in metres and centimetres, and found the perimeter of shapes.

Heinemann Mathematics 6 begins by consolidating and extending this work. A simple word formula for calculating the perimeter of a square is introduced.

The children are introduced to the context of the Animal Protection and Education Centre (APEC).

## Introductory activities

### 1 Snake lengths *(metres and centimetres)*

Revise the equivalence of 100 centimetres and 1 metre. Provide some initial practice in converting from metres and centimetres to centimetres, and vice versa. Some children may find lengths with a zero in either the units or the tens place more difficult. Encourage the children to read each measurement aloud, thus emphasizing the place value of the digits. For example,

### 2 APEC differences *(subtraction)*

Revise the method of subtraction of lengths and provide some initial examples, again in the context of APEC. For example, draw on the chalkboard

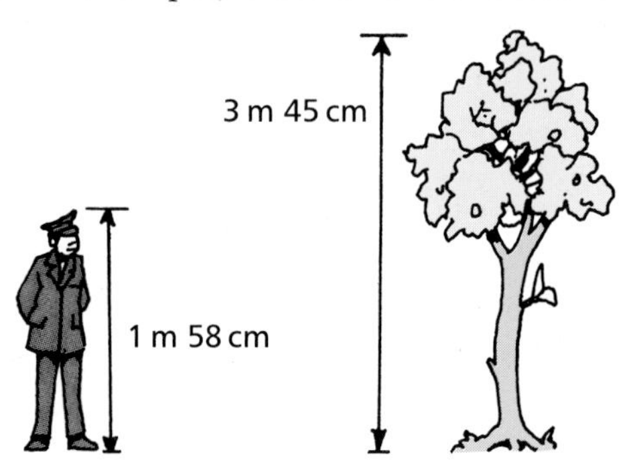

and ask the children how to find the difference between the heights. They should convert the heights to centimetres before subtracting,

3 m 45 cm ⟶ 345 cm
1 m 58 cm ⟶ −158 cm
187 cm ⟶ **1 m 87 cm**

The difference in height is **1 m 87 cm.**

**At this point the children could try Textbook page 69.**

## 3 Perimeters of enclosures

- Discuss with the children the meaning of the word 'perimeter'. They should remember this as the length of the boundary line of a shape. Discuss how to find the perimeter of this rabbit enclosure.

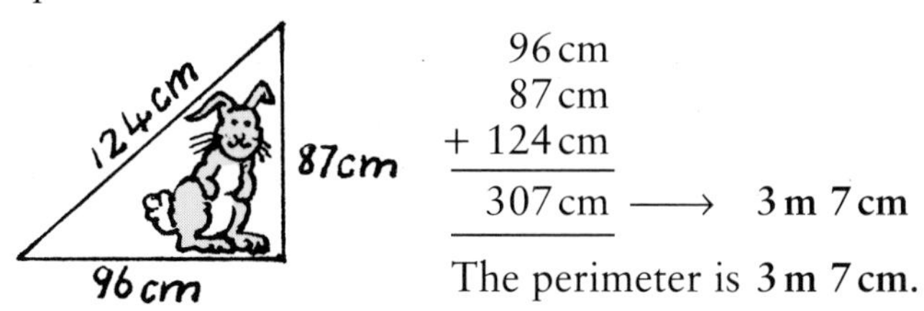


96 cm
87 cm
+ 124 cm
307 cm ⟶ 3 m 7 cm

The perimeter is 3 m 7 cm.

- Draw other shapes on the chalkboard and ask the children to calculate the perimeter of each.

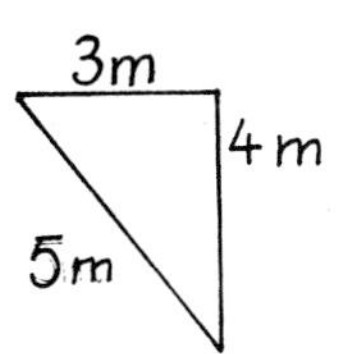


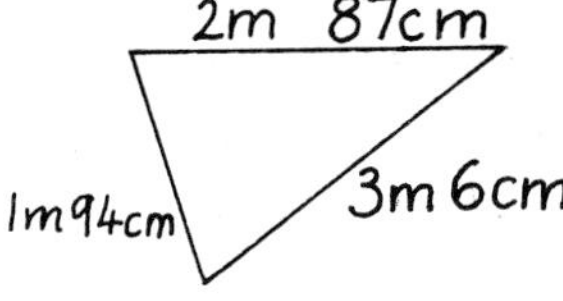


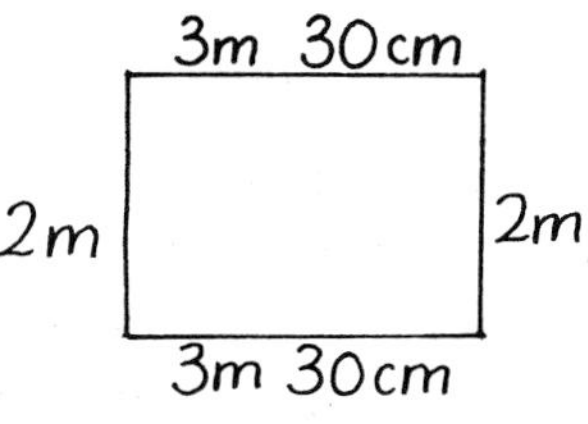


N4a/4 SSM4a/4→5
AS/C3 ME/C5
N/4i M/4a

## Textbook page 69 *Length: m and cm, subtraction*

Discuss the scenario of APEC caring for birds of prey. Some assistance may be required in question 6 to help the children decide how best to measure their armspans – for example, by stretching out their arms against a wall and having a partner mark the position of the fingertips with chalk. The measurement between the two marks can then be found.

UA2d,3b/4 N4a/4 SSM4c/4
PSE AS/C3 FE/D1 PFS/D1
P/4bd N/4i A/4d M/4abc

## Textbook page 70 *Length: addition, perimeter, investigation*
## Workbook page 17

Introduce the context on Textbook page 70 by discussing the different types of animal which may be at APEC and the types of facility they would require – for example, perches for birds; long branches, ropes and tyres for monkeys; and fences and ponds.

Discuss the children's answers to question 3. Making a table on the chalkboard may help them to see patterns, and so derive a word formula for the perimeter of a square.

| length of a side | perimeter |
|---|---|
| 1 cm | 4 cm |
| 2 cm | 8 cm |
| 3 cm | 12 cm |
| 4 cm | 16 cm |

**Problem solving**

R22

On Workbook page 17, in questions 2 and 3 the children could draw additional shapes on sheets of centimetre square dotty paper to help them reach conclusions.

In question 3, the children should discuss their answers. The shape with the shortest possible perimeter is

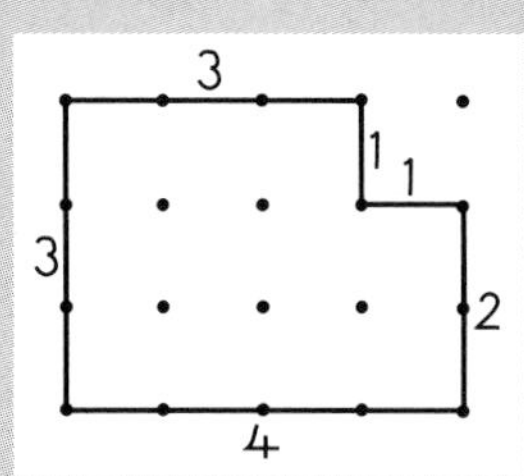


Area = 11 $cm^2$

Perimeter = 14 cm

Here are two possible answers to question 4.

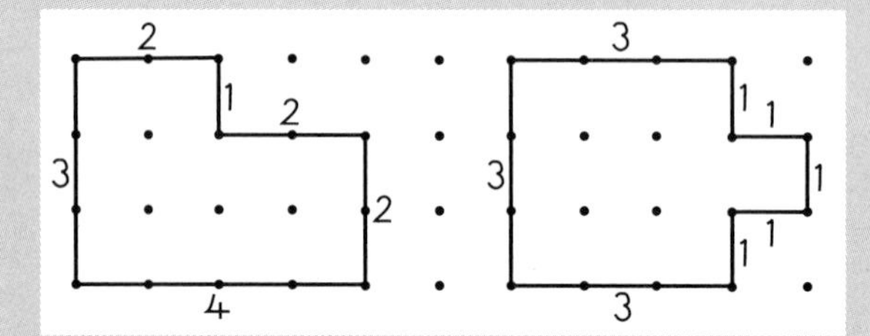


Area = 10 $cm^2$

Perimeter = 14 cm

## SCALE

N4a/4
MD/D1 PFS/E3
N/4i M/5a

Scale drawings are now introduced. Using their skills in measuring to the nearest centimetre, the children measure scale drawings and calculate true lengths using scales of the types 1 cm to 3 cm, 1 cm to 40 cm and 1 cm to 3 m.

The APEC context is developed by considering the 'true' lengths of fish and dinosaurs.

## Introductory activities

### 1 Scale models

Make a collection of models, including, for example, cars, aeroplanes, trains, animals and buildings. Explain that some models are made to **scale**: that is, they are just like the 'real thing', only smaller. Discuss a particular model, such as a model chair.

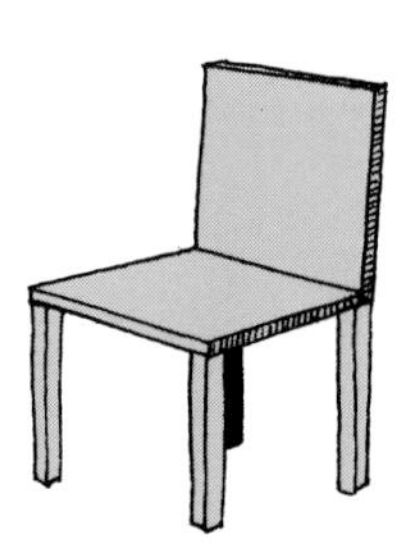

Explain that 1 cm on the model represents 10 cm on the real object. This means that the real chair is

10 times as long

10 times as broad and

10 times as high.

## 2 Finding the true length

- Display a drawing of a fish, say 20 cm long, on a piece of card.

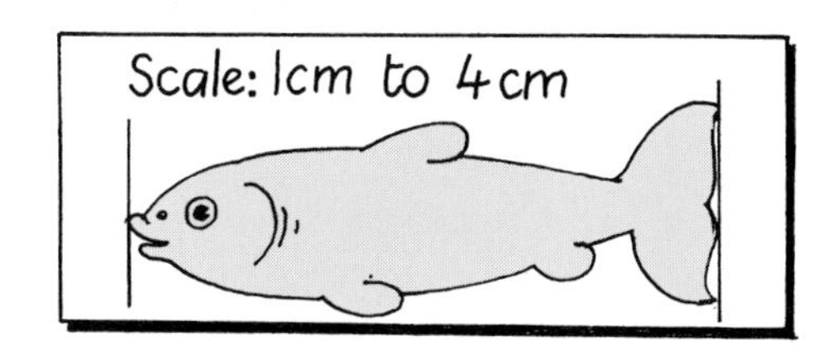


Discuss the scale – each centimetre on this scale drawing represents 4 cm on the real fish. Ask the children how they could find the 'true' length of the real fish (measure the drawing in centimetres and multiply by 4).

A child should measure the length between the lines. Discuss the calculation of 'true length':

True length = 20 × 4 cm = 80 cm

- Consider other scale drawings using different scales where the real objects are relatively large.

For example (measured length 7 cm),

True length of snake
= 7 × 50 cm
= 350 cm
= **3 m 50 cm**

For example (measured length 10 cm),

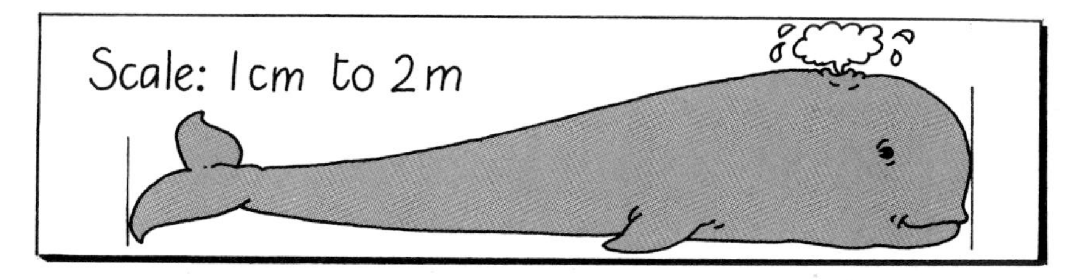


True length of whale
= 10 × 2 m
= **20 m**

N4a/4
MD/D1 PFS/E3
N/4i M/5a

## Textbook pages 71 and 72 *Length: scale*

On Textbook page 71, all the scales are of the types 1 cm to 5 cm (multiply by a single digit) or 1 cm to 50 cm (multiply by a multiple of 10). Emphasize that the children should measure the lengths between the vertical lines.

For question 2, the children should have to judge where to measure the length and height of each object.

In question 3, the length of the Mako shark should be given as $4\frac{1}{2}$ metres.

On Textbook page 72, the scales are of the form 1 cm to so many metres, leading to true lengths in metres.

In question 3(b), the children should estimate the height of the classroom to a whole number of metres before comparing it with a dinosaur.

In question 4(a), some children may need guidance to select the maximum breadth (3 cm) of Dino World.

# THE KILOMETRE

N4a/4 SSM4a/4→5
AS/D2 ME/D7
N/4k M/4a

The units of length are extended to include the kilometre. The initial work is set in the context of raising money for APEC through sponsored races.

## Introductory activities

### 1 The kilometre

- Collect and display articles from magazines and newspapers which mention kilometres.

- The children should be familiar with the kilo**gram** and know that 1000 grams = 1 kilogram. Asking how long they think 1 kilo**metre** is should lead to the suggestion 1000 metres.

## 2 Kilometres and metres

- Provide some initial practice in converting from metres to kilometres and metres, and vice versa.

  Some children may find lengths with a zero in the units, tens or hundreds places confusing. Encourage the children to read each measurement aloud, thus emphasizing the place value of the digits. For example,

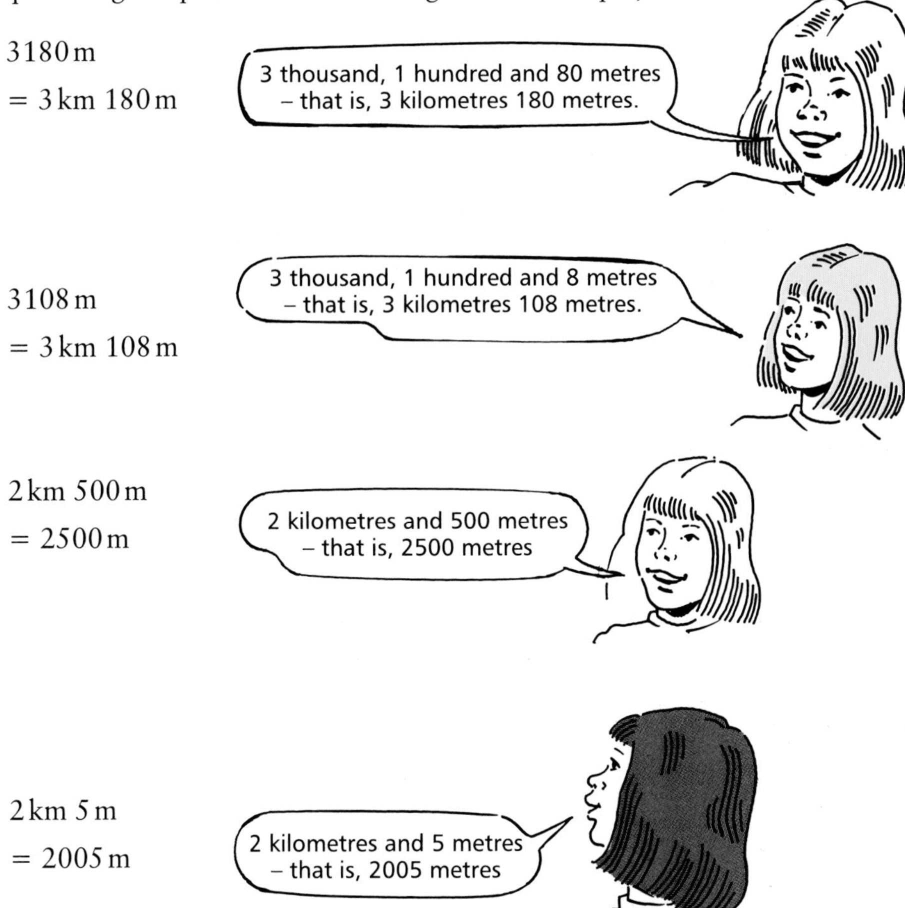


- Include examples such as
  - change to kilometres and metres

    2800 m    2080 m    2008 m
  - change to metres

    50 km 600 m    50 km 60 m    50 km 6 m

**At this point the children could try Textbook page 73.**

## 3 Pathways *(addition and subtraction)*

Draw a map like this on the chalkboard.

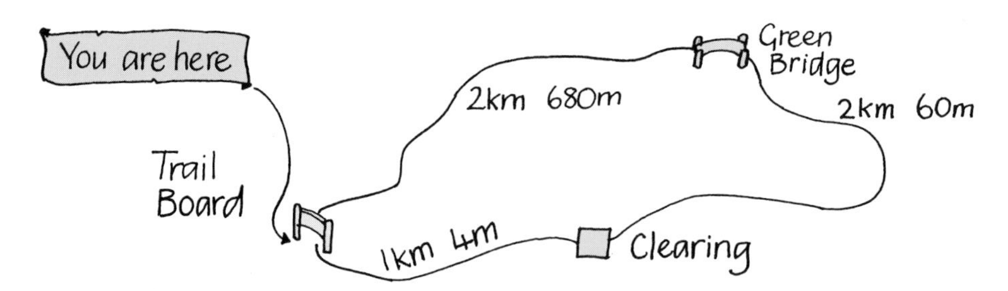


Explain that Kirsty cycles between each place.

Illustrate the method of addition and subtraction of lengths by discussing some examples, such as:

— Find the total distance from the Trail Board to the Green Bridge and then to the Clearing.

| | | | |
|---|---|---|---|
| Trail Board to Green Bridge | 2 km 680 m ⟶ | 2680 m | |
| Green Bridge to Clearing | 2 km 60 m ⟶ | +2060 m | |
| | | 4740 m | ⟶ **4 km 740 m** |

The total distance is **4 km 740 m.**

— Find the difference between the distance Kirsty has to cycle from the Trail Board to Green Bridge and the distance from the Trail Board to the Clearing.

| | | | |
|---|---|---|---|
| Trail Board to Green Bridge | 2 km 680 m ⟶ | 2680 m | |
| Trail Board to Clearing | 1 km 4 m ⟶ | − 1004 m | |
| | | 1676 m | ⟶ **1 km 676 m** |

The difference is **1 km 676 m.**

## Textbook pages 73 and 74 *Length: introducing the kilometre*

On Textbook page 73, in question 1, the children could use a long measuring tape or a trundle (click) wheel.

In question 2(b), the children should not walk 1000 metres, but should calculate how long it would take using their time for 100 m from part (a).

On Textbook page 74, a preliminary discussion to introduce the map would be helpful to familiarize the children with place names and positions.

In question 2, the children may have to be prompted to consider more than one route between places.

In question 3(a), it may be necessary to point out that the route must avoid visiting the same place twice, to be valid.

There are various possible answers to question 5. A partner could check that the route is valid and has a total length between 9 km and 10 km.



UA2a/4 N4a/4 SSM4a/4→5
PSE AS/D1 ME/D1, 7
P/4ab N/4k M/4a

**Problem solving**

**Problem solving**

UA2abc,4d/4
N4a/4
SSM4ab/4
SSM4a/4→5
PSE
AS/D2
MD/D3
ME/C1, 6
ME/D2, 6, 8
P/4a
N/4kl
M/3b
M/4ad

# Weight

## Overview

This section

- revises weighing with 20 g, 10 g and 5 g weights
- revises addition and subtraction of kilograms and grams
- revises and extends reading of scales to include devices calibrated in 5 g, 10 g, 20 g, 25 g, 100 g, 200 g and $\frac{1}{2}$ kg intervals
- provides estimating and weighing activities
- includes multiplication of kilograms and grams by a whole number.

| | Teacher's Notes | Textbook | Workbook | Reinforcement Sheets |
|---|---|---|---|---|
| **APEC: a context for length, weight and volume** | 146 | | | |
| **Kg and g: addition and subtraction** | 160 | 75 | | |
| **Reading scales** | 161 | 76–7 | 24 | 23 |
| **Multiplication, practical work** | 164 | 78 | | |

An extension activity is related to the above section of work as follows:

| | Teacher's Notes | Extension Textbook |
|---|---|---|
| **Measure: imperial units** | 289 | E21 |

Teaching notes for the Extension Textbook are in a separate section at the end of the Teacher's Notes.

# Resources

## Useful materials

- sets of weights which include 20 g, 10 g and 5 g weights
- two-pan balance, scales, spring balance
- a hard-boiled egg
- sand, plastic bags
- other materials suggested within the introductory activities

## Assessment and Resources Pack

### Assessment

*Measure Check-up 2*
Textbook pages 75–8
Workbook page 24
(Scales, kilogram and grams, calculations)

*Round-up 3*
Questions 4(a), 6

### Resources

*Resource Cards*
23, 24 True or false cards (Weighing in kilograms)

# Teaching notes

In Heinemann Mathematics 5 the children used 20 g, 10 g and 5 g weights to weigh light objects. They used and read linear and circular scales, graduated in different intervals, and were introduced to addition and subtraction of kilograms and grams.

In Heinemann Mathematics 6, the work begins by revising the conversion of kilograms and grams to grams, and vice versa, and includes further work on addition and subtraction of kilograms and grams.

The central character is Debbie, who weighs eggs, food for the animals and the animals themselves at the Animal Protection and Education Centre (APEC). For a detailed introduction to this context, which is also used for the Length and Volume sections, see pages 146–7 of these Teacher's Notes.

## Kg AND g: ADDITION AND SUBTRACTION

N4a/4 SSM4a/4→5
AS/D2 ME/C1
N/4k M/4a

### Introductory activities

#### 1 Weighing eggs

Show the children some pictures of eggs from Debbie's surgery in APEC, each clearly showing its weight.

Ask them to put out weights to match the weight of each egg. For example,

Encourage the children to calculate mentally and to use as few weights as possible.

#### 2 Birds' eggs *(addition and subtraction)*

- Show the children pictures of two eggs, labelled as shown.

  Discuss how to find

  — the total weight by changing kilograms and grams to grams before adding. For example,

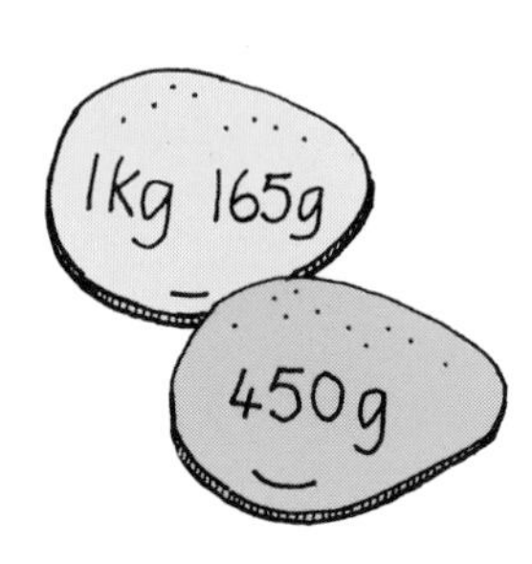


$$\begin{array}{lr} 1\,\text{kg}\ 165\,\text{g} \longrightarrow & 1165\,\text{g} \\ & +\ 450\,\text{g} \\ \hline & 1615\,\text{g} \longrightarrow 1\,\text{kg}\ 615\,\text{g} \end{array}$$

  The total weight is **1 kg 615 g.**

— the difference in weight by changing kilograms and grams to grams before subtracting. For example,

$$1\,\text{kg}\ 165\,\text{g} \longrightarrow \begin{array}{r} 1165\,\text{g} \\ -\ 450\,\text{g} \\ \hline 715\,\text{g} \\ \hline \end{array}$$

The difference in weight is **715 g.**

- Repeat for other examples.

UA4d/4 N4a/4 SSM4ab/4,4a/4→5
AS/D2 ME/C1
N/4k M/4a

## Textbook page 75 *Weight: kg and g, addition and subtraction*

In question 1, the children must choose from the given set of weights, which includes two 20 g weights but only one of each of the others.

In questions 5 and 6, the children should change the weight of each egg to grams before adding or subtracting. The answer, where appropriate, should be converted back to kilograms and grams.

In question 7, use a hard-boiled egg. The weight of a hen's egg is about 70 grams.

# READING SCALES

SSM4ab/4
ME/C6 ME/D2,8
M/3b M/4a

In Heinemann Mathematics 5, the children met scales where an interval represented 10 g, 20 g, 50 g or 100 g. The children are now introduced to scales with intervals of 5 g, 25 g and 200 g.

The APEC context continues with Debbie weighing food for the animals and some of the animals themselves.

## Introductory activities

### 1 Scales

Tell the children that Debbie weighs food for the animals on different scales.

- Draw and label a scale like this on card or on the chalkboard.

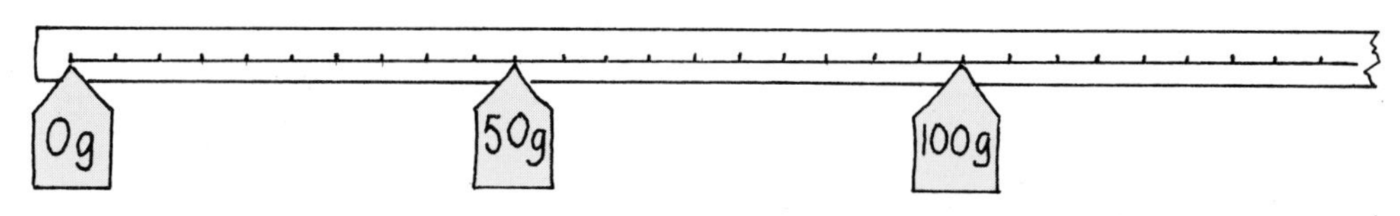


Ask,

'How many intervals are there between 0 g and 50 g?' (10)

'What does one interval represent?' (5 g)

Indicate the weights of some animal foods on the scale and ask the children to read them.

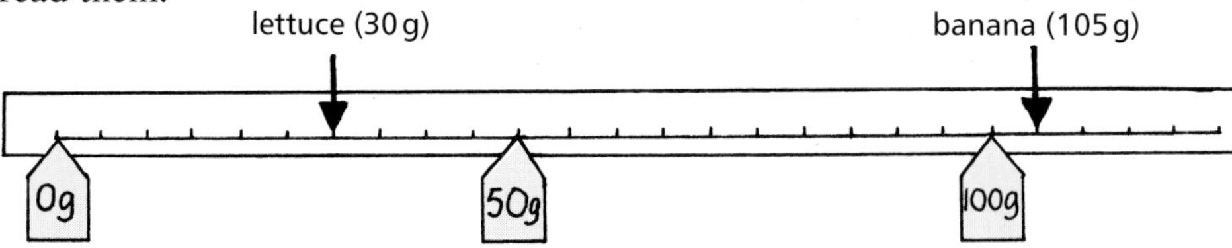


- Repeat for other scales. For example,

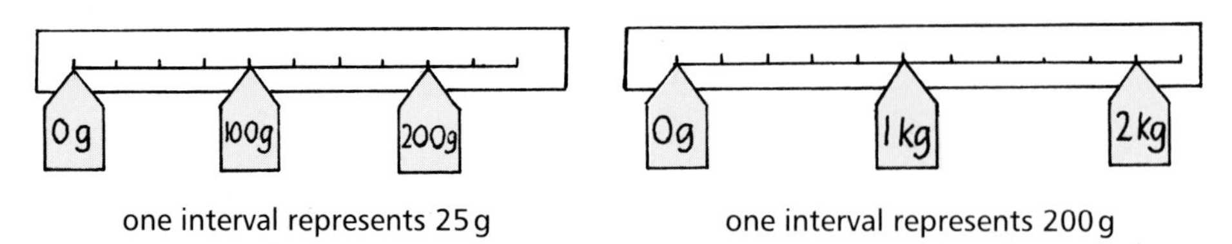


## 2 Spring balances

- Show the children a spring balance. Demonstrate, by attaching an object to the hook of the balance, that the pointer moves **down**, so the scale starts at 0 g at the top.

  Draw and label a vertical scale like this to represent a spring balance. Indicate the weights of some animal foods on the scales.

  Discuss the scale and identify what each interval represents:

  4 intervals represent 100 g

  1 interval represents 25 g

  Ask the children to read the weights of the oats and the turnip.

0g
oats (50 g)
100g
turnip (175 g)
200g

- Repeat for other examples.

## 3 To the nearest mark

- Draw and label a scale like this on card or on the chalkboard.

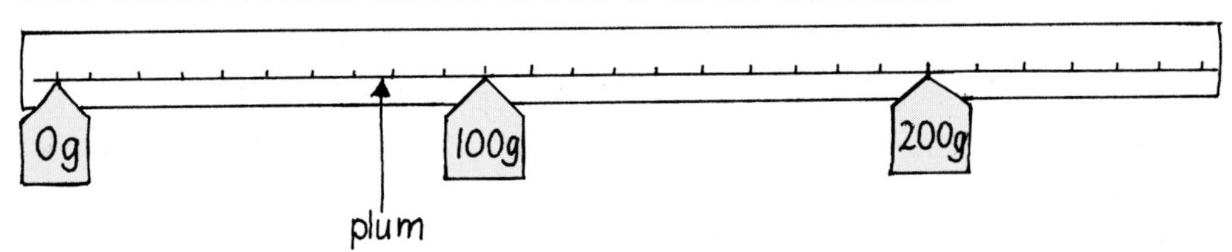


Ask children to estimate the weight of the plum. For example,

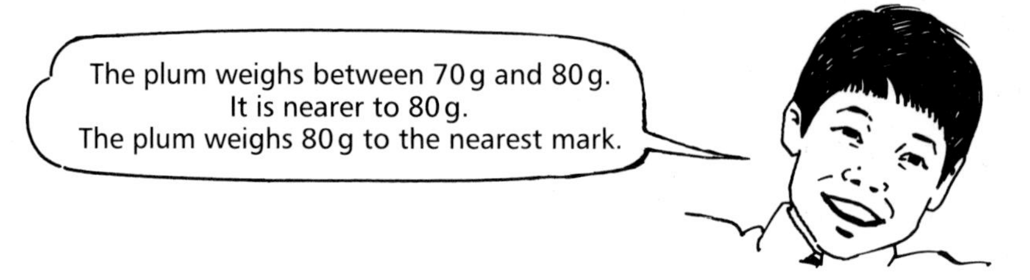


- Repeat for other fruits and other scales. For example,

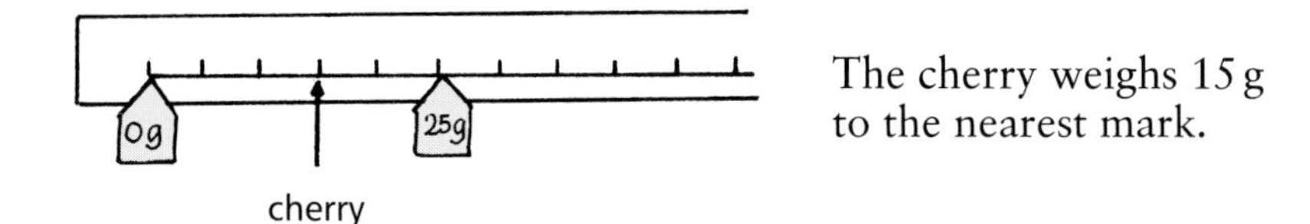


The cherry weighs 15 g to the nearest mark.

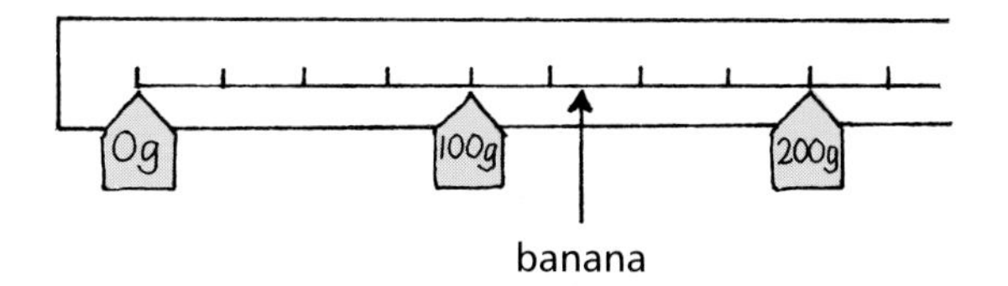


The banana weighs 125 g to the nearest mark.

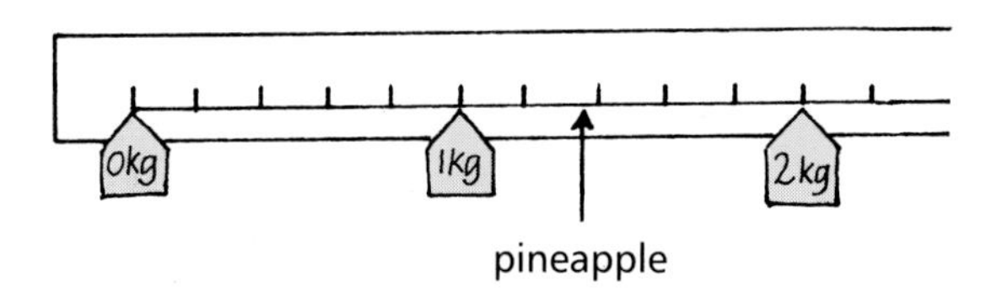


The pineapple weighs 1 kg 400 g to the nearest mark.

## 4 Practical activities

Ask the children to weigh a variety of objects using different types of scale.

The children should be encouraged to

— make sure the pointer is at zero before use

— identify the value of each interval

— record weights to the nearest mark.

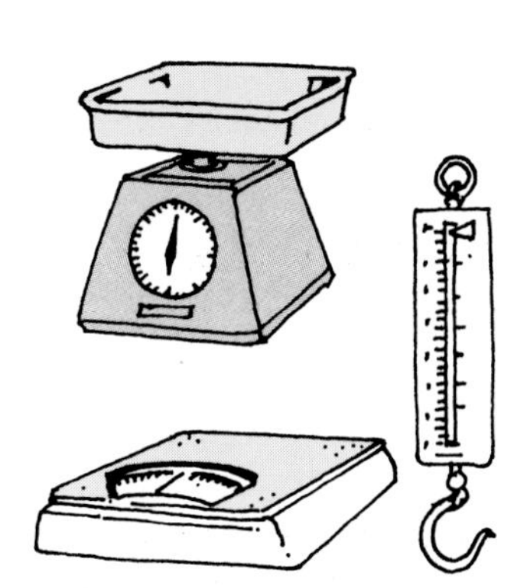

## Textbook pages 76 and 77 *Weight: reading scales*
## Workbook page 24

UA2ac/4 SSM4ab/4
ME/C6 ME/D8
N/3b M/4a

On Textbook page 76, in question 1, the scales have weight intervals of 10 g and 5 g.

In question 2, the circular scale has weight intervals of 20 g, and on the linear scales, the weight intervals represent 5 g, 20 g and 25 g respectively.

On Textbook page 77, in question 1, the spring balances have weight intervals of 200 g and 25 g. The circular scales in question 2 have weight intervals of 100 g and 5 g. The linear scales have weight intervals of 10 g and $\frac{1}{2}$ kg or 500 g.

In question 3, it is expected that the children will use a spring balance to weigh the jacket, and scales to weigh the book.

On Workbook page 24, the children draw pointers to indicate the given weights. It is essential that the children calculate what each interval represents before attempting to draw the pointers. In question 2, if the most suitable scale is chosen, the children should mark

— Benny's weight on the circular scale

— Smudge's weight on the first spring balance

— Patch's weight on the second spring balance.

R23

# MULTIPLICATION, PRACTICAL WORK

N4a/4 SSM4a/4→5
MD/D3 ME/C1
N/4l M/4a

The section is concluded by introducing the multiplication of kilograms and grams by a whole number.

The context continues with the purchase of animal food for APEC.

## Introductory activity

### Total weight

- Display a bag labelled as shown.

  Discuss how to calculate the weight of 7 of these bags.

$$\begin{array}{r} 590\,\text{g} \\ \times 7 \\ \hline 4130\,\text{g} \end{array} \longrightarrow 4\,\text{kg}\ 130\,\text{g}$$

  The total weight of 7 bags is **4 kg 130 g.**

- Repeat for examples such as

$$1\,\text{kg}\ 275\,\text{g} \longrightarrow \begin{array}{r} 1275\,\text{g} \\ \times 3 \\ \hline 3825\,\text{g} \end{array} \longrightarrow 3\,\text{kg}\ 825\,\text{g}$$

  The total weight of 3 bags is **3 kg 825 g.**

UA2abc,4d/4 N4a/4
SSM4b/4 SSM4a/4→5
PSE MD/D3 ME/C1 ME/D2,6,8
P/4a N/4l M/4ad

### Textbook page 78 *Weight: multiplication, practical work*

In question 1, parts (c) and (d), the children are expected to convert to grams before multiplying.

In question 2, the children should find the weight of 5 bags of dog biscuits first (9 kg) and then use this result to find the weight of the other food ($4\frac{1}{2}$ kg of mixed vegetables).

In question 4, the children are expected to choose appropriate objects by estimating then checking their weight. A range of scales should be made available.

**Problem solving**

In question 5, several possible strategies may be used for finding the weights of the given items. For example,

— Chair: Find own weight using bathroom scales.

Hold chair and read scales again.

Calculate the difference between readings to find the weight of the chair.

— Sand: Measure the sand into a 1 litre container.

Pour the sand into the tray of a set of kitchen scales.

Read the weight.

— Water: Weigh a 1 litre container.

Fill the container with water then weigh it again.

Calculate the difference between the two to find the weight of the water.

1 litre of water weighs about 1 kg or 1000 g.

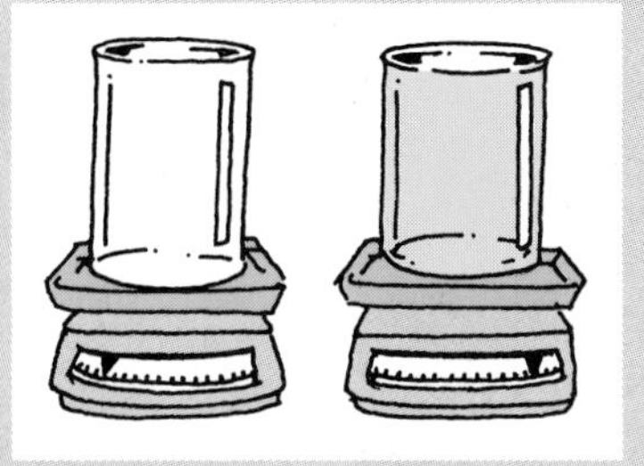

# Lands Beyond

## A context for area and angles

The work on

— Area (Textbook pages 79 and 80, Workbook pages 25–7)

— Angles (Textbook pages 108–10, Workbook page 22)

is set in the context of 'Lands Beyond', a fantasy world. The mathematical activities are developed in a variety of locations, such as the Guardroom and the Groovy Gallery, which are entered by walking through doors scattered across the strange landscape.

### Introducing the context

The context could be introduced by discussing the illustrations on Workbook pages 25 and 26. Lands Beyond and the characters depicted might be compared with other fantasy lands and characters familiar to the children from computer games, television and books – for example, *Alice in Wonderland* and *Rebecca's World.*

Other activities like those suggested below could be developed while the work on the Textbook and Workbook is in progress.

### 1 Creating Lands Beyond

- The children could build a large 3D frieze of Lands Beyond showing plants, animals and landscape using brightly coloured paints, paper, etc. They could also make up detailed factsheets and include information such as size, colour, habitat and special features. These factsheets might be used for a database activity using fields such as Name, Height, Weight, Habitat and Special Features.
- Using cereal boxes or card, the children could make opening doors and place these on the landscape.

- Puppets made from card, paper fasteners and string could be created, of the characters on the Workbook pages and of other characters the children think they might meet when they walk through the doors. These could be displayed on or beside the frieze.

## 2 Adventures in Lands Beyond

Working in groups, the children could create a story to tell what happens and who they meet when they walk through their selected door. Their adventure could be shared with the rest of the class

— by telling the story of what happened and using the puppets made in activity 1 to illustrate it

— by making an illustrated book

— by creating a cartoon strip

— using mime and/or movement

— by making a short audio or video-tape of the story.

UA2ac,3ab,4bc/4
SSM4c/4
ME/C3 PFS/D1
M/4bc

# Area

## Overview

This section

- revises measurement of areas of shapes, including irregular shapes, by counting squares and half squares
- investigates finding the area of a rectangle by 'multiplying the number of rows by the number of squares in each row' or vice versa
- applies this method to finding areas of composite shapes formed by joining together a number of rectangles and/or squares
- investigates how rectangles and squares with the same perimeter (cm) can have different areas ($cm^2$).

| | Teacher's Notes | Textbook | Workbook | Reinforcement Sheets |
|---|---|---|---|---|
| **Lands Beyond: a context for area and angles** | 166 | | | |
| **Area by counting squares** | 170 | | 25, 26 | |
| **Area of a rectangle** | 172 | 79 | | |
| **Area of a composite shape** | 174 | 80 | 27 | 24 |

An extension activity is related to the above section of work as follows:

| | Teacher's Notes | Extension Textbook |
|---|---|---|
| **Area: the square metre, $m^2$** | 288 | E18 |

Teaching notes for the Extension Textbook are in a separate section at the end of the Teacher's Notes.

# Resources

## Useful materials

- 1 cm squared paper
- coloured pencils
- other materials suggested within the introductory activities

## Assessment and Resources Pack

### Assessment

*Measure Check-up 3*
Textbook pages 79, 80
Workbook pages 25–7
(Area: irregular shapes, rectangles, composite shapes)

*Round-up 1*
Question 4

### Resources

*Problem Solving Activities*
16 Threes and fours (Drawing shapes of specified areas)
17 It's a grey area (Investigation involving area and shape patterns)

*Resource Cards*
18 APEC plan (Area of a rectangle)

# Teaching notes

In Heinemann Mathematics 5, the children found the areas of shapes, including irregular shapes, by counting squares and half squares, and square centimetres ($cm^2$) and half square centimetres. The square metre ($m^2$) was also introduced.

This section of Heinemann Mathematics 6

— consolidates much of the above work

— investigates a 'quick way' of finding areas of rectangles

— uses this 'quick way' to find areas of composite shapes.

The formula $A = l \times b$ for the area of a rectangle is **not** formally introduced until Heinemann Mathematics 7.

Although the section begins on Textbook page 79, the children are immediately directed from there to Workbook pages 25 and 26.

The work is set within the context of Lands Beyond.

## AREA BY COUNTING SQUARES

UA2c,3ab/4 SSM4c/4
ME/C3
M/4c

As many children have difficulty with the concept of area, some revision of finding areas by counting squares may well be necessary.

### Introductory activities

#### 1 Emblems *(counting squares and half squares)*

- Discuss the strange emblems found on the shields of the characters from Lands Beyond.
- Draw an emblem as shown on an available squared grid – for example, on the chalkboard, squared paper or an overhead projector transparency. Discuss how to find the area of the emblem in square units.

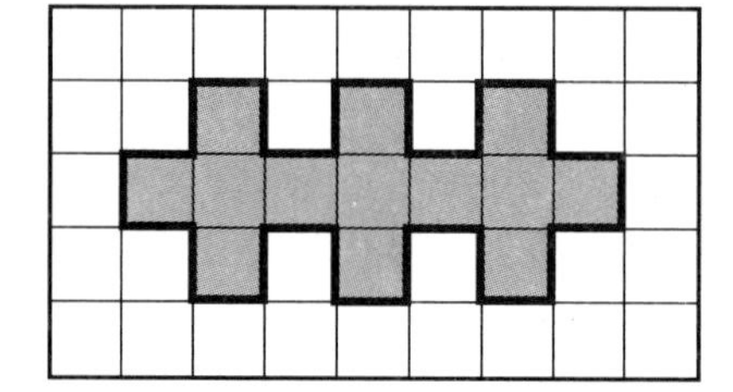

Area = 13 squares

- Repeat for other emblems, including some with half squares. For example,

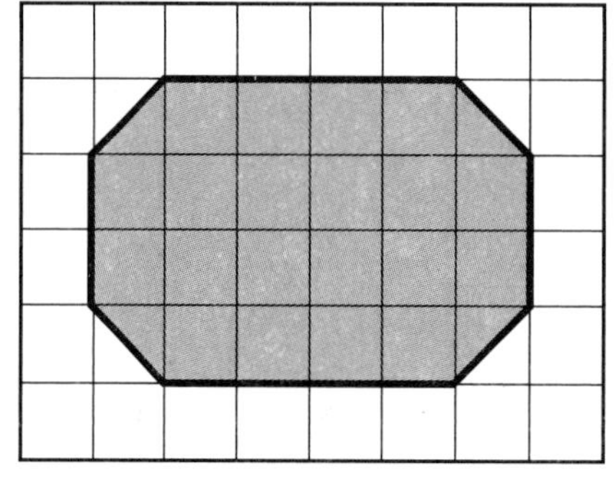

Area = 22 squares

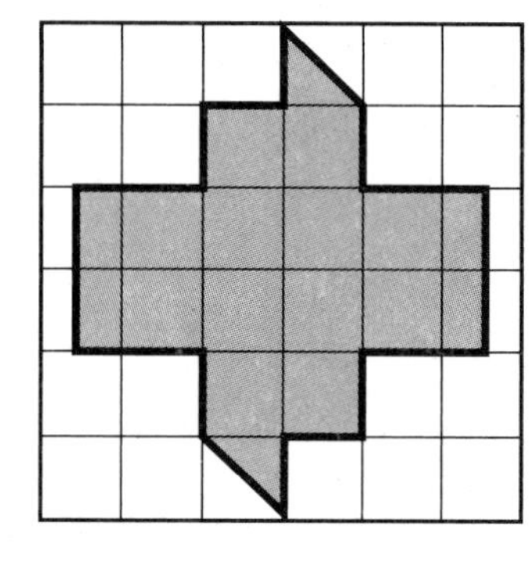

Area = 15 squares

## 2 More emblems *(irregular shapes)*

- Introduce an emblem with some curved edges as shown. Discuss how to deal with the part squares which are not half squares.

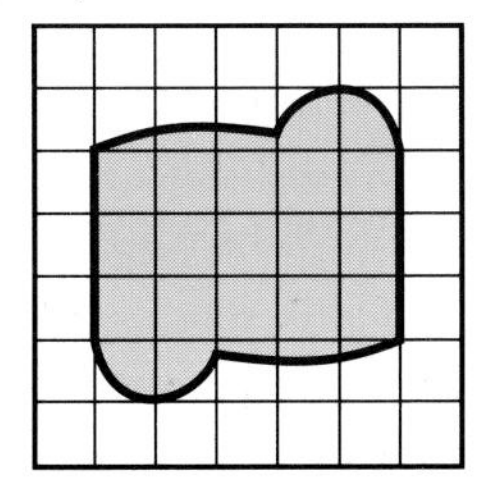

Area about 19 squares

- Repeat for other emblems. For example,

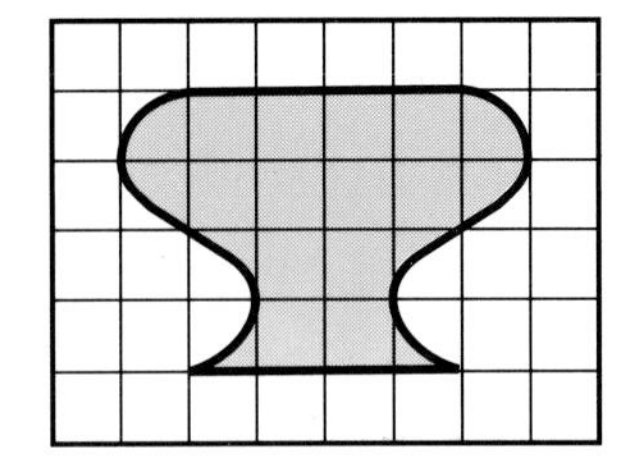

Area about 16 squares

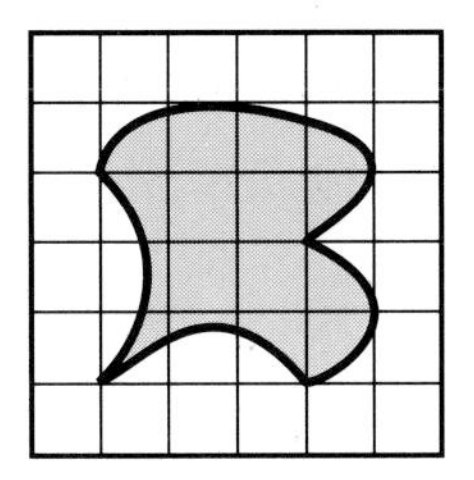

Area about 11 squares

Emphasize that these areas are approximate and must be stated as 'about so many squares'.

- **Note:** The method for dealing with the part squares in irregular shapes should be the one previously used by the children. The method favoured in the Teacher's Notes for Heinemann Mathematics 5 was

  'count the "bits" which are half squares or greater as whole squares and ignore the "bits" of squares smaller than a half square'

as shown below.

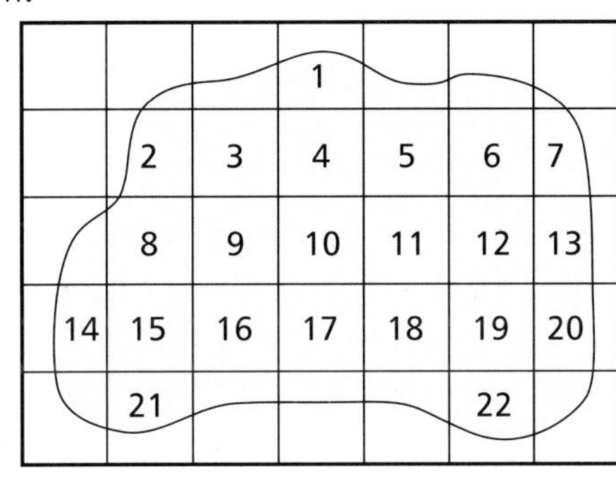


Area about 22cm²

Other valid methods discussed on pages 156 and 157 of the Heinemann Mathematics 5 notes are

— count all the 'bits' as a whole squares and then halve this number

— match up 'bits' of squares to make whole squares.

UA2c,3ab,4bc/4 SSM4c/4
ME/C3
M/4c

## Workbook pages 25 and 26

### *Area: irregular shapes, rectangles*

The Lands Beyond context should be discussed if it has not previously been introduced. The work is concerned with finding areas of shapes by counting squares.

On Workbook page 25, the areas are measured in square centimetres and a majority of the shapes are irregular.

In question 1, some children may need to be reminded that the areas of the final three shapes should be recorded as, for example, 'about 12 cm²' and read as 'about 12 square centimetres'.

Question 2 provides an important type of activity. Although straightforward if the children choose a design with straight edges only, it is much more difficult if they decide upon a design with a curved edge. In the latter case, areas from 25 $cm^2$ to 35 $cm^2$ should be accepted.

On Workbook page 26, the areas are measured in 'squares'. The shapes are rectangles, and the children explore a 'quick way' of finding the areas before any direct teaching takes place.

In question 1, point out that the rows go 'across' the doors. Some children will find the 'area of door in squares' by counting individual squares, while others may realize that they can multiply the 'number of rows altogether' and the 'number of squares in each row'.

In question 2, the children should make use of the pattern obtained in the table in question 1. There should be some teacher-led discussion of their written suggestions.

## AREA OF A RECTANGLE

UA3ab/4 SSM4c/4
ME/C3
M/4c

The 'quick' method of multiplying the 'number of rows' and the 'number of squares in each row' is applied to find areas of rectangles drawn on a square grid. A design has been drawn on each rectangle to obscure some of the squares and thus prevent the children from counting individual squares. The areas are measured in 'squares' and in square metres.

The Lands Beyond context continues with a visit to the Guardroom, where strange characters, carrying rectangular shields with emblems, are encountered.

## Introductory activities

### 1 Doors in Lands Beyond *(areas of rectangles on square grids)*

Refer to the doors on Workbook page 26. Ask the children to describe a 'quick way' of finding the area of a rectangle. Consolidate this using two or three examples.

Jewel Room

'How many rows are there?

How many squares in each row?

How many squares altogether?

How did you find your answer?

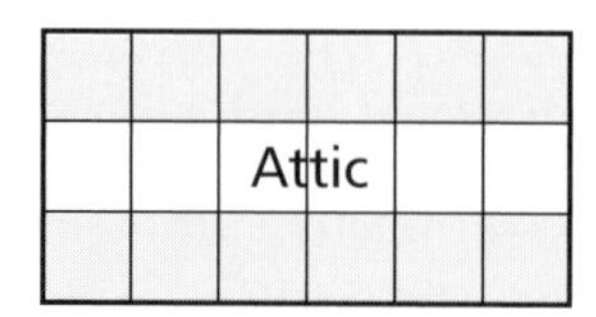


What is the area in squares of • the Jewel Room door • the Attic door?'

## 2 Jewel Room *(areas of rectangles with some squares obscured)*

- 'Large rectangular brooches and pendants with designs are found in the Jewel Room. Find the area in squares of each brooch and pendant.'

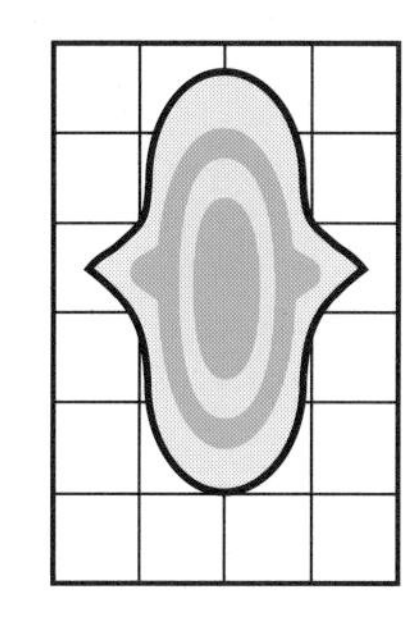

Discuss how the 'quick way' can be used even though some of the squares are obscured and so cannot be counted individually.

- Emphasize an efficient form of language to describe the 'quick way'. For example,
  — 'multiply the number of rows by the number of squares in one row' (or vice versa)
  — 'count the number of squares in one row and the number of rows, then multiply'.

## 3 Jewel Room floor *(using square metres)*

- Remind the children of the size of a square metre by making a metre square using four metre sticks. Revise the name of the unit, the square metre, and its symbol, $m^2$.

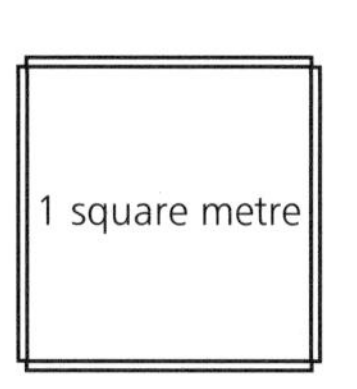


- Draw a rectangle to represent the tiled Jewel Room floor, which is partly covered by a rug. Ask the children to find the area of the floor, $12\,m^2$ (3 rows each with 4 squares). Emphasize that the area is 12 **square metres** and that this is written as $\mathbf{12\,m^2}$.

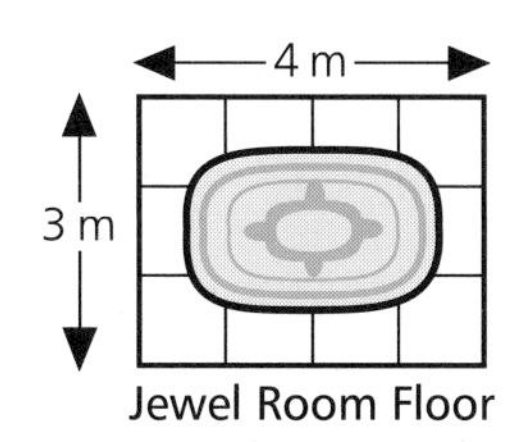

Jewel Room Floor

UA2a,3ab/4 SSM4c/4
ME/C3
M/4c

## Textbook page 79 *Area: rectangles*

In question 1, the examples are graded in that all the squares around the perimeters of the rectangles are visible in parts (a) to (d), whereas fewer squares are visible around the perimeters in parts (e) to (g). The children might be asked to show the multiplications used when they record their answers. For example, in part (a),
Area = $9 \times 6 = 54$ squares.

In question 2(b), the answer should be given as '$20\,m^2$' and not just '20 squares'.

# AREA OF A COMPOSITE SHAPE

UA2c,3ab/4 SSM4c/4
ME/C3
M/4c

The method of finding the area of a rectangle by multiplying the number of squares in one row and the number of rows is now applied to finding the areas of composite shapes which consist of a number of rectangles and/or squares.

An investigation of the areas of rectangles with the same perimeter is included. The Lands Beyond context continues with a visit to the Groovy Gallery, which houses some rather unusual paintings.

## Introductory activities

### 1 Building up shapes

- Give the children squared paper, working individually or in pairs. Ask them to
  - — draw and cut out these shapes
  - — find the area of each shape and record it on the shape.

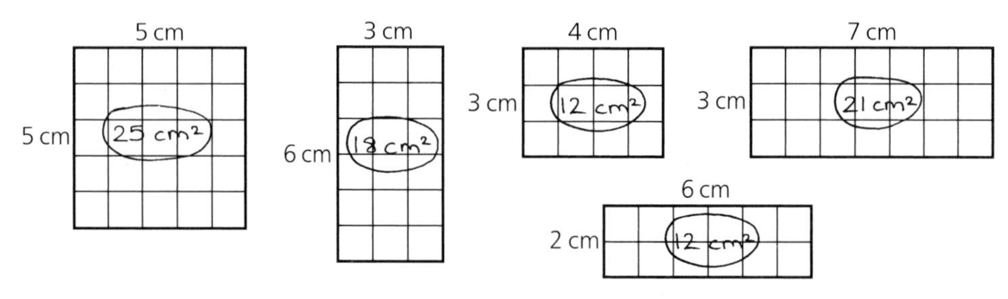


- Ask the children to make a single rectangle by fitting together the two rectangles with areas of $21\,cm^2$ and $18\,cm^2$. Discuss how the area of the single rectangle, $39\,cm^2$, can be found by adding the areas of the smaller rectangles, $21\,cm^2 + 18\,cm^2$. This can be checked by the children directly, ie by multiplying the number of rows, 3, and the number of squares in one row, 13.

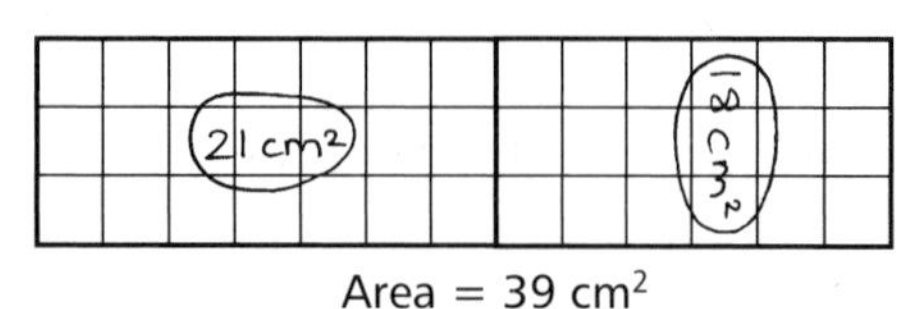

Area = 39 cm²

- Through discussion, establish that the area of any shape made from the two smaller rectangles has an area equal to the total of the two areas.

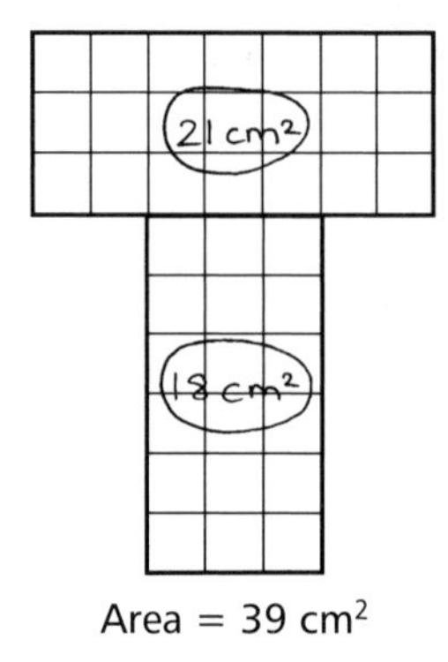

Area = 39 cm²

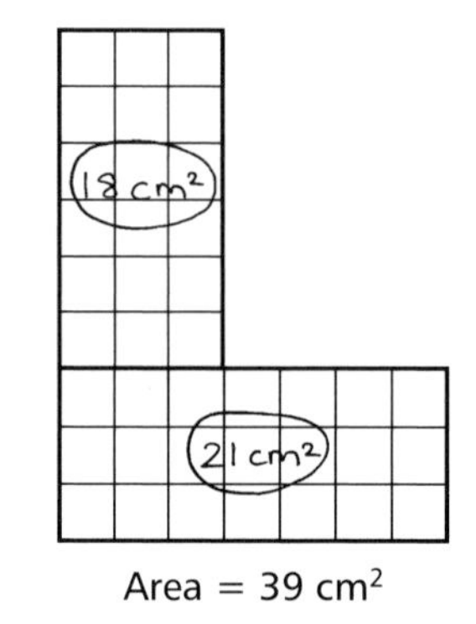

Area = 39 cm²

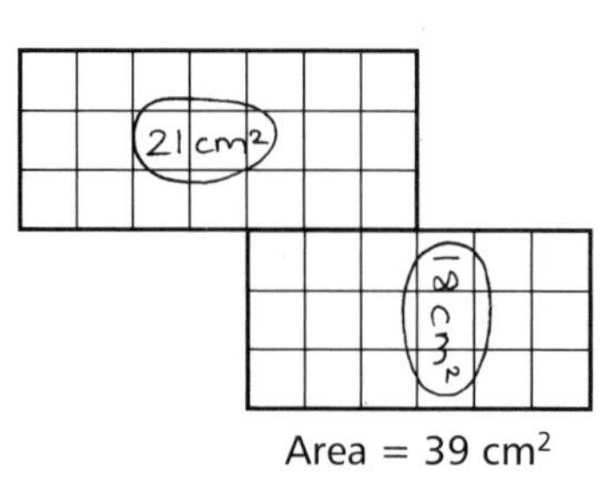

Area = 39 cm²

## 2 Groovy Gallery shapes

The starting rectangles from the above activity can be used to build more complex composite shapes. Ask the children to

— make a single rectangle from **three** of the starting rectangles, find its area by addition and check by multiplying, this time $3 \times 17$.

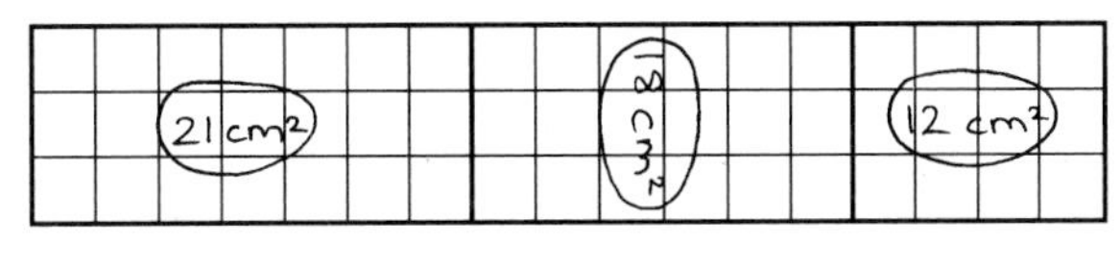


— make a single rectangle using all **five** of the starting rectangles, find its area and check as before:

$$\text{Area} = 12 + 25 + 12 + 18 + 21 = 88\,\text{cm}^2$$

$$\text{Area} = 8 \times 11 = 88\,\text{cm}^2$$

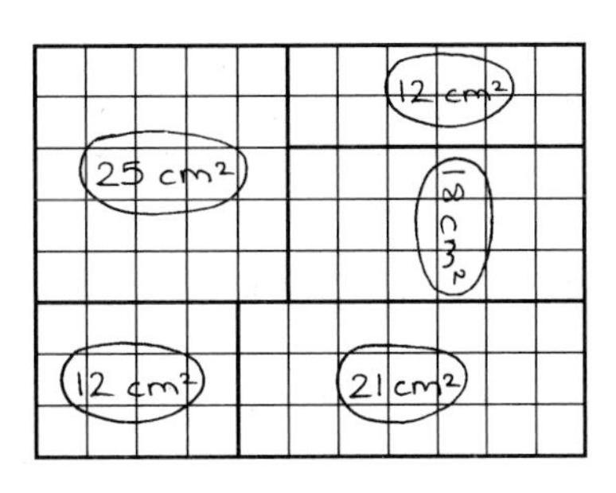


— make a design for the Groovy Gallery (Textbook page 80) using all **five** shapes, and stick this design in their jotters. For example,

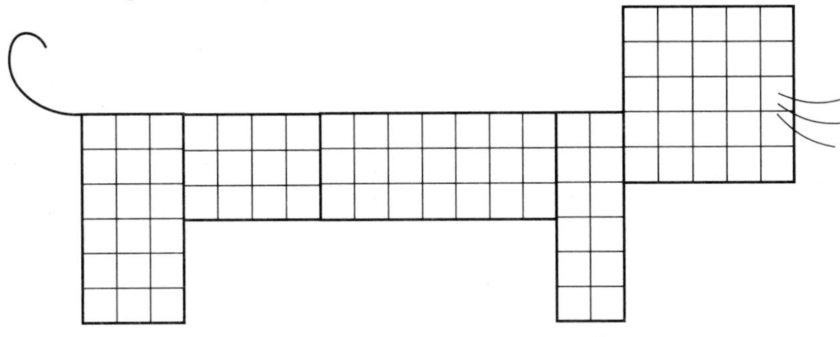

Area of cat from Lands Beyond = 88 cm²

With this design it is not possible to check the area using a single multiplication.

## 3 Groovy Gallery designs

Show the children some designs consisting of a number of rectangles and/or squares as shown below. Discuss how to subdivide the designs so that the areas can easily be found. Some designs should have sufficient small squares obscured to ensure that the children cannot find the areas simply by counting individual squares.

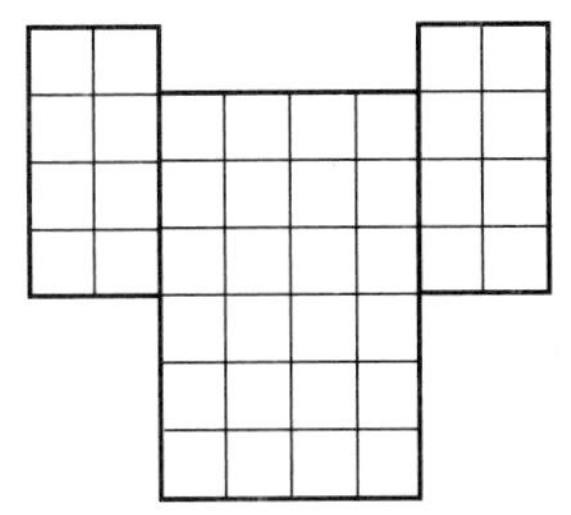

Area = 8 + 24 + 8 = 40 squares

Area = 18 + 16 + 24 = 58 squares

**At this point the children could try Textbook page 80.**

## 4 Splitting the designs

- Draw a design made from rectangles and/or squares on an available squared grid – for example, on the chalkboard, squared paper or an overhead projector transparency. The perimeter lines of these rectangles and/or squares should **not** be highlighted.

  Ask the children where they would draw lines to divide the design into rectangles or squares, so that the area can be found by adding the areas of these constituent shapes.

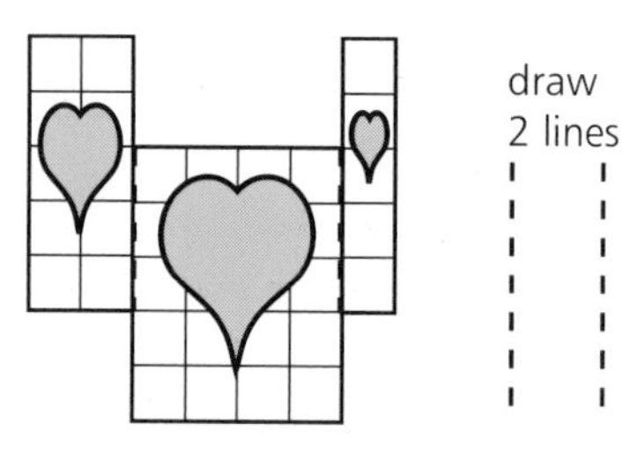


Area = 10 + 20 + 5 = 35 squares

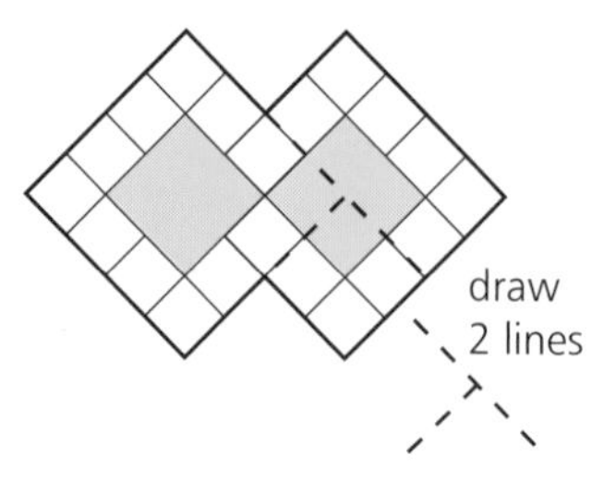


Area = 16 + 8 + 4 = 28 squares

- Some children might suggest other ways of drawing lines to subdivide the design. These should be discussed and the children encouraged, for efficiency, to draw as few lines as possible. For example,

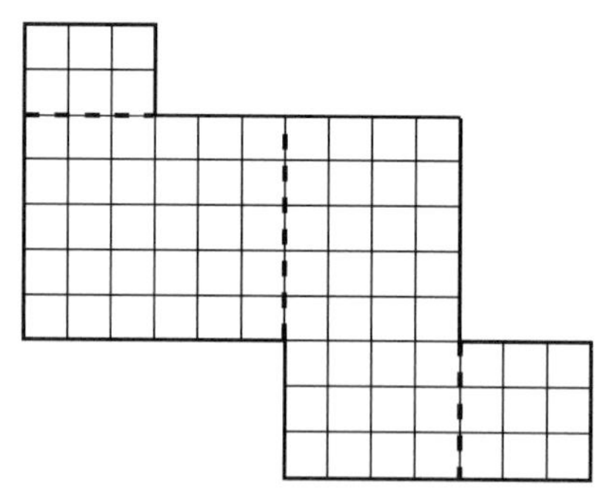

3 lines (3 rectangles and 1 square)

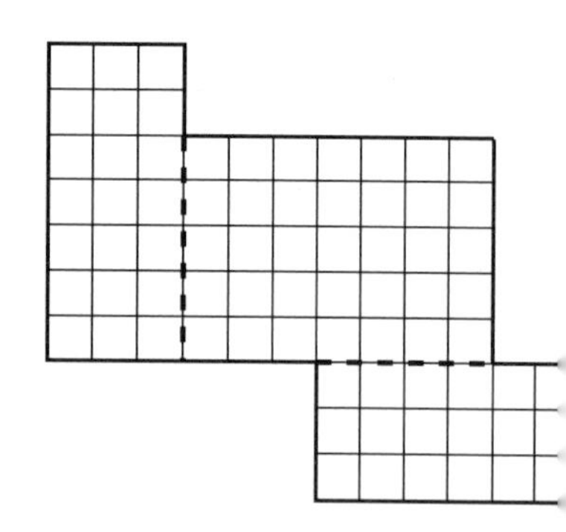

2 lines (3 rectangles)

UA2c,3ab,4b/4 SSM4c/4
ME/C3 PFS/D1
M/4bc

## Textbook page 80 *Area: composite shapes, perimeter*
## Workbook page 27

If the Groovy Gallery is new to the children, it should be introduced. One of the doors in Lands Beyond leads to the gallery, where there are strange designs and paintings.

On Textbook page 80, the designs have been divided into rectangles and/or squares to help the children to find the total areas in 'squares'. Encourage the children to use the 'quick way' of finding the areas of the constituent rectangles and squares. It is, in fact, difficult to do otherwise for those designs which have pictures superimposed on them.

On Workbook page 27, in question 1, the children have first to divide the designs before finding the areas. They should note the area of each constituent shape in turn at the side of the frame, find the total and insert this in the answer space.

In questions 2 and 3, the children should notice that the perimeters of the paintings are all the same but the areas are different. This result should be discussed and linked with the result of a related investigation which appears in the Length section on Workbook page 17, if this has already been completed. The result is that shapes with the same area can have different perimeters.

R24

# Additional activity

## Greatest area

- Ask the children to draw on centimetre squared paper different rectangles or squares, each having a perimeter of 16 centimetres. Challenge them to find the shape with the greatest possible area.
- Repeat for rectangles or squares with a perimeter of 36 centimetres.

UA2ad/4
SSMabc/4
SSM4a/4→5
ME/C6 ME/D3
M/4ac

# Volume

## Overview

This section

- introduces the millilitre, the notation ml and the relationship 1 litre = 1000 millilitres
- introduces marking and reading scales in millilitres (ml)
- revises volumes of cuboids in cubic centimetres ($cm^3$)
- introduces 1 litre as $1000\,cm^3$ and $1\,ml = 1\,cm^3$.

| | Teacher's Notes | Textbook | Workbook | Reinforcement Sheets |
|---|---|---|---|---|
| **APEC: a context for length, weight and volume** | 146 | | | |
| **Millilitres: reading scales** | 180 | 81, 82 | 28 | 25 |
| **Volumes of cuboids, $cm^3$** | 183 | 83 | | |
| **1 litre = $1000\,cm^3$** | 184 | 84 | | |

An extension activity is related to the above section of work as follows:

| | Teacher's Notes | Extension Textbook |
|---|---|---|
| **Volume: cubic centimetres, $cm^3$** | 289 | E19 |

Teaching notes for the Extension Textbook are in a separate section at the end of the Teacher's Notes.

# Resources

## Useful materials

- teaspoon, cup and litre jug
- measuring jars
- containers with volumes less than 1 litre
- centimetre cubes (interlocking cubes would be advantageous)
- a plastic cube container which holds 1 litre
- plastic or wooden 'flats' each with a volume of 100 $cm^3$
- centimetre squared paper
- other materials suggested within the introductory activities

## Assessment and Resources Pack

### Assessment

*Measure Check-up 4*
Textbook pages 81–4
Workbook page 28
(Millilitres and cubic centimetres, scale)

*Round-up 3*
Question 4(b)

### Resources

*Problem Solving Activities*
18 True brew (volumes in millilitres)

*Resource Cards*
23, 24 True or false cards
(Practical work)

# Teaching notes

In Heinemann Mathematics 5, the children were introduced to the quarter litre ($\frac{1}{4}l$) and the cubic centimetre ($cm^3$). This section begins by introducing the millilitre.

The context is the Animal Protection and Education Centre (APEC) which was first introduced for work on length and was also for work on weight. For a detailed introduction to this context, see pages 146 and 147 of these Teacher's Notes.

## MILLILITRES: READING SCALES

SSM4b/4
ME/C6 ME/D3
M/4a

The children should revise the litre, the half litre and the quarter litre. They can then be introduced to the millilitre as a much more accurate unit for measuring volumes.

## Introductory activities

### 1 Medicine bottles at APEC *(the millilitre)*

- Show the children three measuring jugs – a litre, a half-litre and a quarter-litre. They should have met these volumes and be able to explain that

  — two half litres are the same as one litre

  — two quarter litres are the same as one half litre and

  — four quarter litres are the same as one litre.

- Now produce two bottles to represent medicine bottles from APEC. These should contain medicines (coloured water) with volumes of 350 ml and 400 ml.

  Ask two children to find the volumes of the medicines. Using the quarter-litre jug, the children should find that both medicines are more than $\frac{1}{4}l$. Using the half-litre jug, the children should find that the medicines are less than $\frac{1}{2}l$. It is therefore difficult to state the volume of either medicine using the given measuring jugs.

- Discuss the need for another unit to be able to find a volume more accurately. Let the children look at a litre jug which has a scale divided into **1000 millilitres,** and discuss that 1 millilitre is a very small unit of volume. If a hollow centimetre cube is available, this can be used to show the quantity of water which is 1 millilitre. Discuss the word 'millilitre' so that the children realize that it means one thousandth part of a litre.

- The children should look at containers which are marked with a given number of millilitres, such as the medicine spoon, 5 ml, and a medicine bottle with a volume of 200 ml.

## 2 Dispensing medicines at APEC *(reading scales)*

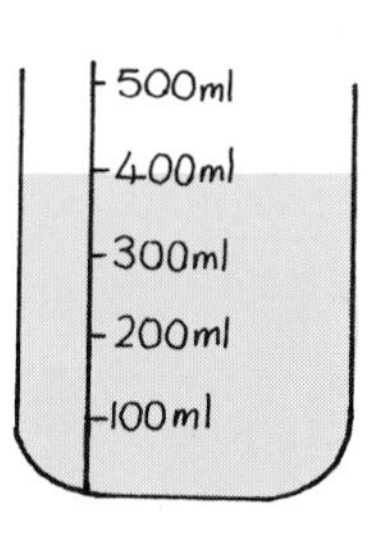


- A large drawing of a scale could be used to investigate
  - — the maximum volume which can be read
  - — the volume represented by each interval (100 ml).

Use coloured paper, held in position by Blu-tack, to represent liquid, so that different volumes can be read using the scale. At this stage, each volume, represented by the top edge of the paper, is aligned with a mark and read to that mark.

- Repeat the above activity for scales with intervals of 50 ml and 10 ml.

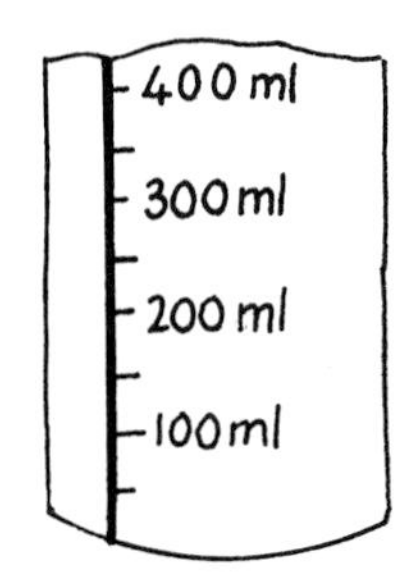


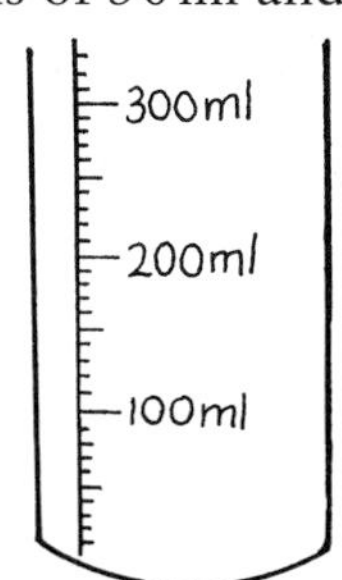


- Let the children look at different measuring jugs and jars to identify the scale intervals. These should include scales with intervals of 100 ml, 50 ml, 20 ml and 10 ml.
- Now let the children use an appropriate measuring jug to read the volume of the half-litre jug as 500 ml, and the quarter-litre jug as 250 ml, and to find the volumes of the two medicines introduced in Activity 1. They should also find the volumes of some other containers used at APEC for giving medicines – for example, a tumbler, a mug, a cup and a tablespoon. It is essential that each container has a **marked** level, so that the children can be asked to find volumes which align with a mark on the appropriate scale.
- Finally, the children can be asked to pour out different volumes of 'medicine'. For example,

  'Pour out 260 ml of Bear Beverage.'

**At this point the children could try Textbook page 81 and Workbook page 28.**

## 3 Medicines in stock *(reading to the nearest mark)*

- Use the scale diagram with intervals of 100 ml which was prepared for activity 2. Show the level representing medicine not aligned to a mark, and discuss how the volume might be expressed as

  '200 ml **to the nearest 100 ml**'

  The children may also wish to give an estimate of the volume as 'about 180 ml'.

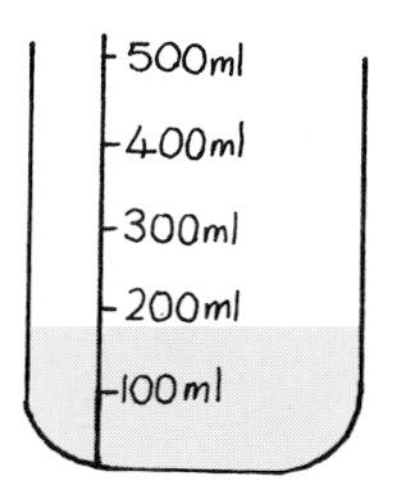


Move the level to give the children practice with other readings to the nearest 100 ml.

Insert the 50 ml marks on the scale again, and ask the children how a reading might be made this time – for example, 'about 240 ml' and '250 ml to the nearest 50 ml'.

- Use the scale with the 50 ml intervals. Establish that levels would be read

  **'to the nearest 50 ml'.**

  Discuss an example like the one illustrated, so that it is read as '250 ml to the nearest 50 ml'.

  300 ml
  200 ml
  100 ml

  Repeat for scales with other intervals – 250 ml, 20 ml and 10 ml.
- Ask the children to find the volume of prepared 'medicines'. These could be labelled with names such as Ostrich Ooze, Lion Lotion and Panther Potion. For each, the children should select an appropriate measuring jug or jar, identify the scale interval, and then find the volume of the 'medicine' expressed to the nearest mark.

UA2ad/4 SSM4ab/4
ME/C6 ME/D3
M/4a

## Textbook pages 81 and 82 *Volume: millilitres, reading scales*
## Workbook page 28

The APEC context should be discussed before the children carry out the work on these pages, if this has not already been done during the introductory activities.

On Textbook page 81, the panel at the top should be discussed to consolidate that 1 litre is the same volume as 1000 millilitres.

In question 4, encourage the children to align their answers so that the total volume of the mix is easily added:

| | |
|---|---|
| water | 260 ml |
| plant juice | 310 ml |
| tonic | 180 ml |
| factor X | + 250 ml |
| total volume | 1000 ml or l litre |

On Workbook page 28, in question 1, the children need to write the numbers on the scales as neatly as possible, since they must be able to read them in question 2. For question 4, the children require a selection of containers. Labels would be useful for them to name each feed.

On Textbook page 82, for question 4, sets of three containers could be made ready for use by pairs of children. Sets could then be exchanged, so that each pair of children finds six volumes. Two pairs of children should then compare their answers.

R25

# VOLUMES OF CUBOIDS, $cm^3$

SSM4c/4
ME/D3
M/4c

Work on the cubic centimetre which was introduced in Heinemann Mathematics 5 is now extended. The number of centimetre cubes in a layer is found, and then this number is multiplied by the number of layers to find the volume of a cuboid.

The APEC context continues with the vet dispensing vitamin cubes.

## Introductory activity

### 1 Stacking vitamins *(building cuboids)*

- Show the children a cuboid with one layer of centimetre cubes like this.

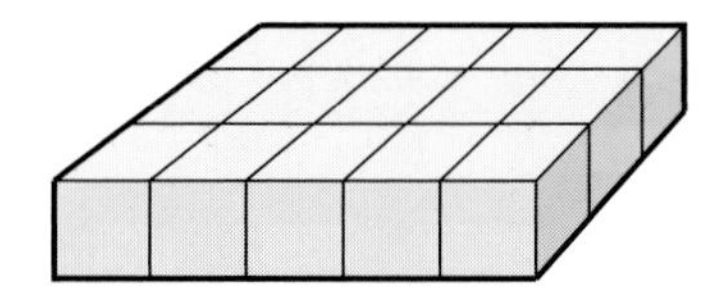

Ask the children to find the volume of the layer. ($15\,cm^3$)

Build a second identical layer on top of the first.

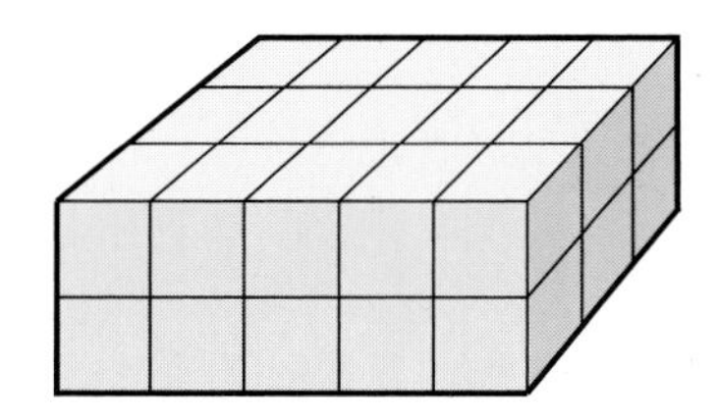

Ask the children how they will find the volume of the cuboid. The steps could be recorded as

Volume of one layer is $15\,cm^3$.

Number of layers is 2.

So volume of the cuboid is $2 \times 15\,cm^3 = 30\,cm^3$.

Ask the children to build a third layer on the cuboid and find the total volume as

$3 \times 15\,cm^3 = 45\,cm^3$

Generalize the method as

'to find the volume of a cuboid, multiply the volume of one layer by the number of layers'.

- Ask the children to find the volumes of these cuboids in the same way:

3 cm by 6 cm by 2 cm

5 cm by 2 cm by 4 cm

- The children could find the volumes of small boxes such as a matchbox and a drawing pin box. They should be encouraged to fit one layer of cubes, and then to build up cubes in one corner to find out how many layers there would be. It is not necessary to fill each box with cubes.

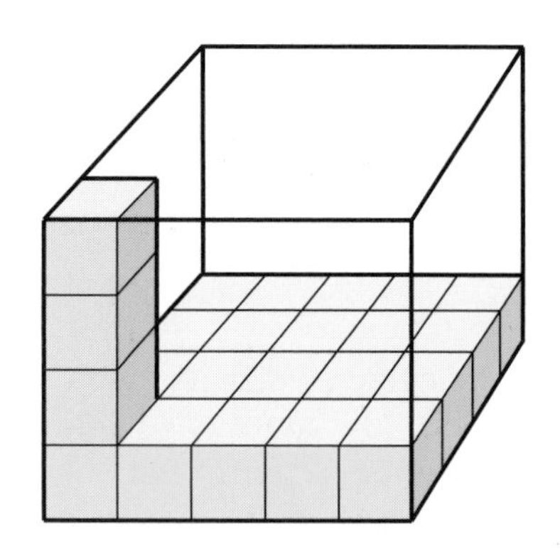

SSM4c/4
ME/D3
M/4c

### Textbook page 83 *Volume: cubic centimetre, cm³*

For question 4, a selection of small boxes is required. These could be made up in sets of three for a pair of children to use.

SSM4c/4
ME/D3
M/4c

## Additional activity

### Finding the dimensions

Ask the children to make cuboids with volumes of $36\,cm^3$, $40\,cm^3$ and $48\,cm^3$, and then record the volume of each layer and the number of layers for each cuboid.

# 1 LITRE = $1000\,cm^3$

The children are given the experience of finding that $1000\,cm^3$ fill a box which has the volume of 1 litre, and so establish that $1\,cm^3$ is equivalent to 1 ml.

SSM4c/4 SSM4a/4→5
ME/D3
M/4c

## Introductory activity

### Special boxes

- Ask the children to fill a plastic box which has a volume of $\frac{1}{4}$ litre with centimetre cubes. 'Longs' from base-ten materials may be useful.

  The children should fill one layer and then find how many of these layers will fill the box. They should find the volume to be $250\,cm^3$.

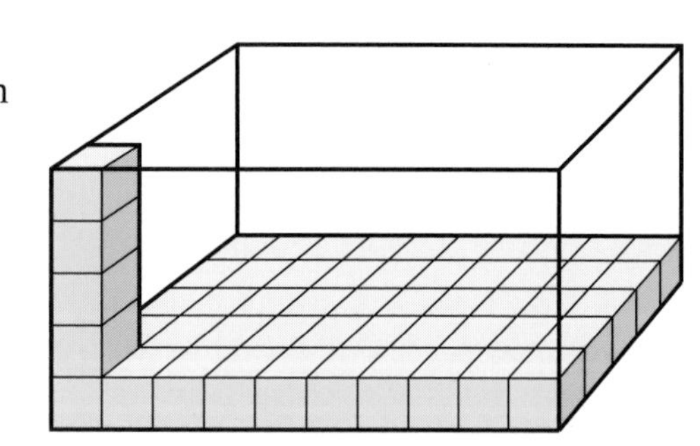

  The children should then find the volume of the box by filling it with water. They should find this to be 250 ml or $\frac{1}{4}$ litre. So $250\,cm^3 = 250\,ml$.

- The children could fill another container and find its volume as $500\,cm^3$ and as 500 ml or $\frac{1}{2}$ litre ($500\,cm^3 = 500\,ml$). The relationship between the cubic centimetre and the millilitre can be discussed now or after the children have carried out the work of Textbook page 84.

SSM4c/4 SSM4a/4→5
ME/D3
M/4c

## Textbook page 84 *Volume: 1 litre = 1000 cm³*

In question 1(b), the children could find the first two answers by repeatedly adding 20. For example,

20 + 20 + 20 makes 60, so that is 3 layers.

60 + 20 + 20 makes 100, so that is 5 layers.

For the last part, some may see 1000 as 10 × 100, which is 10 × 5 layers or 50 layers. Others may be able to find all three answers by dividing by 20.

In question 2, the 'flats' are pieces from base-ten materials that are 10 cm by 10 cm by 1 cm, so each has a volume of $100\,cm^3$.

In question 4, the 'special cube' is a plastic litre cube and is filled by ten of the 'flats'.

# Channel 6TV

## A context for time

The work on Time on Textbook pages 85–91 and Workbook pages 29 and 30 is related to the activities of Class 6 who, having watched the maths programme *Timewarp* on schools television, find out more about how programmes are made by visiting the Channel 6TV studios.

### Introducing the context

The activities suggested below are intended to encourage the children to discuss what goes on inside a modern television studio.

### 1 Television companies

Ask the children to suggest the name of a television company and any programmes it makes or broadcasts. Record their suggestions. For example,

| Scottish | Taggart |
|---|---|
| BBC | Match of the Day |
| Grundy | Neighbours |
| ViaCom | Roseanne |
| Channel4 | Film on 4 |
| Granada | Coronation Street |

### 2 Television schedules

Give groups of children a page from a TV guide. Ask them to note the types of programme broadcast on that day. Discuss possible categories, such as

— soaps

— documentaries

— sport, news

— game shows

— films, cartoons.

Discuss each group's findings.

### 3 6TV studio

- Ask the children to find out about the people who work in a television studio, such as writer, costume designer, make-up artist, director, actor and camera operator.

  An information card could be made for each one.

- Ask them to create a studio background and then add equipment, people and the information cards.

- Pictures of different programmes and presenters could be collected from TV guides and added to the studio picture.

## 4 Time diary

The children could create an illustrated time diary, showing events and times from getting out of bed to entering their classroom. For example,

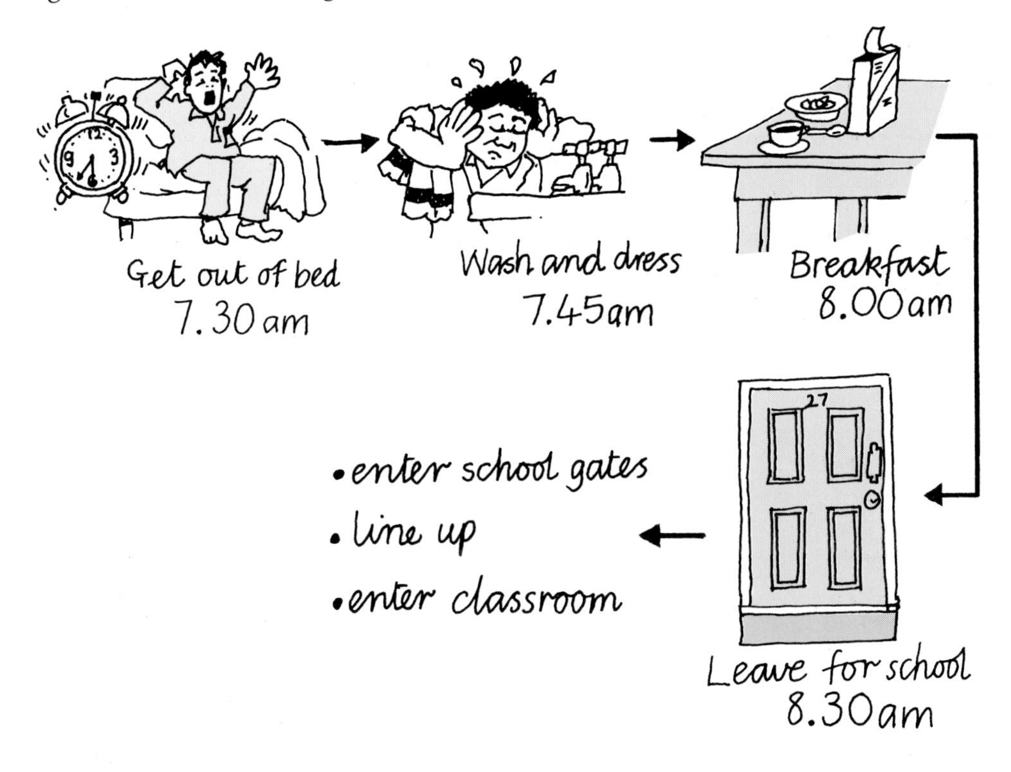

UA2ac,3ab,4d/4
N4a/4
SSM4ab/4
PSE
T/C1,2,3
T/D1,2,3
P/3a
P/4bd
M/3b
M/4a

# Time

## Overview

This section

- revises reading and writing times to the nearest minute from analogue and digital displays, using the notation 'am' and 'pm'
- consolidates and extends durations in minutes and in hours and minutes, with 'bridging' of an hour
- consolidates and extends counting on and counting back given durations in hours and in minutes, with 'bridging' of an hour
- introduces the 24-hour notation, including applications and simple durations
- introduces the second.



Extension activities related to the above section of work are as follows:



Teaching notes for the Extension Textbook are in a separate section at the end of the Teacher's Notes.

# Resources

## Useful materials

- analogue and digital stopwatches (minutes and seconds)
- TV programme guides or listings
- cubes, dice
- other materials suggested within the introductory activities

## Assessment and Resources Pack

### Assessment

*Measure Check-up 5*
Textbook pages 85–8
(Time: 12-hour notation, durations)

*Measure Check-up 6*
Textbook pages 89, 90
Workbook page 29
(Time: 24-hour notation, applications)

*Round-up 2*
Questions 3, 6, 7

### Resources

*Problem Solving Activities*
19 Time diary (12-hour notation, durations)

*Resource Cards*
19–22 Channel 6TV (Times in words and in 24-hour notation)
23, 24 True or false cards (Practical work)

# Teaching notes

In Heinemann Mathematics 5 the work on time included

- — reading times to the nearest minute
- — the notation 'am' and 'pm'
- — durations in minutes and in hours and minutes, and those which 'bridge' an hour
- — calendar work

Heinemann Mathematics 6 revises and extends this work with the exception of calendar work. Calendar work is provided in the 'Other activity' on Textbook page 92.

## 12-HOUR NOTATION: AM AND PM

UA2c,3a/4 SSM4ab/4
T/C1,2
M/3b M/4a

The work begins by revising

- — recording and writing times from analogue and digital displays, using 'past' and 'to'
- — the notation 'am' and 'pm'.

The context begins with Class 6 watching the TV programme *Timewarp*.

## Introductory activities

### 1 Time cards *(analogue and digital times)*

- Prepare a set of cards showing analogue displays, digital displays and times written in words.

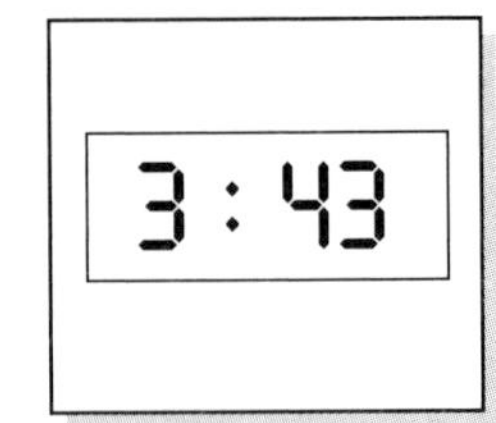

Shuffle the cards and ask the children to sort the cards into sets of three which show the same times.

- The shuffled cards are dealt so that each player holds three cards. The remaining cards are placed in a pile face down. Each player, in turn, takes either the card on the top of the pile or a card from the discard pile. The player then discards one card.

  The winner is the first player to hold three matching cards. For example,

Measure
Time
6TV

## 2 6TV programmes *(am and pm)*

- List the names of six programmes that the children would like to see on 6TV.
- Say a time, in words, for each programme, and ask the children to write the time using 'am' or 'pm'.

| | |
|---|---|
| Cartoon time | 2.50 pm |
| Football | 8.02 am |
| Weather | |
| Fun on 6 | |
| Music man | |
| Schools quiz | |

- Ask the children to write the programmes and times in the correct sequence.

UA3a/4 SSM4ab/4
T/C1,2
M/3b M/4a

### Textbook page 85 *Time: 12-hour clock, am, pm*

Class 6 is watching a schools programme called *Timewarp*, broadcast by 6TV.

In question 1, emphasize that the children write the time using 'past' or 'to', and also using am or pm notation.

# DURATIONS IN MINUTES AND HOURS AND MINUTES

UA3a/4 N4a/4 SSM4ab/4
T/C1,2,3 T/D2
M/3b M/4a

In Heinemann Mathematics 5, durations in minutes and hours and minutes were calculated for times like these:

from 1.00 pm till 1.35 pm (minutes)

from 1.20 pm till 1.45 pm (minutes)

from 1.20 pm till 2.05 pm (bridging the hour)

from 1.20 pm till 2.45 pm (hours and minutes)

from 11.15 am till 12.45 pm (bridging noon)

These types of example are now consolidated and longer durations are included – for example, between 8.20 am and 3.05 pm.
The context continues with Class 6 planning a trip to 6TV studios.

# Introductory activities

## 1 Filming times *(durations in minutes)*

On the chalkboard make a list of filming times for a day at one of the 6TV studios.

Ask questions such as

'How long does each filming session last?'

Studio 1

| Day Monday | filming from |
|---|---|
| Timewarp: | 9.00am to 9.40pm |
| Sing-a-long: | 10.10am to 10.55am |
| Around Europe: | 11.15am to 12noon |
| Storytime: | 1.30pm to 2.05pm |

For some children it may be necessary to demonstrate these durations by moving the hands of an analogue clock from the starting time to the finishing time and counting the minutes. Alternatively, a time line might be used.

## 2 Visiting times *(durations in hours and minutes)*

- Make up a chart showing the times during the week when 6TV studios are open to the public.

Ask questions such as

'How long are the studios open to the public each day?'

Channel 6 TV
Visiting times

| | |
|---|---|
| Monday | 9.30am to 11.45am |
| Tuesday | 10.30am to 12noon |
| Wednesday | 10.30am to 2.15pm |
| Thursday | 9.30am to 4.10pm |
| Friday | 1.25pm to 6.15pm |

For the durations on Wednesday, Thursday and Friday, the children could use two clocks to help them calculate the time interval. For example,

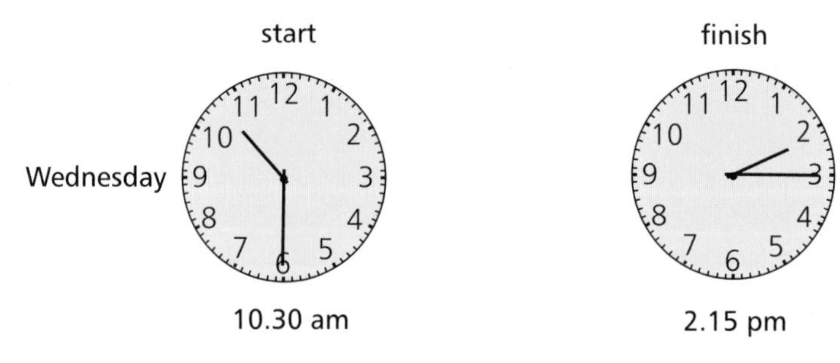


Move round the hands of the first clock while discussing a way of finding the duration.

'Ten thirty to one thirty is **3 hours.**

One thirty to two fifteen is **45 minutes.**

The studios are open for **3 hours 45 minutes.**'

- Repeat for other examples by changing the visiting times for each day.

## Textbook page 86 *Time: durations*

On this page, the children are planning their journey to the TV studios by interpreting the data contained in the three timetables.

Some children may wish to use the clockface on the city centre illustrations to help them answer questions 1 to 6.

Discuss Jo's method by finding how long his journey takes. Questions 4, 5 and 6 provide practice in this method. In question 6, some children may not realize that they have to refer once again to the timetables at the top of the page to find the answers.

**Measure**
Time
6TV

UA3a/4 N4a/4 SSM4ab/4
T/C1,2,3 T/D2
M/3b M/4a

**R26**

# DURATIONS: COUNTING ON, COUNTING BACK

UA3a/4 N4a/4 SSM4ab/4
T/C1,2,3 T/D2
M/3b M/4a

In Heinemann Mathematics 5, the children were given durations in either hours **or** minutes and were asked to find

- — finishing times by counting on
- — starting times by counting back.

This work is now revised and extended to durations in hours **and** minutes.

The context continues with the children visiting 6TV studios. They find out from 'studio schedules' about the time and activities involved in setting up the studios for filming, lighting, sound, costumes and so on.

Two of the children also interview some of the people they meet at the studios.

## Introductory activities

### 1 Pre-filming schedules *(counting on to find finishing time)*

- Prepare a set of large cards showing the length of time each pre-filming task takes. Alternatively, draw these on the chalkboard.

| Lighting<br>2 hours | Make -up<br>25 minutes | Camera call<br>45 minutes |
|---|---|---|

Set a starting time on a real clock or a clock with geared hands.

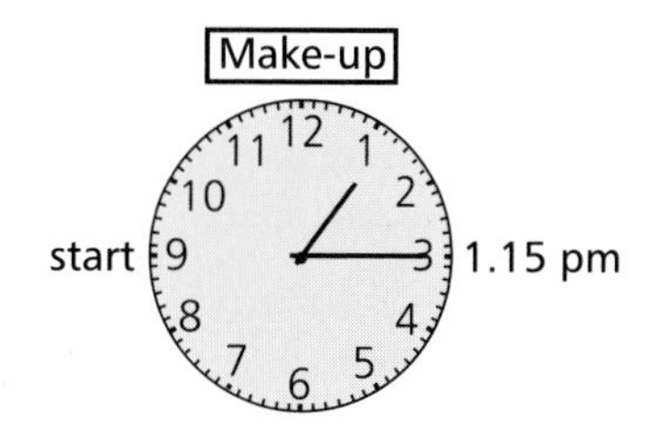


Ask when the actor should finish being made-up. The children should count in five-minute intervals to find the finish time of 1.40 pm.

- The activity can be repeated for different starting times and durations which require 'bridging the hour' to find the finishing times. For example,

  'When would the actors finish "camera call"?'

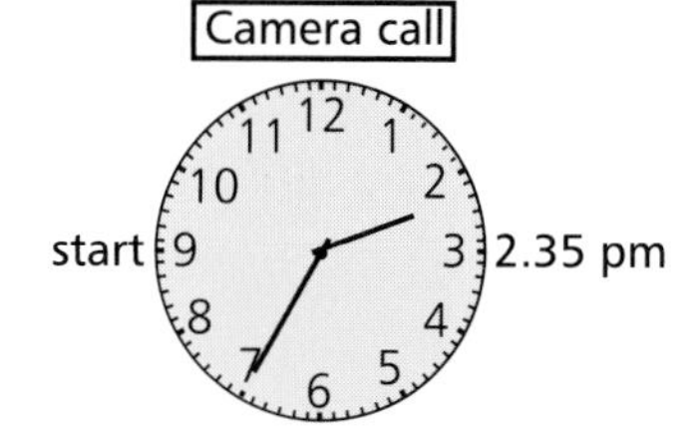


The children should count on to 3 o'clock (25 minutes) and then another 20 minutes to give a finish time of 3.20 pm.

- Introduce further cards where the durations involve hours and minutes.

| Costume fitting<br>1 hour 15 minutes | Rehearsal<br>2 hours 35 minutes |
|---|---|

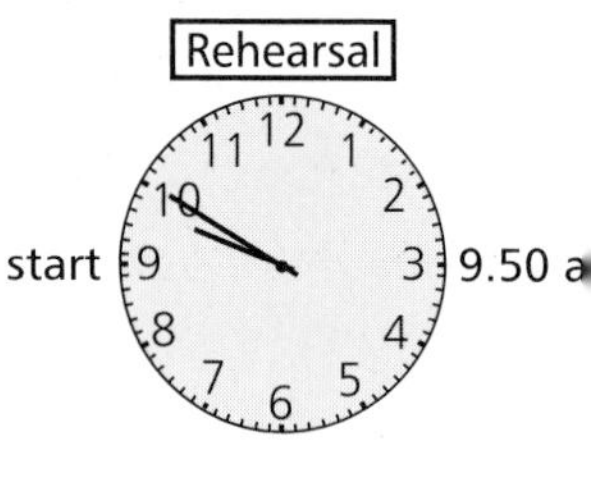


'When should the rehearsal finish?'

The children should count on the hours 9.50, 10.50, 11.50 am, then count on 10 minutes to 12 noon, then count on the remaining 25 minutes to give a finishing time of 12.25 pm.

**Several examples with various starting times should be discussed in order to introduce this new teaching point.**

Some children will need access to a clock to visualize the counting on process. Some may require to move the hands of the clock to find the finishing times. Alternatively, a time line could be used.

**At this point the children could try Textbook page 87.**

## 2 Outside broadcasts *(counting back to find a starting time)*

Prepare a set of large cards like these, or write these times on the chalkboard.

| Horse racing<br>20 minutes | Tennis<br>35 minutes | Cricket<br>2 hours 10 minutes | Athletics<br>1 hour 45 minutes |
|---|---|---|---|

Set a finishing time on an analogue clock, choose a card, and find when the outside broadcast should start. Grade the examples as for the 'pre-filming schedules' activities, i.e. counting back

- — minutes only
- — minutes involving bridging the hour
- — hours and minutes

For example, ask the children when the athletics outside broadcast should start.

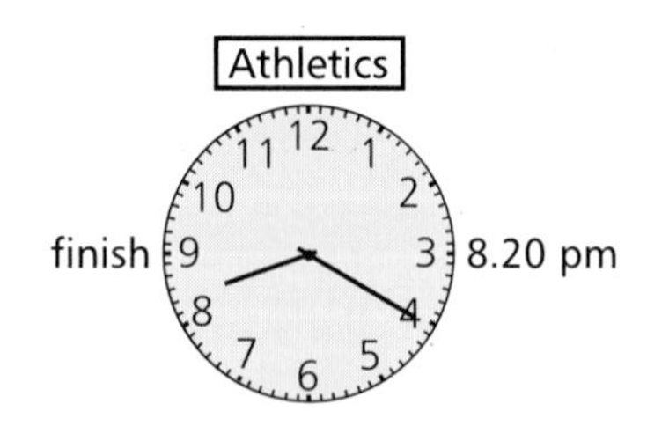


The children should count back 1 hour, to give 7.20 pm, and then a further 45 minutes (including 'bridging the hour') to give 6.35 pm.

Several examples with varied finishing times should be discussed in order to reinforce this quite difficult technique.

Access to a clockface or time line would be helpful, and indeed is essential for many children.

## Textbook pages 87 and 88
### *Time: durations, counting on and back*

On Textbook page 87, the examples are graded as follows.

In question 1, the durations involve whole numbers of hours or minutes where no 'bridging' of an hour is required.

In questions 2 and 3, the minute durations require the hour to be bridged to find the finishing time.

In question 4, the durations involve hours and minutes. 'Bridging the hour' is required in questions 4(e) to 4(j).

In all questions, children may need to use the clockface at the top of the page to help find the finish times.

The examples on Textbook page 88 are graded in a similar fashion. Some children may need guidance in interpreting the information in question 2.

Measure
Time
6TV

UA2ac,3a/4 N4a/4 SSM4ab/4
T/C1,2,3 T/D2
M/3b M/4a

R27

# 24-HOUR NOTATION: APPLICATIONS

The 24-hour notation and writing times in the form 04.35 and 21.18 are now introduced.

The context continues with 6TV broadcasting for 24 hours a day. Four children visit Lynchester airport to help film the arrival of musicians attending the Lynchester Pop Festival.

UA3ab/4 SSM4ab/4
T/D1
M/4e

## 1 Introducing 24-hour notation

- Elicit that there are 12 hours of am time and 12 hours of pm time, and that there are 24 hours in a complete day.

  Tell the children that 6TV broadcasts through the day time and night time – that is, for 24 hours.

  Ask the children to give examples of programmes that are broadcast before noon and after noon.

- Draw a time line on the chalkboard.

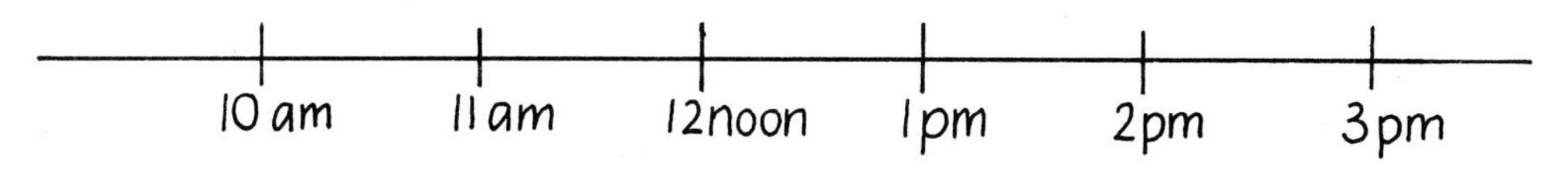


Add the new 24-hour notation by showing equivalences, as follows:

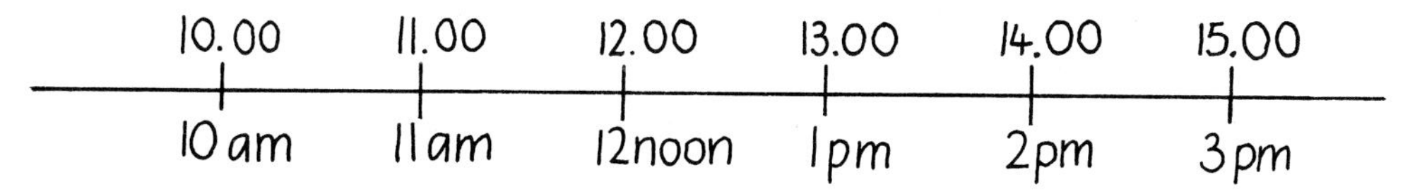


Discuss further examples such as

7 am → 07.00    5 pm → 17.00    2 am → 02.00    11 pm → 23.00

- Explain that when writing a 24-hour time
  - — four digits are always used
  - — the first two digits show the hours 17 : 00
  - — the last two digits show the minutes 17 : 00

  When hours only are recorded, the last two digits will be **zeros**.

- Ask the children to explain why the 24-hour notation is used in places like railway and bus stations, at airports and in timetables.

  The children should appreciate that this notation helps prevent people making errors about times or trains. For example, 5 o'clock can be 5 am or 5 pm, whereas 17.00 can only be a pm time.

- Some 24-hour times can be difficult to say – for example, those involving whole hours only. Discuss such examples and explain how to 'say' 24-hour times. For example,

  05.00 → 'oh five oh oh' or 'zero five zero zero'

  16.00 → 'sixteen zero zero' or 'sixteen oh oh'

  14.36 → 'fourteen thirty-six'

  At this stage it is best to avoid using expressions such as 'oh five hundred hours' 'or sixteen hundred hours'. While these are often used by adults, such usage can be misleading for children, who might mistakenly assume that there are 100 minutes in 1 hour.

## 2 Changing between 12-hour and 24-hour times

Draw this number line on the chalkboard.

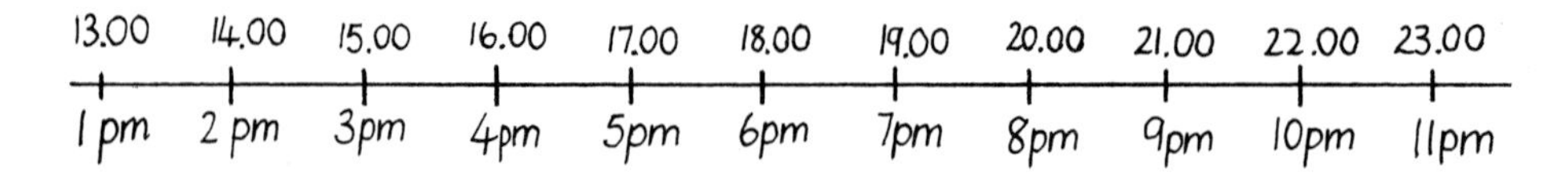


Ask the children if they can find a simple way of changing pm time to 24-hour time:

5 pm (add 12 hours) → 17.00

11 pm (add 12 hours) → 23.00

The reverse process should be discussed:

14.00 (subtract 12 hours) → 2 pm

17.00 (subtract 12 hours) → 5 pm

stressing again that

1 am is 01.00    11 am is 11.00    07.00 is 7 am    09.00 is 9 am.

## 3 Time cards *(using 24-hour notation times)*

Prepare a yellow set of cards showing hourly times from 1 am to 11 pm

1am | 2 am . . . . . . . 12noon | 1pm . . . 11pm

and a blue set of cards showing their 24-hour equivalents.

01.00 | 02.00 . . . . . . . 12.00 | 13.00 . . . 23.00

These cards can now be used

— as a matching activity

— to play 'Snap'

— to play 'Pairs'. Put all the cards face down. Pick a yellow and a blue card. If they match, the child keeps them; if not, the cards are replaced. The winner is the child who collects most pairs.

Although no examples appear on Textbook pages 89 and 90, it may be worthwhile to show and explain how midnight is recorded using 24-hour notation.

midnight is **either** 24.00 (**end** of a day)

**or** 00.00 (**start** of a day)

On digital displays, midnight is always shown as 00 : 00 .

## 4 Time cards *(24-hour times involving minutes)*

Having established the notation and associated language for the 24-hour clock, extending to hours and minutes should be reasonably straightforward. For example,

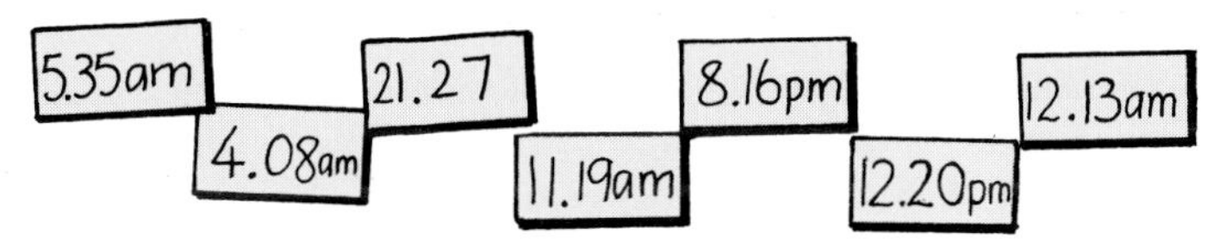


5.35 am is 05.35, spoken as 'oh five thirty-five'

9.27 pm is 21.27, spoken as 'twenty-one twenty-seven'.

Give orally a 12-hour am or pm time. For example,

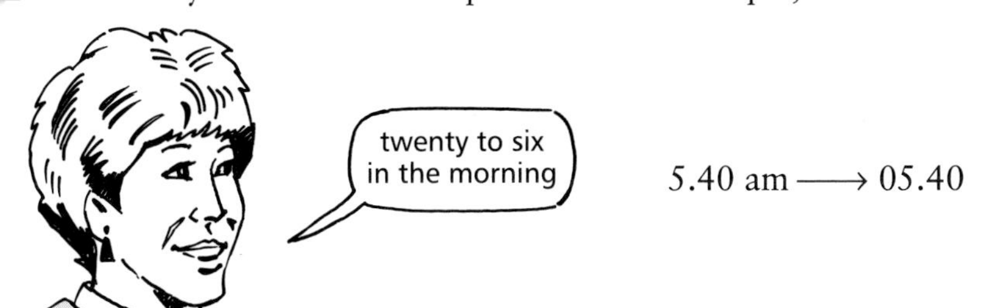


Ask children to say and record times in 12-hour and 24-hour notation.

Measure
Time
6TV

UA3a,4d/4 SSM4ab/4
T/D1
M/4e

## Textbook pages 89 and 90

### *Time: the 24-hour clock, applications*

Textbook page 89 introduces 24-hour times through the context of television programmes broadcast day and night by 6TV.

Questions 1 and 2 deal with the conversion of times given in hours only.

In question 1, emphasize that the children should use four digits to record times, such as 6 am → 06.00 and 9 pm → 21.00.

In question 2, the time 12.00 should be recorded as 12 noon or noon.

Questions 3 and 4 extend the work to include times in hours and minutes.

Children may have difficulty with questions 5(e) and 5(f). 12.50 am is 50 minutes after midnight and is written as 00.50 in 24-hour notation. 00.18 is 18 minutes after midnight and is written as 12.18 am.

On Textbook page 90, in questions 1, 2 and 3, the children are given experience of matching 12-hour and 24-hour times.

In questions 5, 6 and 7, the children record flights that arrive **after** or **before** a given time. The answers can be given either as places or as times.

Questions 6 and 7 are of the same type but involve arrivals.

Question 8 requires the children to use both the TV timetable and the flight arrival times. Dan Ely is on time for *Pop Quiz* as he arrives at the studio ten minutes before the start time of 6.15 pm or 18.15.

# 24-HOUR NOTATION: DURATIONS

UA3ab/4 N4a/4 SSM4b/4
T/D1,2
M/4e

Work using the 24-hour clock notation is concluded with mental calculations involving durations in hours, in minutes and in hours and minutes. To keep the work at a simple level, examples which involve 'bridging' an hour in minutes – for example, from 14.35 to 15.20 – are not included.

The 6TV context continues with Class 6 considering programme schedules.

## Introductory activity

### TV programmes *(durations using 24-hour notation)*

- Display information about a TV programme on the chalkboard as shown. Ask the children to say, in 24-hour time, when *Quest* finishes. Repeat for other programmes, starting on the hour and lasting 30 minutes, $1\frac{1}{2}$ hours and 2 hours.

- Change the chalkboard display as shown. Discuss finding mentally how long *Superhero* lasts. Some children may count on:

  14.35 to 14.45 (10 min)

  14.45 to 15.00 (15 min)

  *Superhero* lasts 10 + 15 → 25 minutes.

Others may subtract, 60 – 35 → 25 minutes.

Repeat for other programmes, choosing start and finish times which do not require 'bridging' an hour in minutes. Include examples such as: start 20.00, finish 21.40.

- Change the display again as shown. Ask the children to say, in 24-hour time, when *Cartoon Time* begins. Repeat for other programmes, again choosing finish times and programme lengths which avoid 'bridging' difficulties. For example,

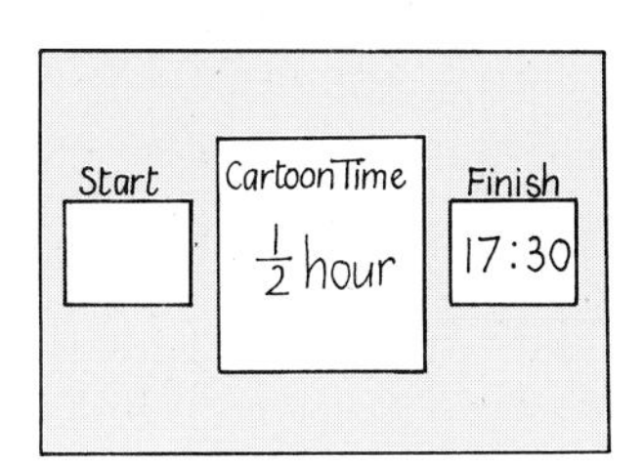


— finishes 20.50, lasts 40 min

— finishes 21.00, lasts $1\frac{1}{2}$ hours.

Measure
Time
6TV

UA3ab/4 N4a/4 SSM4b/4
PSE T/D1,2
M/4e

## Workbook page 29 *Time: 24-hour clock durations*

Discuss the scenario briefly – at 6TV Antonia plans the sequence of each day's programmes. Ensure that the children understand the meaning of 'schedule'.

In question 1, the children use the programme lengths and titles to complete Antonia's schedule for Thursday afternoon. At an appropriate moment, discuss the fact that the 'Finish time' for one programme is also the 'Start time' for the next programme, since this is important information for question 2.

**Problem solving**

Question 2 requires the children to complete a similar schedule for Thursday evening's programmes. This time they have to make entries in the 'Start', 'Programme' and 'Finish' columns. Different approaches are possible, but all involve logical reasoning. For example,

> *Pop Toppers* lasts half an hour and finishes at 19.30 – it must begin at 19.00. Thus *Meet the Mob* finishes at 19.00 and, since it lasts 40 min, begins at 18.20. The final programme lasts 1 hour 45 minutes, so it must be *Snooker*. The third programme must be *Showtime* – it begins at 19.30, finishes at 21.00 and so lasts $1\frac{1}{2}$ hours, which is the given length.

# THE SECOND

UA3a/4 SSM4ab/4
T/D3
M/4ad

The second is now introduced. Illustrations of times on stopwatches are provided for the children to interpret. Teachers without access to stopwatches should use whatever timing devices are available to enable the children to carry out the essential timing activities. Most digital watches have a stopwatch function.

The context continues with the children back in school and again viewing the *Timewarp* programme.

# Introductory activities

## 1 Introducing the second

Discuss occasions which highlight the need for a unit of time less than 1 minute such as the timing of a 100 m or 200 m race, and the countdown for a rocket launch. Many children will already know that this unit is the second, and that there are 60 seconds in 1 minute. They will also have heard people use phrases such as 'be with you in a second' or 'hang on a second'.

## 2 Around the class *(reading a stopwatch)*

- Use a large card stopwatch or one drawn on the chalkboard. Discuss the watch and explain how it works.
- Select and give times for some activities that the children might do in the classroom and which take less than 1 minute – for example,
  - — counting aloud quickly to 60
  - — fastening a coat
  - — saying a multiplication table.

  Illustrate each time on the stopwatch and ask the children to say how long the activity 'took' in seconds.
- Introduce some other activities which take longer than 1 minute – for example,
  - — reading a poem or short story
  - — completing a worksheet
  - — giving out jotters or textbooks.

  Use an example such as

  'How long did it take to read the short story?'

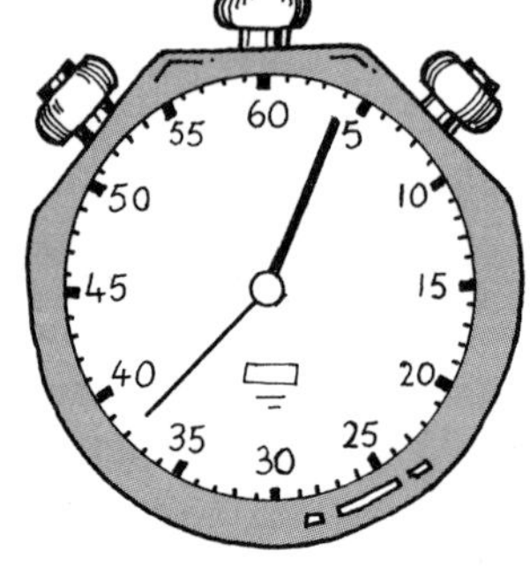


  Explain that, in order to find out how long it took, they must
  - — look at the minute hand, which is pointing between 4 and 5
  - — look at the second hand, which is pointing to 38 seconds.

  It took 4 minutes and 38 seconds to read the short story.

**At this point the children could try Textbook page 91.**

## 3 Varieties of stopwatches

Although only one type of stopwatch is illustrated on Textbook page 91, the children should be aware of other types and how to use them.

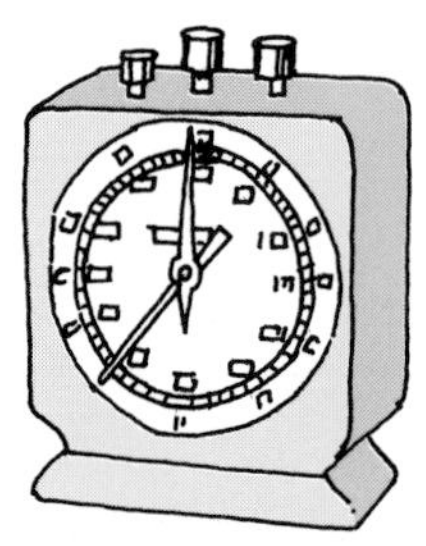

Stopclock

Digital stopwatch

Watch with a seconds hand

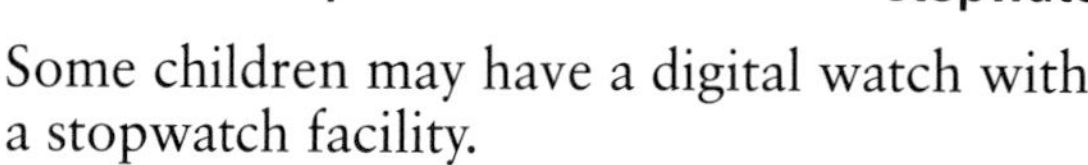

Some children may have a digital watch with a stopwatch facility.

There are many different digital displays available, but a simplified version, like the one illustrated, is useful for demonstration purposes.

10 : 05 00

## 4 Mini-Olympics *(using a stopwatch)*

Tell the children that the stopwatch should be re-set to zero before each activity.

Select five or six practical activities that the children could do in the hall or in the playground. For example,

Decide how long each activity will last, say 30 seconds, 1 minute, and 2 minutes 30 seconds.

In pairs, the children take turns to try each activity, one performing and counting, the other timing.

The results – for example, the number of skips in 40 seconds – should be recorded and discussed to find who is the Mini-Olympic Champion.

Name:

| activity | time | number |
|---|---|---|
| skip | 40 seconds | |
| bat and ball | 30 seconds | |
| netball shot | 1minute | |

## 5 Tracks *(using a stopwatch)*

- Ask the children to mark out a long track in the playground or hall – for example, an 'oval' track or a zig-zag track like these:

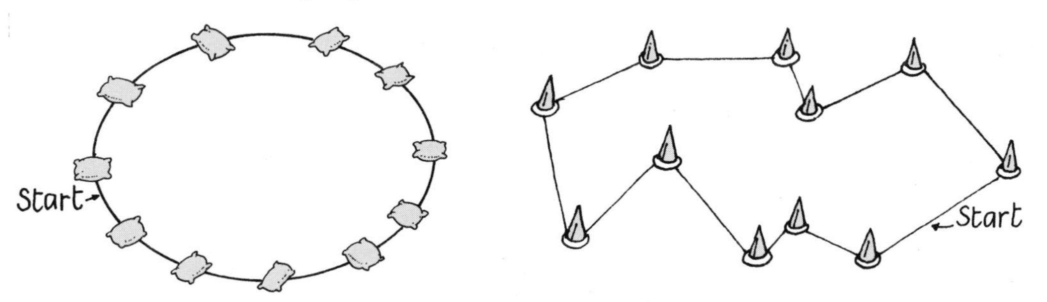


Cones, bean bags or chalk could be used as markers.

Working in pairs, one child could be asked to walk twice round the track, while the other child uses the stopwatch to find the time taken. The results could be recorded and discussed.

- This activity could be extended by asking the children to estimate how long they will take to complete the task. It could also be repeated for other activities in the classroom, to give further practice in using a stopwatch.

UA3a/4 SSM4ab/4
T/D3
M/4a

## Textbook page 91 *Time: introducing the second*

Before the children attempt this work, they should know about seconds and have experience of using a stopwatch as suggested in the introductory activities.

In question 1, as the times are all shorter than 1 minute, the children need only concentrate on reading the seconds hand: that is, the red hand.

In question 2, the children must read the number of minutes indicated by the black hand as well as the number of seconds indicated by the red hand.

UA2a,3a/4 SSM4ab/4
T/D3
M/4a

## Workbook page 30 *Time: practical timing*

This page contains practical activities where the children work in pairs and time each other performing certain tasks.

In question 3, after they have had some experience of timing activities, the children are given the opportunity to estimate times in seconds before they measure them. A simple way of helping the children to estimate the time in seconds is to tell them to say,

'one thousand and one, one thousand and two . . .'

This gives a reasonably accurate way of determining time in seconds. There may be other phrases known to the teacher or children, such as

'one elephant, two elephant . . .'

which could also be used. Again the children should work in pairs, one estimating and the other checking actual times with a stopwatch.

UA2a/4 SSM4a/4
PSE T/C4
P/4a M/3b M/4a

## Textbook page 92 *Other activity: calendar*

- The calendar activities on this page can be attempted at any time as they are **not** dependent on the Time section on Textbook pages 85–91.
- Children who experience difficulty with question 1 could use any calendar to help them. In questions 2 and 3, they should use the given calendar for September and October.

**Problem solving**

- In question 4, some children might use a guess and check strategy. Others might use a more systematic approach, based on the information that Gran is six times Roddy's age. For example,

— Try 10 for Roddy's age. Gran would be 60. Roddy is only 50 years younger.

— Try 11 for Roddy's age. Gran would be 66. Roddy is only 55 years younger.

— Try 12 for Roddy's age. Gran would be 72. Roddy is 60 years younger.

## Textbook page 93 *Other activity: timetables*

UA2a/4 SSM4a/4
PSE T/D2
P/4a M/4e

**Problem solving**

- This activity involves interpretation of bus, train and ferry timetables. It is best attempted after the children have completed the work on durations involving the 24-hour clock on Workbook page 90.
- Discuss the map to ensure that the children are familiar with the routes which have buses **only**, those which have a bus **and** a train service, and the two ferry crossings to Calport.
- In question 4, the children have to work backwards from the 12 o'clock appointment in Calport.
  - If they travel via Garton, the only suitable ferry is the one arriving in Calport at 1100. It leaves Garton at 0930. To catch this ferry they would need to take the 0820 bus from Fairfield to Garton.
  - If they travel via Portsea, the appropriate ferry arrives in Calport at 1130. It leaves Portsea at 1045. They could catch this ferry by taking the 0845 bus from Fairfield to Portsea.

  Taking the train offers no advantage. The latest time to leave Fairfield is 0845.
- The buses, trains and ferries all run at regular intervals. Able children could be told that the services continue at these regular intervals, and further questions could be posed. For example,

  'What is the latest you could leave Fairfield for an appointment in Calport at 1515?'

The Shape part of Heinemann Mathematics 6 is concerned with Shape, Position and Movement and has four sections, each with an Overview and accompanying notes.



Shape

SSM2c,3b/4
PM/D3
S/D1,2
S/3c
S/4d

# Co-ordinates

## Overview

This section

- introduces the use of co-ordinates such as (3, 4) to locate a point
- introduces the terms 'co-ordinates', 'horizontal axis', 'vertical axis' and 'origin'
- includes identifying and plotting points to complete pictures, some of which are symmetrical.

| | Teacher's Notes | Textbook | Workbook | Reinforcement Sheets |
|---|---|---|---|---|
| **Co-ordinates, line symmetry** | 207 | 94 | 31, 32 | |

Extension activities related to the above section of work are as follows:

| | Teacher's Notes | Extension Textbook |
|---|---|---|
| **Co-ordinates: transformations** | 291 | E24 |
| **Other activity: co-ordinate patterns** | 292 | E25 |

Teaching notes for the Extension Textbook are in a separate section at the end of the Teacher's Notes.

## Resources

### Useful materials

- squared paper
- coloured pencils or pens
- other materials suggested within the introductory activities

### Assessment and Resources Pack

Assessment

*Shape Check-up 1*
Textbook page 94
Workbook pages 31, 32 (co-ordinates, line symmetry)

*Round-up 1*
Question 9

Resources

*Problem Solving Activities*
22 Shapes on grids (Making shapes and writing the co-ordinates of each vertex)
23 Four sides game (Making quadrilaterals)

*Resource Cards*
25 Archaeological dig (Co-ordinates)

## CO-ORDINATES, LINE SYMMETRY

SSM2c,3b/4
PM/D3 S/D1,2
S/3c S/4d

In Heinemann Mathematics 5, the children used numbered columns and rows to provide grid references for **regions**. For example, the shaded region is **4, 2**.

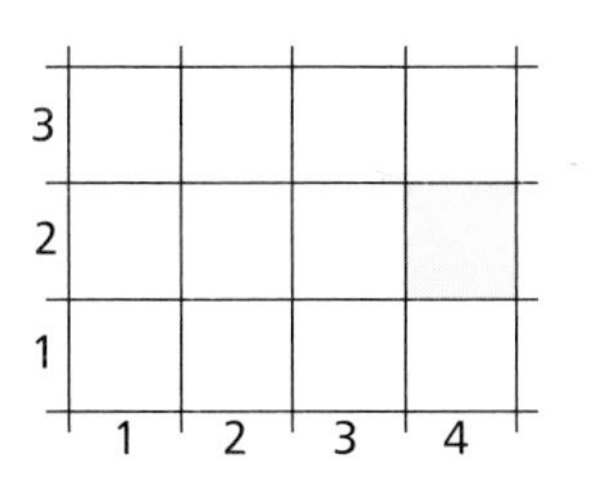


In this section, the use of co-ordinates to plot and locate **points** is introduced. For example, the co-ordinates of point **A** are given using the notation (3, 2).

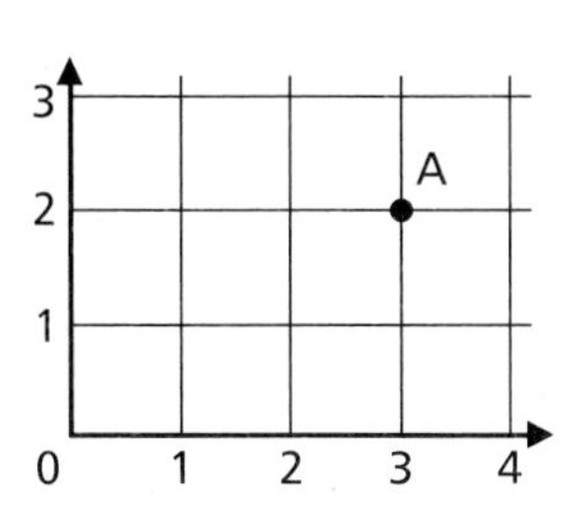


The work on co-ordinates is set in the context of space travel.

## Introductory activities

### 1 Meteor storm *(locating/plotting co-ordinates)*

- Use a chalkboard or overhead projector to display the positions of meteors on a squared grid.

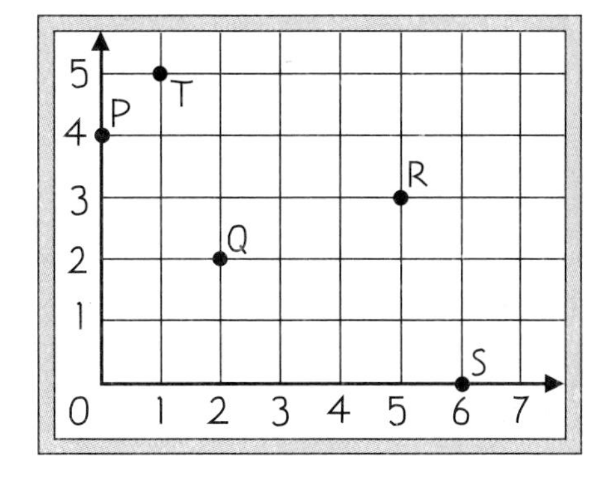


- Emphasize that the position of meteor **R** is at the meeting point of the lines '5 along' and '3 up'.
- Tell the children that the **co-ordinates** of point **R** are written as (5, 3).
- Now ask them to give the co-ordinates of meteors **T** (1, 5) and **Q** (2, 2).

  Emphasize the use in the notation of brackets and a comma.
- Ask the children to give the co-ordinates of meteors **S** and **P**. They may need to be prompted to give these as (6, 0) and (0, 4) respectively. Tell them that **S** lies on the **horizontal axis**, while **P** is on the **vertical axis**.

  Discuss the point (0, 0) where the co-ordinates 'start', and introduce its special name, the **origin**.
- Tell the children that three other meteors have been traced as follows:

  U ⟶ (4, 1)

  V ⟶ (7, 5)

  W ⟶ (3, 0)

  Ask them to show where these points should be plotted on the grid.

**At this point the children could try Textbook page 94.**

## 2 Moon mountains *(plotting and joining points)*

Display a grid like this.

Ask different children, in turn, to mark the following points:

(0, 2) (1, 4) (2, 1) (3, 5)

(4, 3) (5, 0) (6, 4)

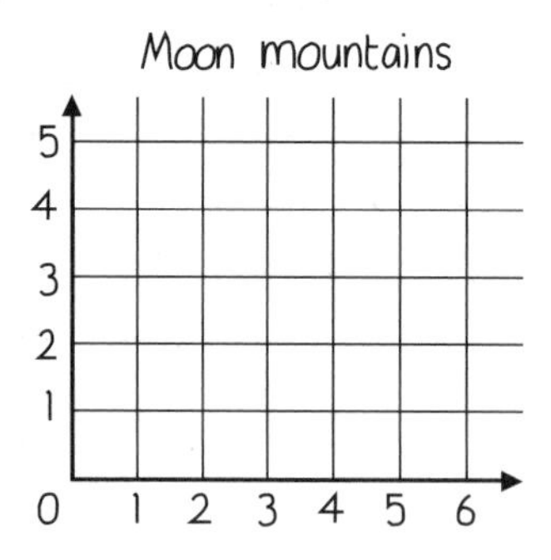


Each new point marked should be joined to the previous one with a straight line, thus revealing the outline of the mountains.

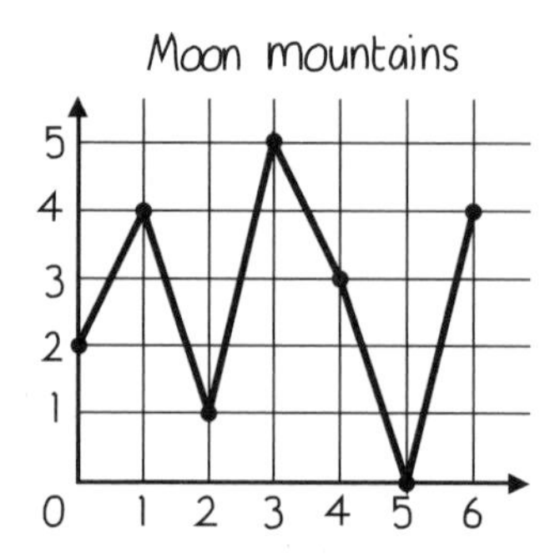


## 3 Control tower

*(identifying co-ordinates and plotting and joining points to complete a symmetrical picture)*

- Ask different children, in turn, to mark the following points on a large 8 × 8 grid:

  (4, 2) (1, 2) (1, 0) (0, 0) (0, 4)

  (1, 4) (1, 6) (2, 6) (2, 7) (4, 8)

  Each new point marked should be joined to the previous one by a straight line, resulting in the diagram shown.

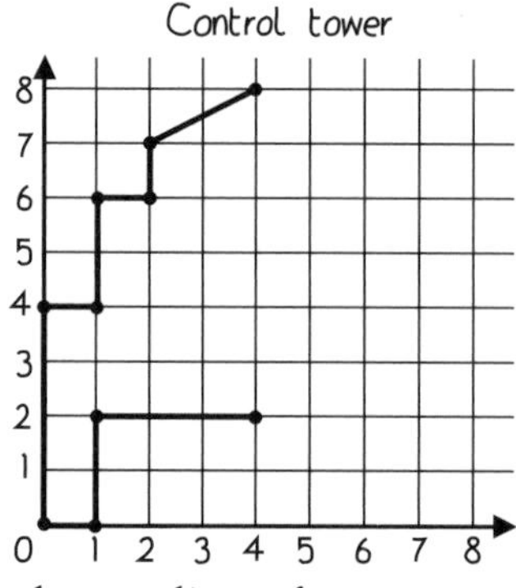


- Tell the children that the control tower is to have a line of symmetry. Ask them to suggest the co-ordinates of points to complete the other half of the picture so that it is symmetrical. The correct points, starting from (4, 8), are:

  (6, 7) (6, 6) (7, 6) (7, 4) (8, 4) (8, 0) (7, 0) (7, 2) and back to (4, 2)

  Emphasize the importance of giving the co-ordinates of the starting point to complete the control tower.

  Finally, ask a child to join up the new points to complete the symmetrical picture.

  When the children are working independently on a similar activity, they should join each new point to the previous one as it is plotted.

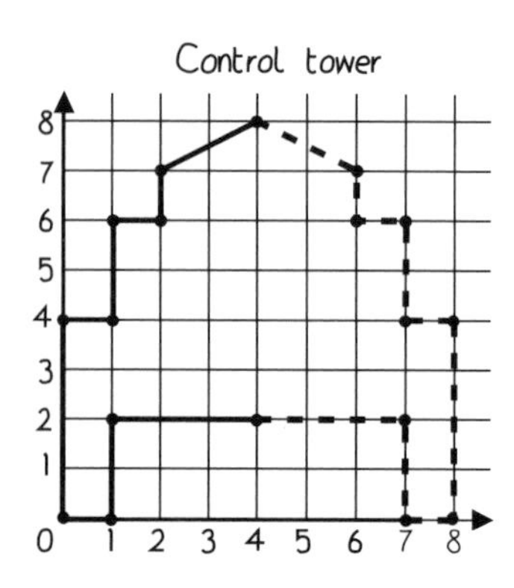

## Textbook page 94 Workbook page 31 *Co-ordinates*

SSM3b/4
PM/D3
S/4d

Discuss the 'Spaceflight control' context used on Textbook page 94.

In questions 1, 2 and 3, some children may have more difficulty with co-ordinates of points which lie on either the horizontal or vertical axis, and also those which have equal co-ordinates, such as (2, 2).

On Workbook page 31, in question 1, emphasize the importance of joining the points **in order**. The children might find it easier to draw a line to each new point as it is marked, rather than marking all of the points and then joining them.

In question 2, some children may forget to return to the starting point to complete a picture.

## Workbook page 32 *Co-ordinates: line symmetry*

SSM2c,3b/4
PM/D3 S/D1,2
S/3c S/4d

In question 1(b), the children need to ensure that when they colour the 'badge' it retains **all four** of its lines of symmetry.

In question 2(c), the designs are completed using the following co-ordinates:

cruiser: (3, 9) (4, 7) (4, 6) (6, 4) (6, 2) (4, 4) (4, 2) (5, 1) (4, 0)
and back to (3, 1)

station: (7, 4) (8, 5) (8, 6) (9, 7) (10, 7) (11, 8) (12, 7) (13, 7) (14, 6) (14, 5)
and back to (15, 4)

UA3a/4
SSM2abc,3b/4
**PSE**
**RS/C2**
**RS/D1,2,3,4,5**
**RS/E1**
**PM/D3**
**S/D1**
**S/E1**
P/4ab
S/3abc
S/4abd
S/5ac

# 2D shape

## Overview

This section

- introduces the parallelogram and investigates equal sides, right angles and lines of symmetry of four-sided shapes
- introduces the term 'diagonal'
- introduces equilateral and isosceles triangles
- gives practice in drawing shapes on co-ordinate grids
- introduces the term 'congruent'
- introduces drawing circles using a pair of compasses, and the terms 'radius', 'diameter' and 'circumference'
- introduces rotational symmetry.

| | Teacher's Notes | Textbook | Workbook | Reinforcement Sheets |
|---|---|---|---|---|
| **Four-sided shapes** | 212 | 95, 96 | 18, 20 | |
| **Triangles** | 215 | 97 | 20 | |
| **Drawing shapes on grids** | 217 | | 33 | 29 |
| **Congruent shapes** | 219 | 98 | 18 | |
| **Circles** | 221 | 99, 100 | | |
| **Rotational symmetry** | 223 | 101, 102 | | |
| *Other activity* | *225* | *103* | | |

Extension activities related to the above section of work are as follows:

| | Teacher's Notes | Extension Textbook |
|---|---|---|
| **2D shape: properties** | 292 | E26 |
| **Other activity: translation** | 278 | E1 |
| **Other activity: 2D shape** | 285 | E12 |

Teaching notes for the Extension Textbook are in a separate section at the end of the Teacher's Notes.

# Resources

## Useful materials

- wood, plastic or metal strips and fasteners, for making shapes
- paper, card or plastic shapes
- squared paper
- sheets of A4 paper
- coloured paper, card, coloured pens/pencils, scissors
- pairs of compasses
- other materials suggested within the introductory activities

## Assessment and Resources Pack

### Assessment

*Shape Check-up 2*
Textbook pages 95–8
Workbook pages 18, 20, 33 (4-sided shapes, triangles, congruency, symmetry)

*Round-up 1*
Question 6

*Round-up 2*
Questions 4, 6

### Resources

*Problem Solving Activities*
20 Shape shuffle (Making hexagons)
21 Two diagonals (Making shapes within a hexagon)
22 Shapes on grids (Making quadrilaterals)
23 Four sides game (Making quadrilaterals)

*Resource Cards*
26 Tangrams (Making congruent shapes)
26, 30, 31 Strip patterns (Transformations)
27–9 Shape Link (Rotational symmetry)
32 Shape template (For making large demonstration shapes)

# Teaching notes

In Heinemann Mathematics 5, the work on tiling involved squares, rectangles, triangles, pentagons and hexagons, and introduced briefly rhombuses and kites. The children found lines of symmetry for some of these shapes by folding, and met shapes with more than two lines of symmetry for the first time.

This work is now developed through activities which highlight simple properties of different types of four-sided shapes. The parallelogram is introduced, and the children construct four-sided shapes including the rhombus and kite with strips. They explore equal sides and right angles, and find lines of symmetry by folding paper shapes. The term 'diagonal' is introduced.

## FOUR-SIDED SHAPES

SSM2ac/4
RS/D1,2,3 RS/E1 S/D1
S/3abc S/4b

### Introductory activities

#### 1 What shape is it?

- Use a set of large paper shapes like these.

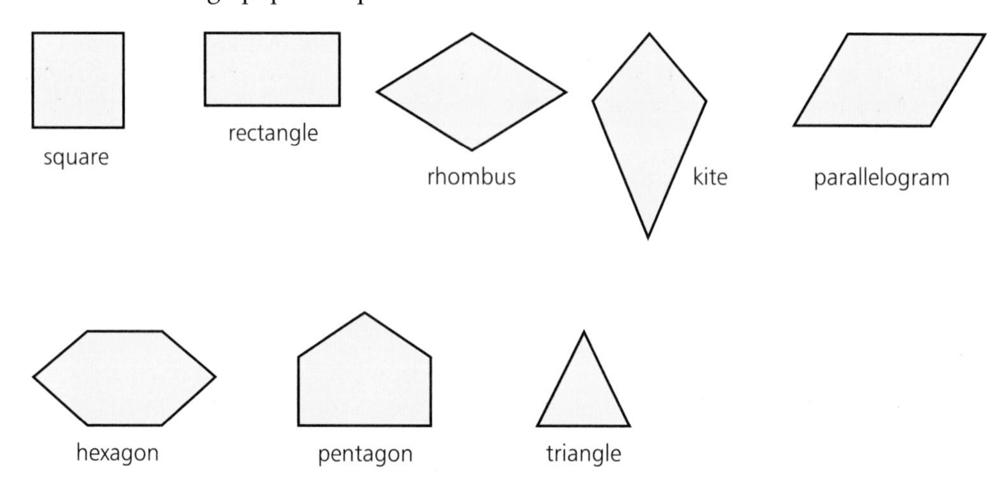


The shapes can be drawn using the template from Resource Card 32.

Although they do not appear in the Textbook or Workbook activities, a trapezium and also a quadrilateral with no 'special' properties could be included.

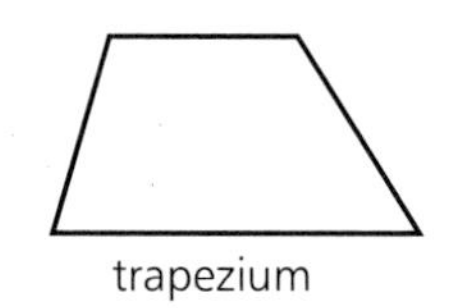

- Put the shapes in an envelope or bag. Ask a child to pick a shape and name it. The children have previously met all the coloured shapes above, except for the parallelogram. They may require some help with the rhombus and kite. Attach each shape, as it is selected, to a chalkboard with Blu-tack, and write its name. Give practice in saying and spelling particularly difficult names, such as 'rhombus' and 'parallelogram'.

- Discuss the sides of each four-sided shape as it is selected, asking questions such as

  'How many sides are there?'

  'Are any of them equal in length?'

  'Are they all equal?'

  The children should point to those sides they think are equal and then check using a paper strip or a ruler.

  Ask about the right angles of the shape. If necessary, the children can check with a right angle tester.

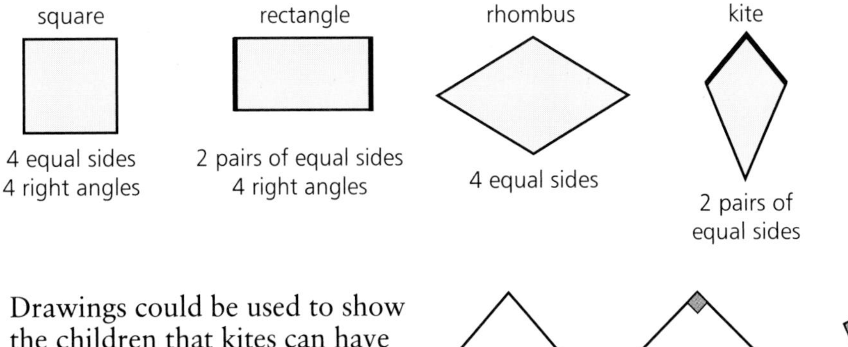


Drawings could be used to show the children that kites can have 0, 1 or 2 right angles.

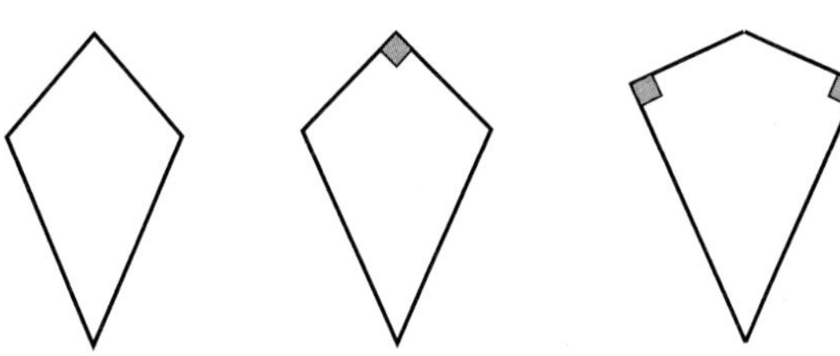

**At this point, the children could try Textbook page 95.**

## 2 Lines of symmetry

Use two of the large paper shapes. Ask the children how many lines of symmetry each shape has. Ask children to fold each shape to confirm that 'the two-halves match', and to draw the lines of symmetry.

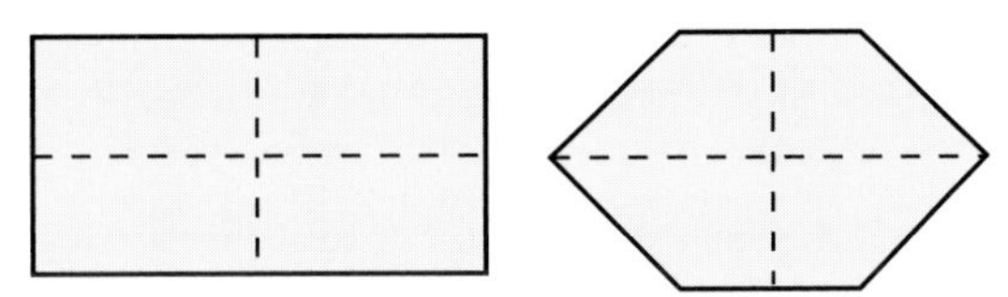

## 3 Diagonals

- Draw a rectangle on the chalkboard and insert one of its diagonals.

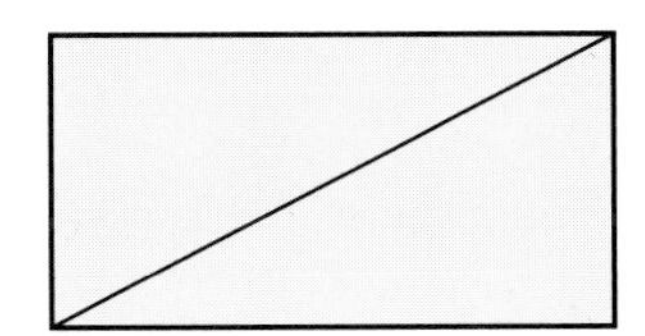

Tell the children that this line is a diagonal. Ask them to describe a diagonal, establishing that it joins two **corners** but is not a side of the shape.

Ask if the rectangle has any more diagonals, and let a child draw the other one.

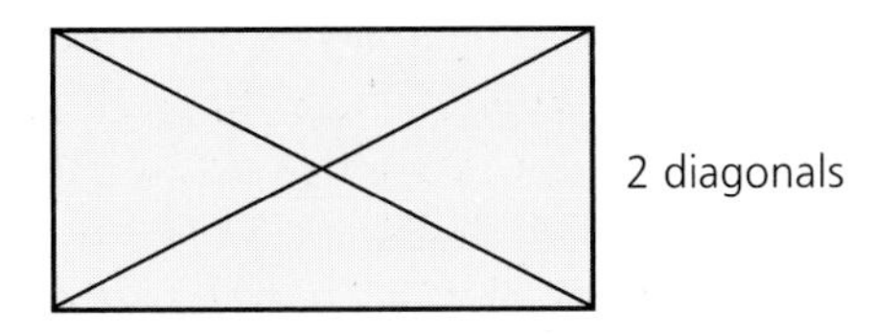

Repeat for other four-sided shapes. Each has 2 diagonals.

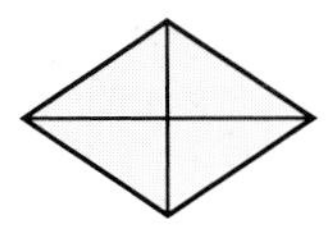
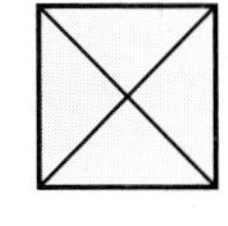
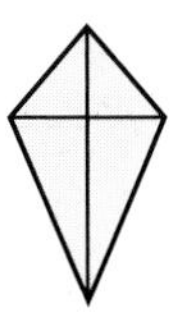

- Make a large drawing of a pentagon. Ask the children how many diagonals it has. Put 'blobs' on the corners and have the children draw diagonals using coloured chalk.

The pentagon has 5 diagonals.

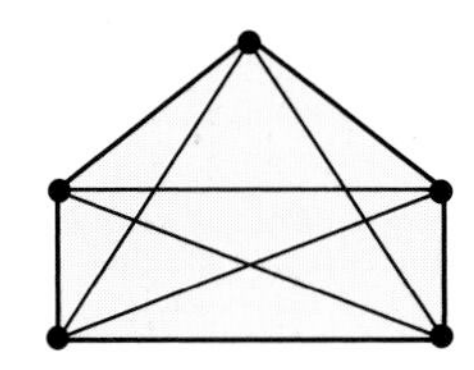

## 4 Lines of symmetry and diagonals

- Use a paper rectangle. Ask if a diagonal of a rectangle is also a line of symmetry.

  Ask a child to fold to show that the two halves do **not** match.

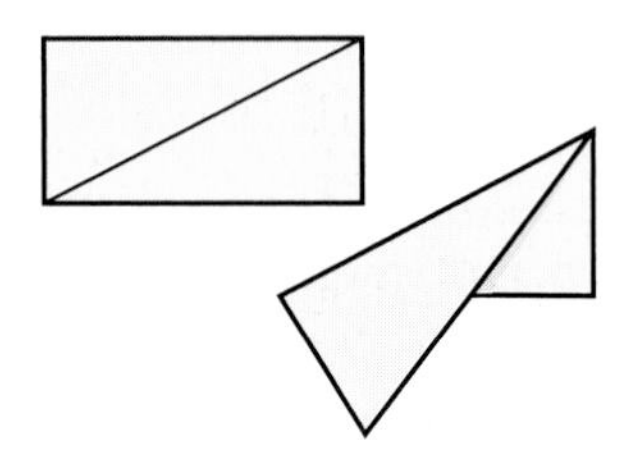

- Use a paper hexagon with its two lines of symmetry drawn.

  Ask if these lines of symmetry are also diagonals. One is a diagonal (it joins two corners), but the other is not.

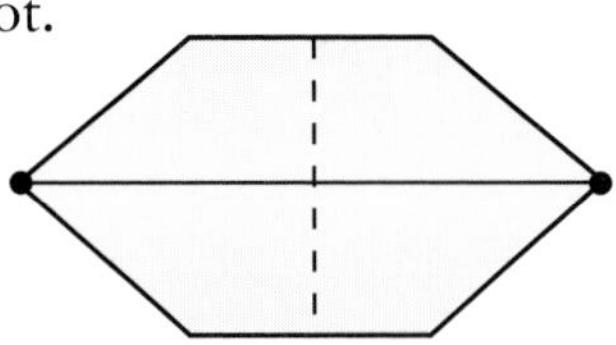

UA3a/4 SSM2b/4
RS/D1 RS/E1
S/3ab S/4b

### Textbook page 95 *2D shape: rhombus, parallelogram, kite*
### Workbook page 18

On Textbook page 95, in questions 1 to 3, a supply of strips or straws of two different lengths is required. All of the shapes can be made, in turn, from a set like this

with suitable bolts or fasteners or pipe cleaners to join the corners. It is important to discuss the children's conclusions in questions 1(c), 2(c) and 3(b). Make sure that they appreciate the following, possibly expressed in other ways:

Q1(c) – Both the square and the rhombus have 4 equal sides.
– The rhombus becomes a square when it also has 4 right angles.

Q2(c) – Each shape has 2 pairs of equal sides.
– The parallelogram becomes a rectangle when it also has 4 right angles.

Q3(b) – The equal sides of the kite are next to each other.
– The equal sides of the parallelogram are opposite each other.

## Textbook page 96 *2D shape: symmetry, diagonals*
## Workbook pages 18 and 20

UA3a/4 SSM2abc/4
RS/D1 RS/E1 S/D1
S/3abc S/4b

In question 2, the children should answer by writing names of the shapes. Stress that the parallelogram has **no** lines of symmetry. If necessary, show that the lines below are **not** lines of symmetry, by folding the shape.

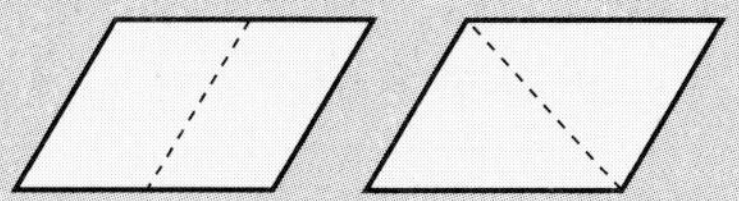

In question 4(a), the children should make a parallelogram like one of these:

In part (b), where one piece is flipped over, they can make a kite.

In question 5, the shapes should be like these:

one of two 'V'-kites

one of three parallelograms

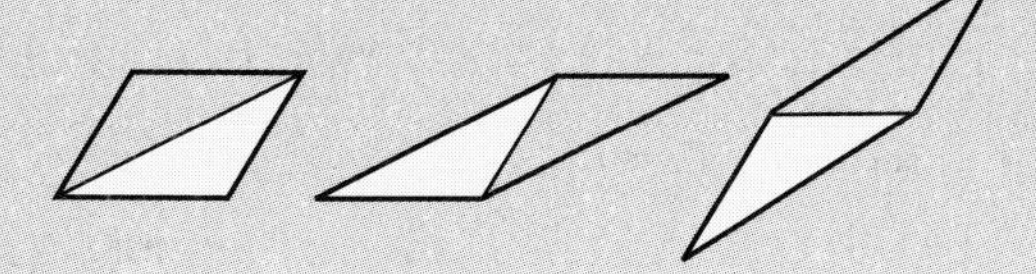

# Additional activity

## Name the shape

SSM2ac/4
PSE RS/D1,2 S/D1
P/4b S/3abc

Hide a 2D shape in an envelope. Tell the children that they have to find out what shape it is. They can ask any question they like about it, but the only answer they will be given is 'Yes' or 'No'.

Allow **four** questions. Award 4 points if the shape is guessed correctly after only one question, 3 points after two questions, and so on.

# TRIANGLES

SSM2ac/4
RS/D1,3 S/D1
S/3ac S/4b

The children have met triangles of various types, including right-angled triangles, in earlier stages of Heinemann Mathematics. The children folded triangles to find their lines of symmetry.

The terms 'isosceles' and 'equilateral' are now introduced to describe triangles with two or three equal sides respectively. The children make these triangles using strips, and fold paper triangles to find their lines of symmetry.

# Introductory activities

## 1 Equilateral triangles

- Show the children a design made from three equilateral triangles stuck on top of each other.

Turn the design around to show that it 'still looks the same'.

Flip the design over so that only the large triangle is visible, and turn it in the same way. Ask the children about the sides of the triangle and establish, by measuring if necessary, that it has **three equal sides**. Introduce the name **equilateral triangle**.

- Show the children a set of triangles containing an equilateral triangle, some isosceles triangles and a scalene triangle. Ask the children to find one which is equilateral. They should try to do this by inspection or by turning it round 'to see if it looks the same'. The sides can be measured as a check.

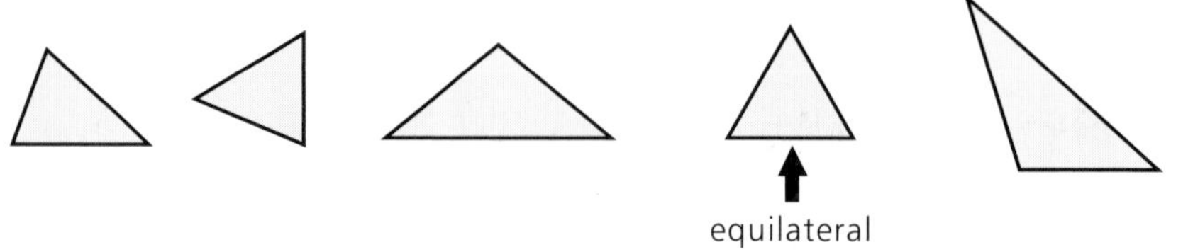


## 2 Isosceles triangles

- Use a pair of equal strips, joined at the corner, with a piece of elastic forming the third side of a triangle. Tell the children that this type of triangle has **two equal sides** and is called an **isosceles triangle.**

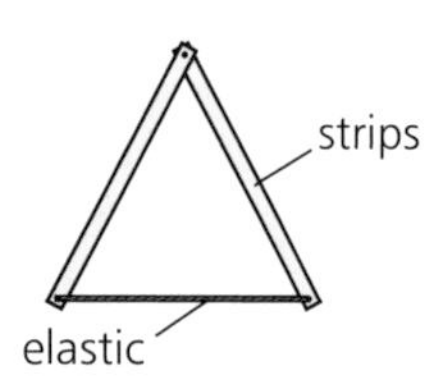


Move the strips to show different isosceles triangles.

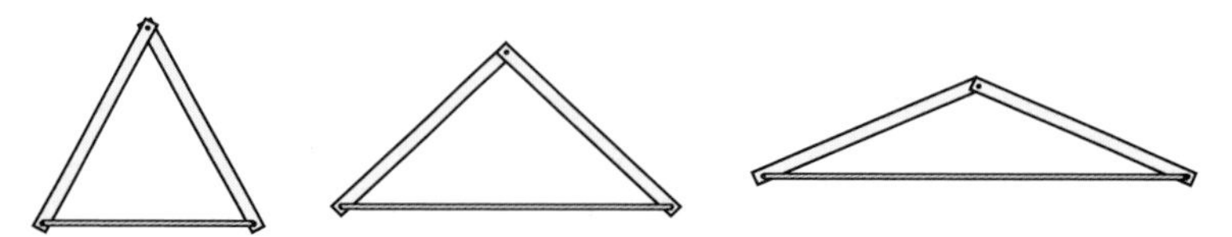

Show that, when the elastic side is the same length as the other two sides, the triangle is equilateral.

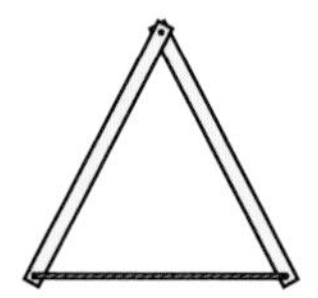

- Use the set of shapes shown in the last part of activity 1. Ask the children to look for the triangles which are **isosceles.**

## Textbook page 97, Workbook page 20

### *2D shape: equilateral and isosceles triangles*

In order to make the triangles shown in questions 1 and 2 at the same time, the children need a set of seven strips of equal length plus one shorter and one longer strip. Alternatively, straws and pipe cleaners might be used.

Discuss the children's completed work for question 4 to stress that an equilateral triangle has 3 lines of symmetry and an isosceles triangle 1 line of symmetry.

In question 5, the two equilateral triangles will make a rhombus.

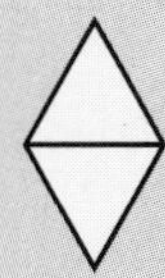

The two isosceles triangles make a parallelogram or a rhombus or a kite.

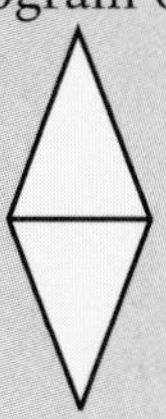

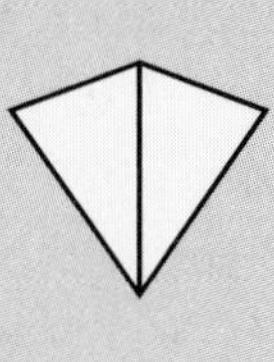

One isosceles and one equilateral triangle produce a kite in this particular case.

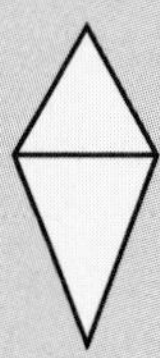

**Shape**
**2D shape**

SSM2abc/4
RS/D1,3 S/D1
S/3ac S/4b

# DRAWING SHAPES ON GRIDS

The work on four-sided shapes and triangles is now reinforced by providing activities on co-ordinate grids. This work also makes use of some of the properties of these shapes.

## Introductory activities

SSM2ab,3b/4
RS/D1,3 PM/D3
S/3a S/4bd

### 1 Shapes within shapes

- Draw a grid on a large sheet of paper or use a 'squared' chalkboard or overhead projector transparency. Ask the children to give the co-ordinates of the three points shown below [(1, 2), (5, 4), (9, 2)].

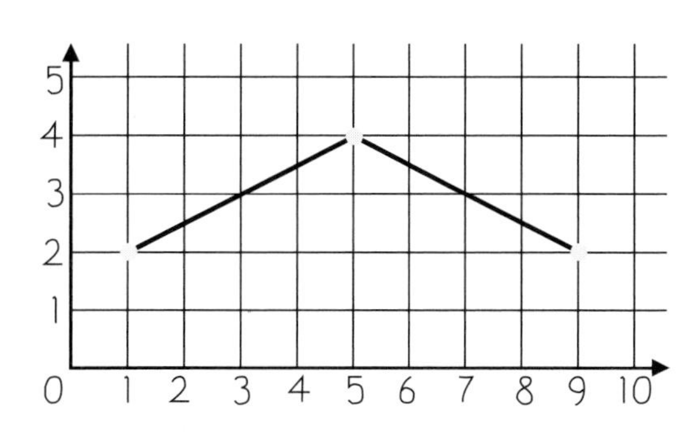

Ask them to choose a fourth point which will give a rhombus when the four points are joined to make a shape [(5, 0)].

- Mark the points in the middle of the sides of the rhombus and ask for their co-ordinates [(3, 3), (7, 3), (7, 1) (3, 1)].

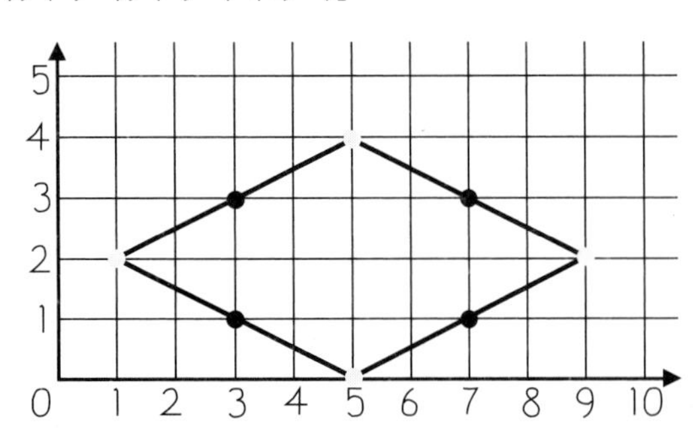

Ask what type of four-sided shape they will make when the points are joined in order (a rectangle).

- Ask a child to mark the points in the middle of each side of the rectangle (shown as crosses below).

  'What four-sided shape will be made when the points are joined in order?' (a rhombus again).

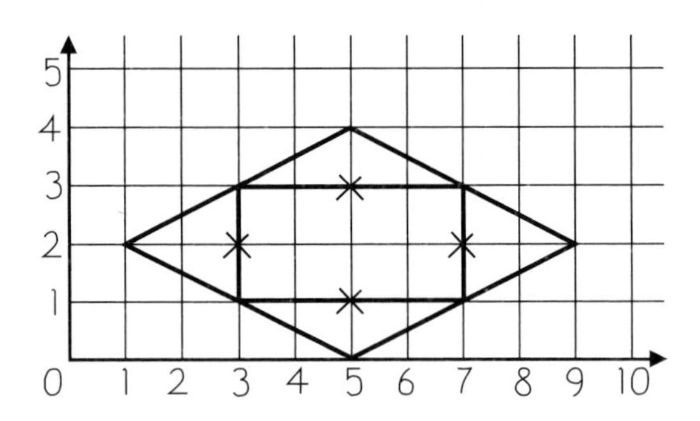

- Some children could follow this up by
  — marking the points (1, 2), (5, 4), (9, 2), (9, 0) on squared paper
  — joining them in order to make a four-sided shape
  — marking the points in the middle of the sides of the shape and joining them in order. What shape do they have? (a parallelogram).

UA3a/4 SSM2ab,3b/4
PSE RS/D1,3 PM/D3
P/4a S/3a S/4bd

## Workbook page 33 *2D shape: drawing shapes*

In question 1, the children should join up the points as they go to ensure they do so in the right order. They could write the names on the shapes themselves or underneath the grid. The shapes are **(a)** parallelogram **(b)** rectangle **(c)** isosceles triangle **(d)** rhombus.

**Problem solving**

In question 2, the first two shapes are

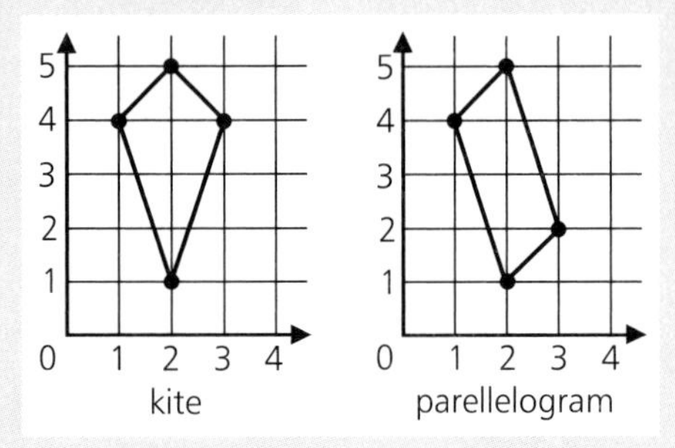

Shape
2D shape

The third shape is more difficult to find. The children may need to check with a right angle tester.

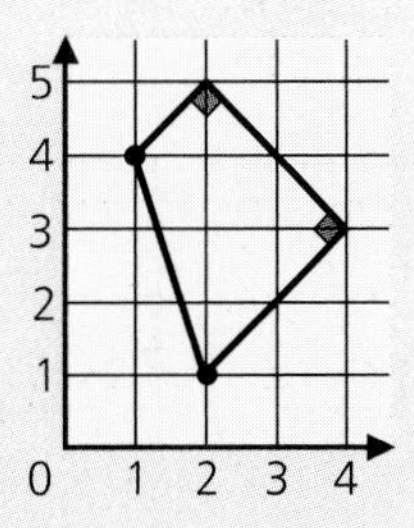


Problem solving

In question 3, some of the possible shapes are

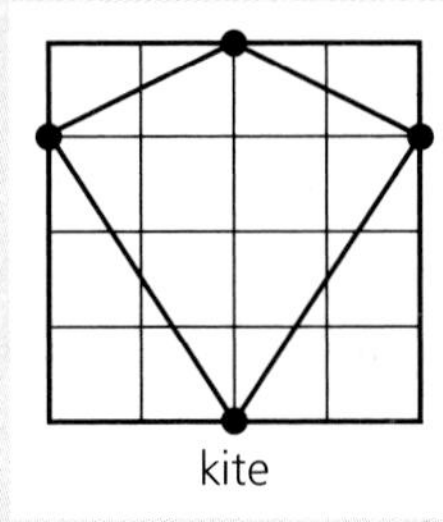
kite

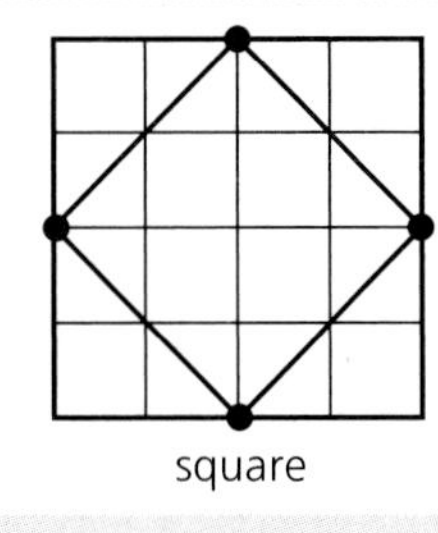
square

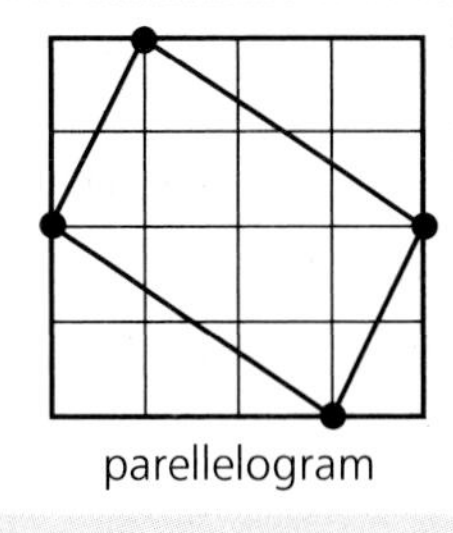
parellelogram

. . . and so on.

R29

## CONGRUENT SHAPES

SSM2bc/4
RS/D1,5
S/3a S/5a

The term 'congruent' is now introduced to describe identical shapes.

## Introductory activities

### 1 Rectangles

- Prepare a set of 7 paper or card rectangles from four A4 sheets as shown, by folding and cutting.

A4 sheets

- Ask the children to spot which pairs of shapes are 'identical' and to check by fitting one on top of the other.

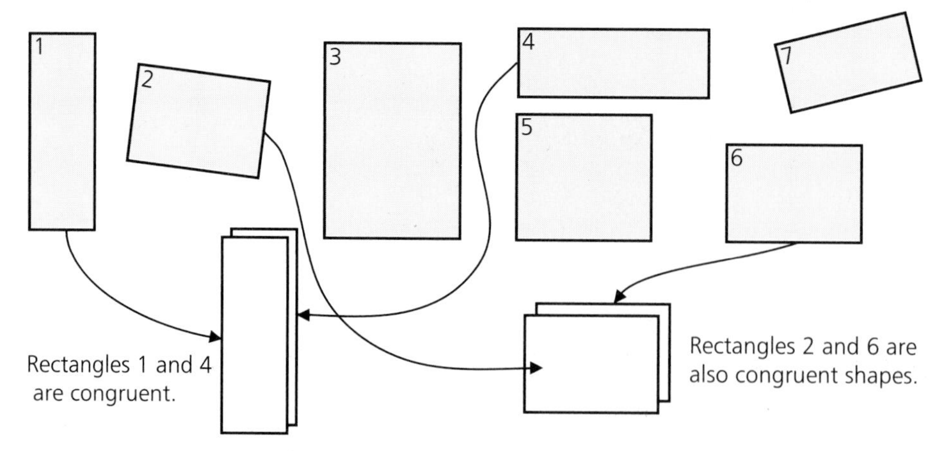


Describe each pair of identical shapes as **congruent** shapes.

- Ask the children to find a pair of shapes which will fit together to make a shape congruent to the largest rectangle. For example,

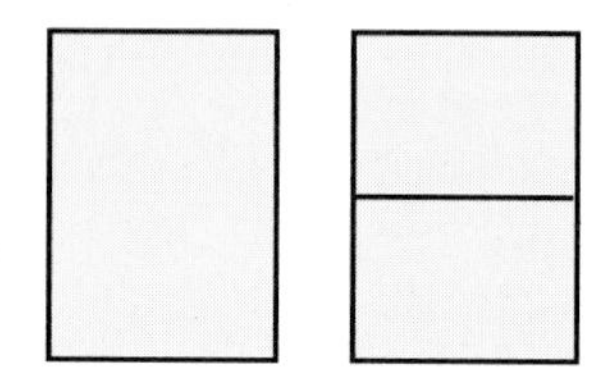

Repeat until there are 4 congruent rectangles (A4 sheets) as shown in the first diagram in this activity.

## 2 Making congruent shapes

- Give some children a shape each, cut from card or paper. For example,

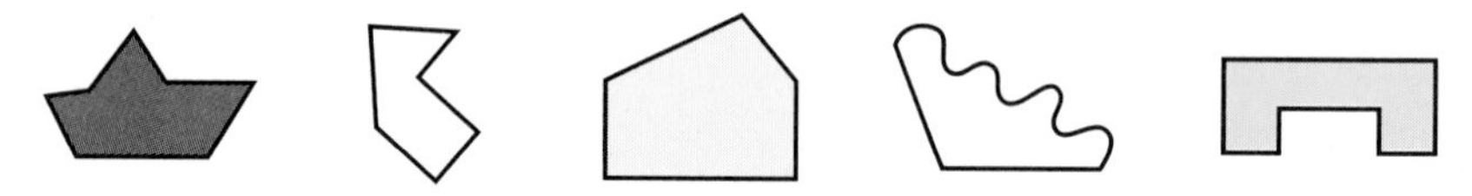

Ask them to make another shape congruent to it. They can do this by tracing the shape or drawing round it.

- Give other children paper to fold and cut to make a pair of congruent shapes.

- Display the pairs of congruent shapes on a wall poster.

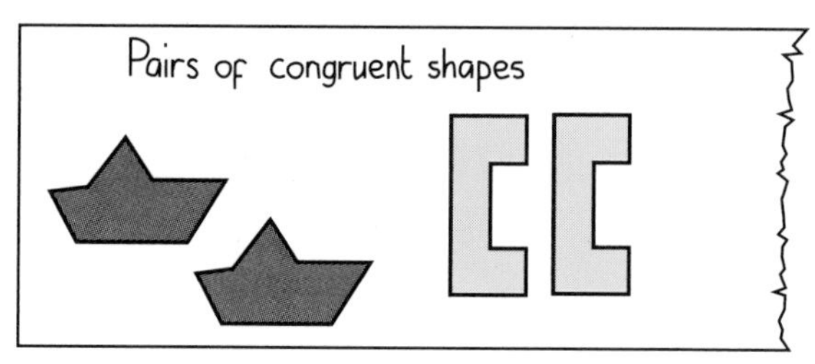

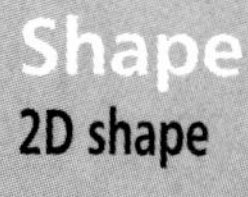


## 3 Tiling patterns of congruent shapes

Supply the children with shape templates and ask them to draw tiling patterns of congruent shapes. For example,

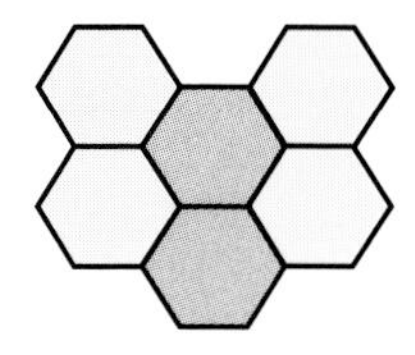
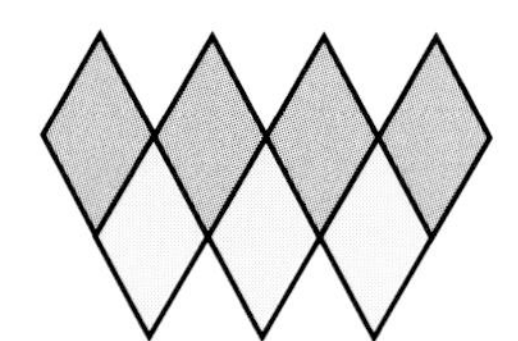
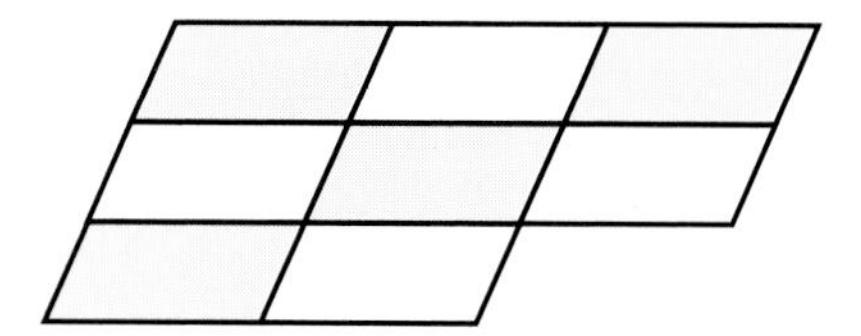

UA3a/4 SSM2bc/4
PSE RS/D1,3 PM/D3
P/4a S/3a S/4d S/5a

## Textbook page 98 *2D shape: congruent shapes*
## Workbook page 18

The children should draw round the tangram pieces carefully, so that congruent shapes are accurate enough to appear 'identical', and also show how the pieces fit. For example, in question 2,

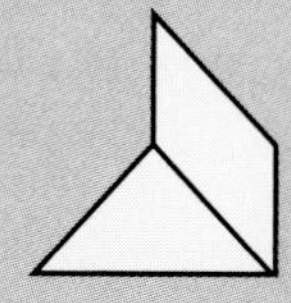
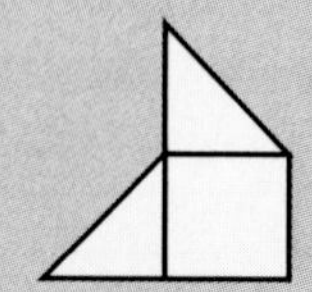

Alternatively, the children could sketch the shapes. The grid on the pieces may help them to use squared paper for this, if desired.

Possible arrangements of pieces in questions 3 to 5 are as follows:

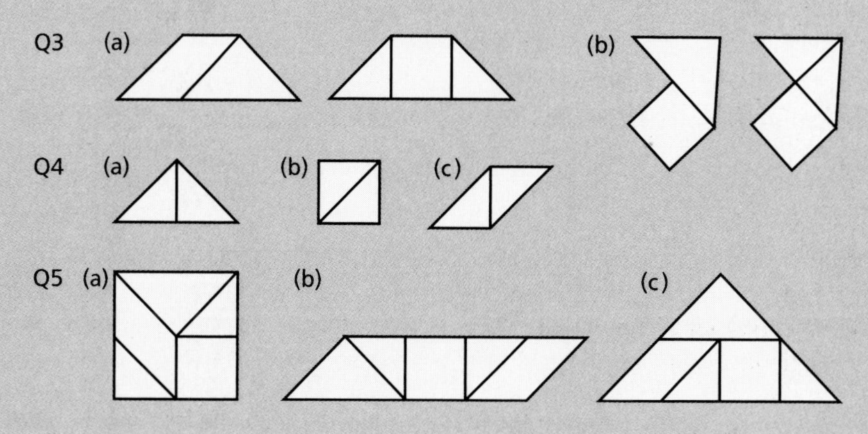


**Problem solving**

# CIRCLES

In Heinemann Mathematics 5, the children constructed circles in a variety of ways

- — by drawing around suitable objects
- — using string and chalk
- — using a plastic strip and a pencil.

They now use a pair of compasses to draw circles, and are introduced to the terms 'radius', 'diameter' and 'circumference'.

SSM2b/4
RS/C2 RS/D4
S/4b

# Introductory activity

## Drawing circles

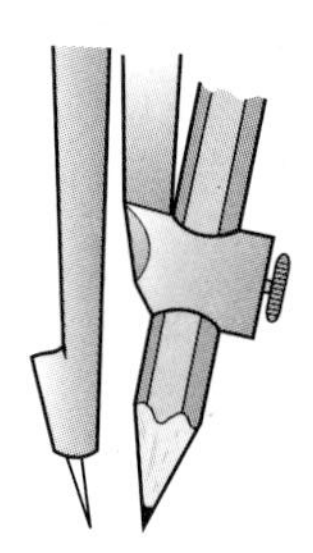

- Show the children how to insert the pencil into the pair of compasses, ensuring that the point of the pencil and the point of the compasses are aligned when the compasses are closed.

  Now demonstrate how to draw a circle, emphasizing that gentle pressure should be applied on the point of the compasses, since too much pressure on the pencil may cause the point of the compasses to move.

  Demonstrate how to set a radius of a given length using a ruler.

  Draw a circle and highlight the following terms: 'radius', 'diameter' and 'circumference'.

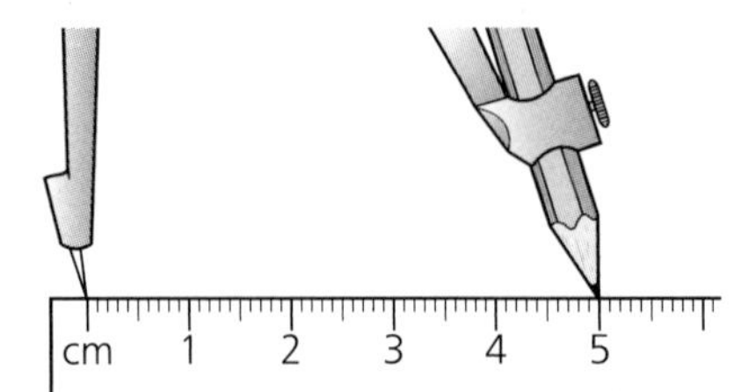


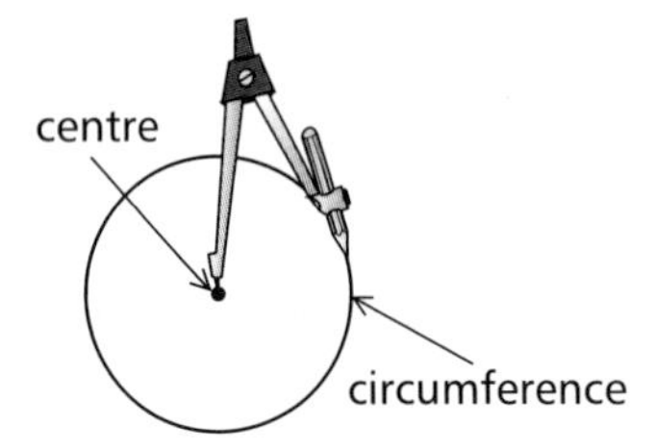


'The distance around the perimeter'

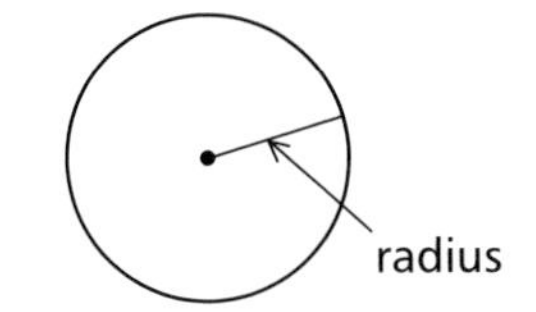


'The distance from centre to circumference'

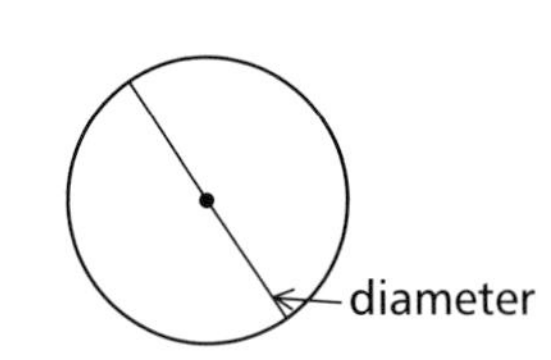


'Twice as long as the radius'

  Give the children practice in drawing circles of given radii.

- Repeat for drawing circles of given diameter where the children have to halve this length to set the correct radius.

SSM2b/4
RS/C2 RS/D4
S/4b

## Textbook pages 99 and 100 *2D shape: circles*

On Textbook page 99, in question 1, the children are not expected to make an exact copy of the design, but simply to draw a variety of different-sized, overlapping circles.

In question 2, the circles drawn by the children could overlap.

In question 4, the children should realize that each dot is used as the centre for two circles with radii 4 cm and 6 cm respectively.

In question 6, some children may require help in starting the design. They could begin by drawing a line 28 cm long with dots marked every 4 cm (ie half of the 8 cm diameter of the larger circle).

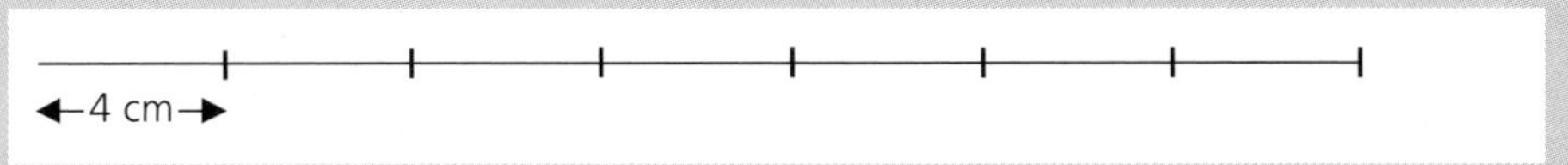


These dots are the centres of the circles. The diameters of the semi-circles are 8 cm and 6 cm. The children should realize that their radii are 4 cm and 3 cm respectively.

# ROTATIONAL SYMMETRY

SSM2c/4
S/E1
S/5c

In Heinemann Mathematics 5, shapes and designs with one and two lines of symmetry were revised, and some with more than two lines of symmetry were introduced.

This section of Heinemann Mathematics 6 concludes with the introduction of rotational symmetry.

## Introductory activities

### 1 Rotating shapes

- Show the children card or paper shapes like these.

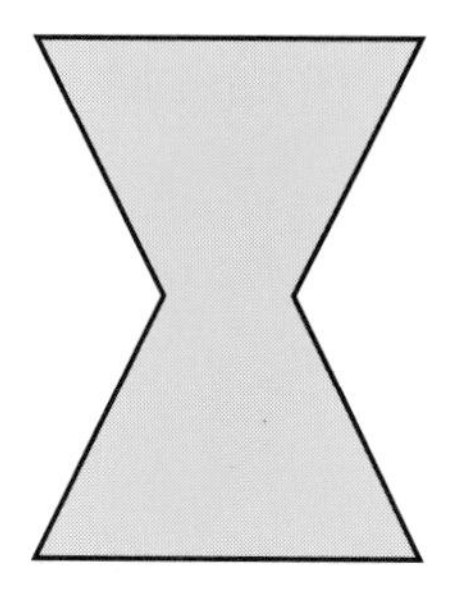

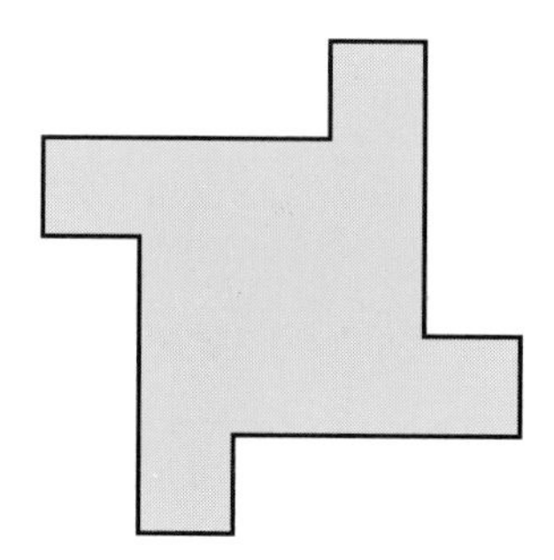

Ask about their lines of symmetry. Check by folding or using a mirror that the first shape has two lines of symmetry and the second one has none.

- Explain that both of these shapes have another type of symmetry called **rotational symmetry**, which they can find out about, not by folding the shape, but by **turning it round.**

Draw round the first shape with chalk to mark its outline.

Mark one corner of the outline and the same corner of the shape with a dot.

Turn the shape round to show that it **fits its outline two times in one full turn.**

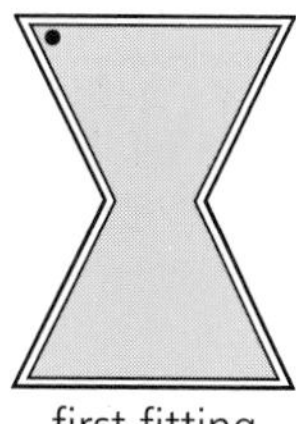

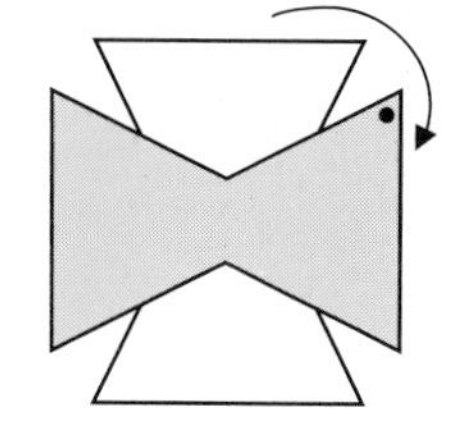

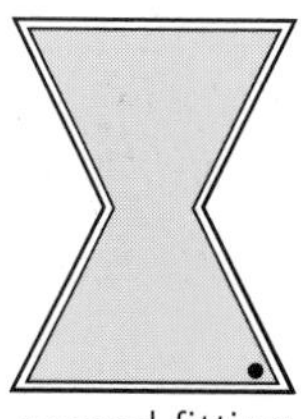

second fitting

Repeat for the other shape to show that it fits its outline **four times** in one full turn.

Introduce the term 'rotational symmetry' to describe the property of a shape 'fitting its outline' as it is turned round.

## 2 Rotating patterns

- Draw a rectangle with a pattern like this and mark a dot in one corner.

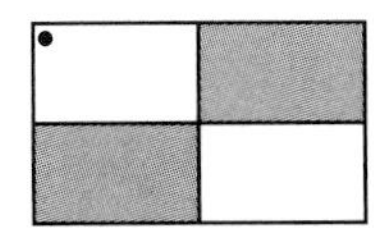

Trace the shape, then rotate the tracing about its centre. Ask the children to say when the tracing matches the original shape.

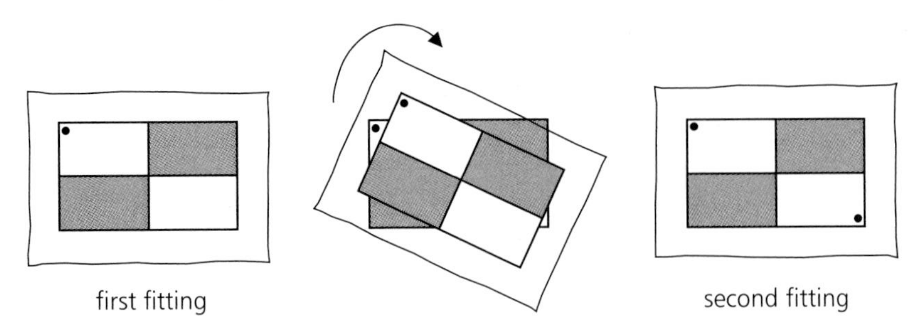


Continue rotating until the tracing returns to its original position. Establish that the shape matches the original or 'fits its outline' **two times** in one complete turn.

- Repeat for shapes such as these.

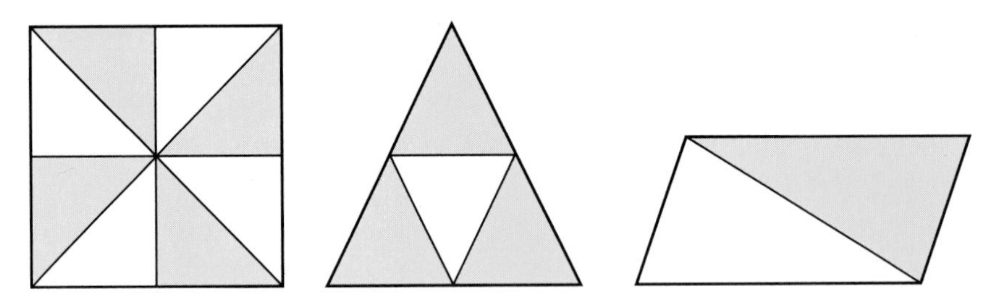

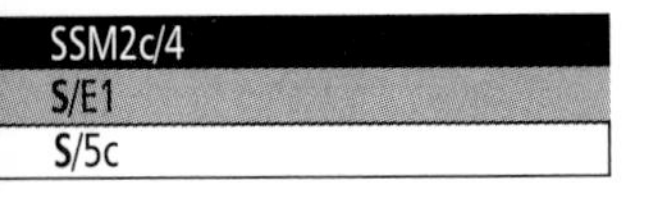


### Textbook pages 101 and 102 *2D shape: rotational symmetry*

On Textbook page 101, in question 2, some children may be able to tell by inspection how many times a shape fits its own outline. This is acceptable, but they should be encouraged to check their answers by tracing and rotating the tracing.

On Textbook page 102, in question 3, the children should realize that the colours as well as the outline need to match when the tracing is rotated.

For some children it may be worthwhile introducing the language 'order of rotational symmetry'. For example,

— a square has the order of rotational symmetry 4 because it fits its outline four times in one complete turn

— an equilateral triangle has the order of rotational symmetry 3 because it fits its outline three times in one complete turn.

UA3a/4 SSM2ac/4
PSE RS/D1,3 S/D1,2
P/4a S/3ac S/4b

## Textbook page 103

### *Other activity: shape properties, problem solving*

- This activity could be attempted at any time after the children have completed the 2D shape section on Textbook pages 95–102.
- In question 1(a), the kite and the parallelogram are being described.
- In question 1(b), giving 'opposite sides equal' as a clue would not identify the rectangle, since it also applies to the rhombus. It would be better to give 'all its angles are right angles' as the identifying clue.
- In question 1(c), possible clues to distinguish the other three shapes are:

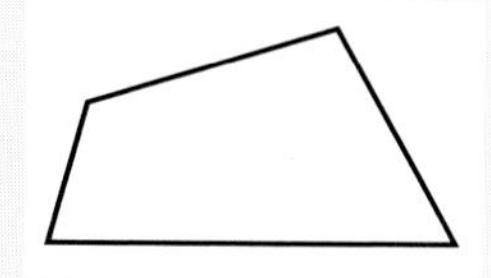

Has no equal sides, a 'basic' four-sided shape (quadrilateral).

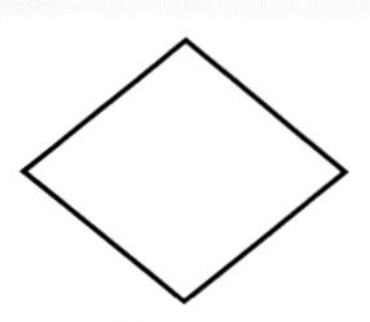

All four sides are equal in length – in this case, a rhombus.

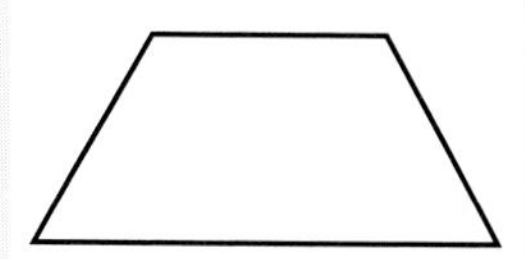

Only one pair of equal sides (trapezium).

There are other possible descriptions.

- In question 2(a), the first description applies to *Shape 3*, the second to *Shape 2*.
- In question 2(b), possible answers are:

Shape 1: The triangle has sides half as long as the square and is inside the square.

Shape 4: The triangle is outside the square and touching one side of the square. The sides of the triangle and the square are the same length.

**Problem solving**

- In question 2(c), possible solutions include:

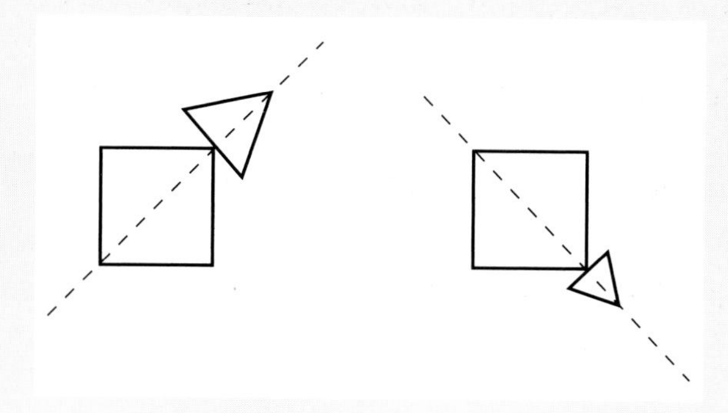

SSM2b/4
RS/C1
RS/D1,6
RS/E4
S/4b

# 3D shape

## Overview

This section

- consolidates nets of cubes and cuboids
- introduces nets and skeleton models of triangular prisms and pyramids.

| | Teacher's Notes | Textbook | Workbook | Reinforcement Sheets |
|---|---|---|---|---|
| **Nets: cubes, cuboids** | 227 | 104 | | |
| **Nets and skeleton models:** | | | | |
| **triangular prisms** | 229 | 105 | | |
| **pyramids** | 231 | 106 | | |
| *Other activity* | *234* | *107* | | |

## Resources

### Useful materials

- model cubes, cuboids, triangular prisms, square and triangular pyramids (plastic, wooden, cartons, etc.)
- construction kits with interlocking plastic faces, such as 'Clixi' and 'Polydron'
- sets of 6 identical plastic or card squares
- centimetre squared paper, card, sticky tape, glue sticks, scissors
- plasticine, Blu-tack, coloured paper, plastic straws
- other materials suggested within the introductory activities

### Assessment and Resources Pack

Assessment

*Round-up 3*
Question 3

Resources

*Problem Solving Activities*
24 Open boxes (Nets)

# Teaching notes



In Heinemann Mathematics 5, the children were introduced to nets and skeleton models of cubes and cuboids.

In Heinemann Mathematics 6, the work on cubes and cuboids is consolidated, and some emphasis is placed on the use of the terms 'faces', 'edges' and 'vertices'. Nets and skeleton models of triangular prisms, square pyramids and triangular pyramids are introduced.

## NETS: CUBES, CUBOIDS

### Introductory activities

SSM2b/4
RS/C1 RS/D1,6
S/4b

#### 1 Nets of a cube

- Cut along the edges of a carton in the shape of cube, or open out a cube made from interlocking squares or from card and sticky tape, to obtain a net.

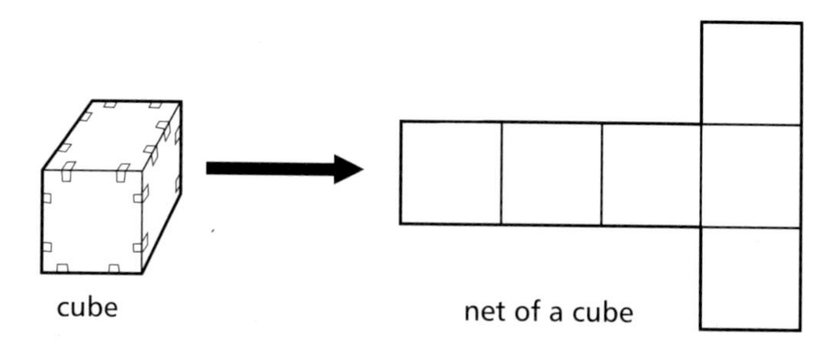


Discuss the net with the children, emphasizing that it consists of 6 square faces all of the same size.

Re-assemble the cube and describe it using the terms 'faces', 'edges' and 'corners'.

Introduce the terms 'vertex' and 'vertices' as alternatives to 'corner' and 'corners' respectively.

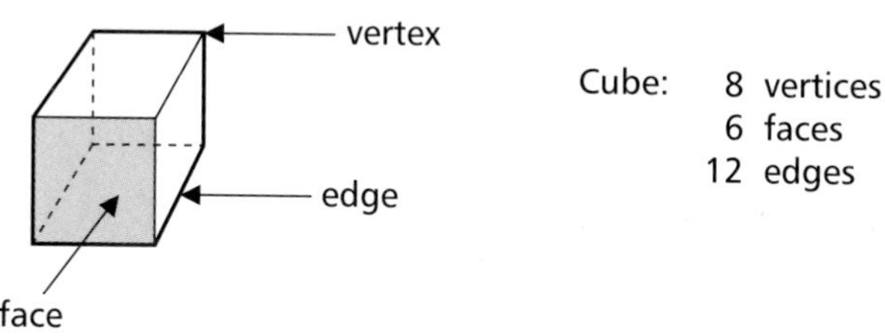


- Use interlocking squares, or card and sticky tape, to make a different net of a cube.

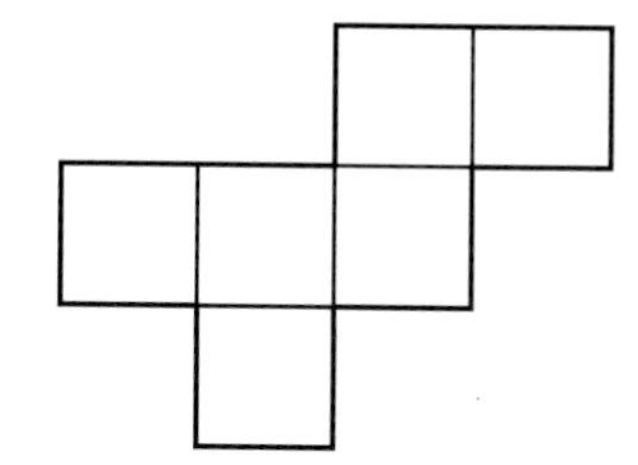

Show by folding that this net also makes a cube. Sketch the net on the chalkboard.

Give 6 identical squares to groups of children. Ask them to make different nets of cubes and to sketch these on centimetre squared paper. Discuss the different nets that the children make, collating their results on the chalkboard or on squared paper. Eleven different nets are possible.

## 2 Opening out a cuboid

Cut along some of the edges of a carton in the shape of a cuboid, or a cuboid made from card and sticky tape, and open it out. Alternatively, open out a cuboid made from interlocking rectangles.

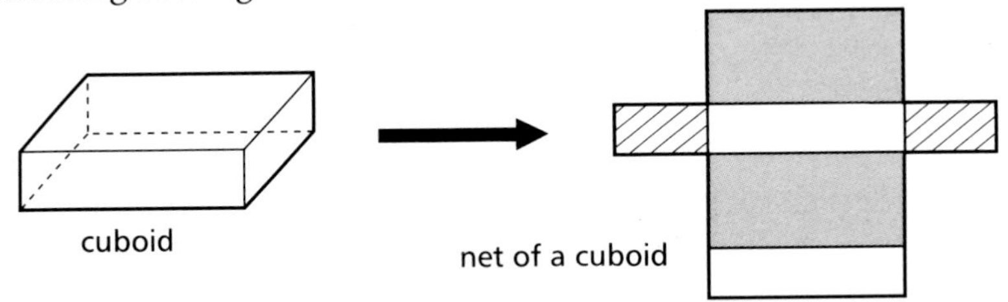


Discuss the net, emphasizing that it consists of 6 rectangular faces that are not all identical. The faces occur as 3 pairs of identical or congruent rectangles. Re-assemble the cuboid and ask the children to use the terms 'faces', 'edges' and 'vertices' to describe it.

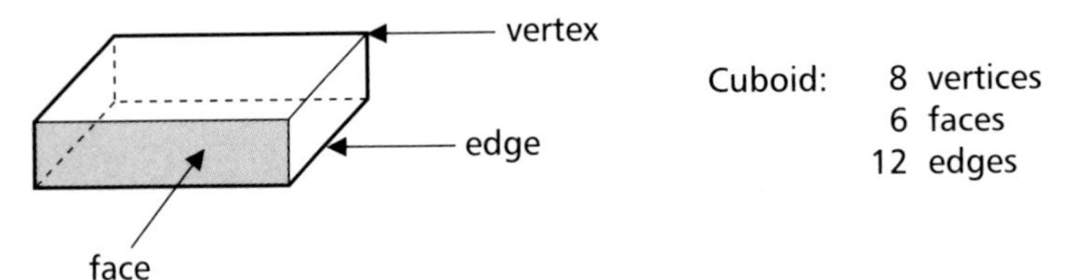


Show that it is possible to have a different arrangement of the 6 rectangles which will still fold to make a cuboid. For example,

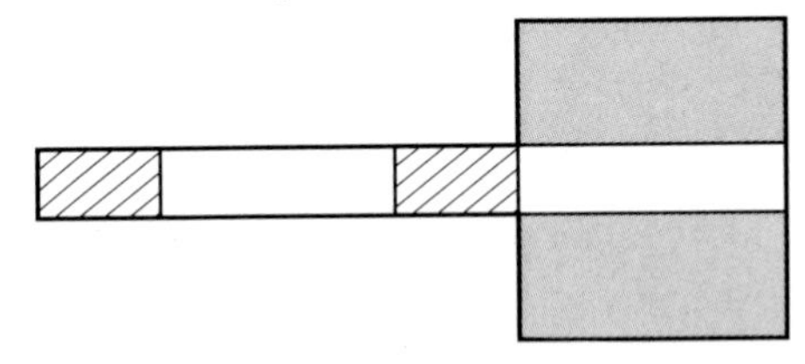

Sketch this different net of a cuboid on the chalkboard. Give groups of children 3 compatible pairs of identical rectangles. Ask them to make different nets of cuboids and to sketch these on squared paper. Discuss the different nets that the children make, collating their results on the chalkboard or on squared paper.

SSM2b/4
RS/C1 RS/D1,6
S/4b

### Textbook page 104 *3D shape: cube, cuboid, nets*

The term 'vertices' should be discussed if it has not already been met as part of the introductory activities.

# NETS AND SKELETON MODELS: TRIANGULAR PRISMS

## Introductory activities

### 1 Triangular prisms

Discuss different triangular prisms, such as a prism made from

— 3 squares and 2 equilateral triangles

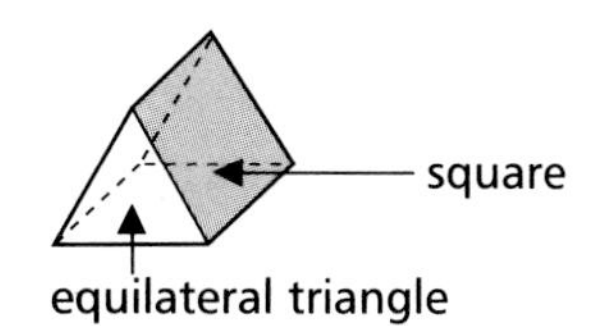


— 1 square, 2 rectangles and 2 isosceles triangles

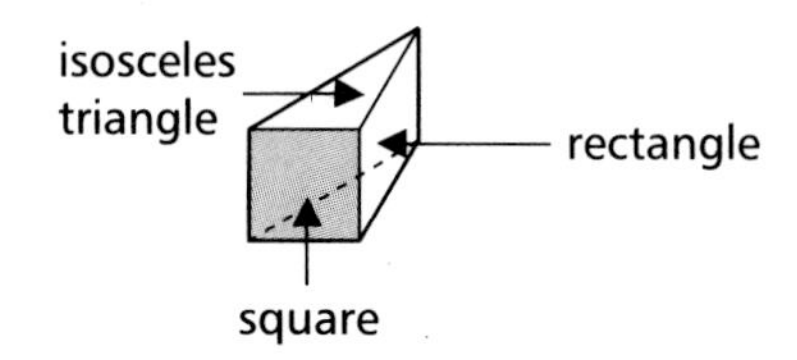


Ask the children to use the terms 'faces', 'edges' and 'vertices' to describe these triangular prisms.

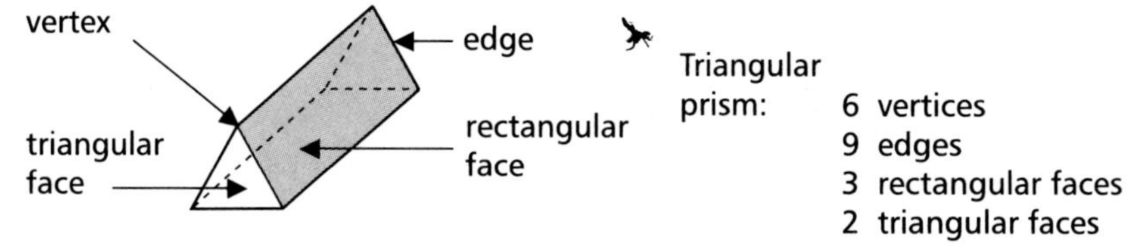


Make sure that the children are familiar with the name **triangular prism**.

### 2 Finding the net of a triangular prism

- Using any of the methods previously described for a cube or cuboid, open out a triangular prism to obtain a net, and sketch this net on the chalkboard.

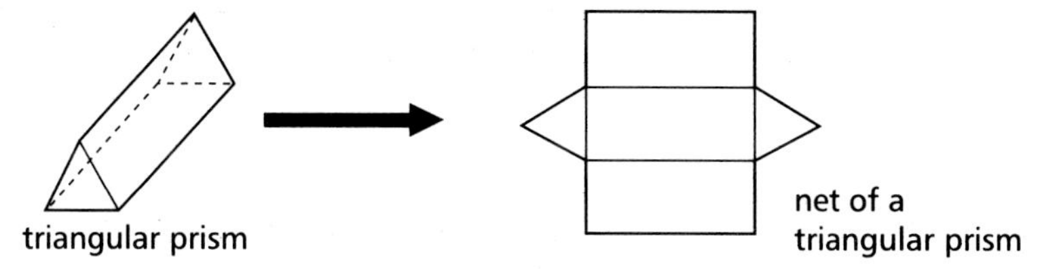


  This net is made from 3 identical or congruent rectangles and 2 identical or congruent equilateral triangles.

- Discuss with the children how to make another net for this triangular prism by moving one or both of the equilateral triangles. For example,

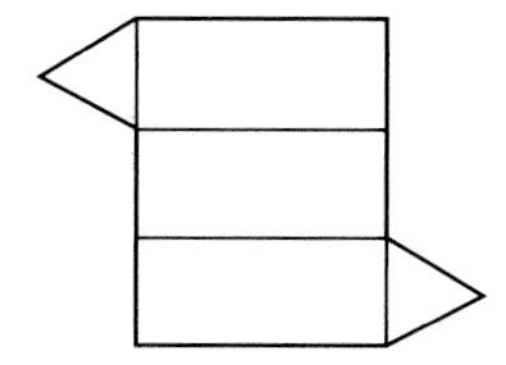

  Fold to show that this new net makes the triangular prism.

## 3 Skeleton model of a triangular prism

While card or plastic nets of 3D shapes emphasize **faces**, skeleton models highlight **edges** and **vertices**.

Show the children a skeleton model of a triangular prism made from straws and blobs of plasticine or Blu-tack.

Discuss the number of straws and the lengths of straws needed to make the triangular prism.

Ask the children how many blobs of plasticine have been used.

SSM2b/4
RS/D1 RS/E4
S/4b

### Textbook page 105

***3D shape: triangular prisms, nets, skeleton models***

In questions 1 and 2, the children make a net of a triangular prism, using plastic or card shapes, by copying the given net. They should use the terms 'faces', 'edges' and 'vertices' to describe the triangular prism constructed.

In question 3, the children make skeleton models of different types of triangular prism using straws and plasticine or Blu-tack.

The children should realize that for

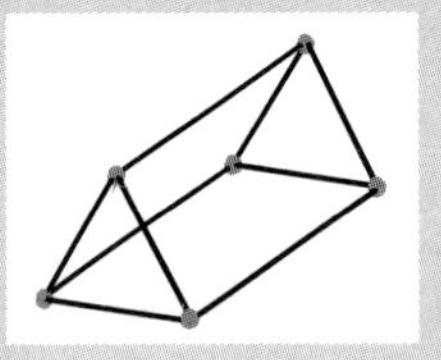

they require 6 short and 3 long straws of equal lengths

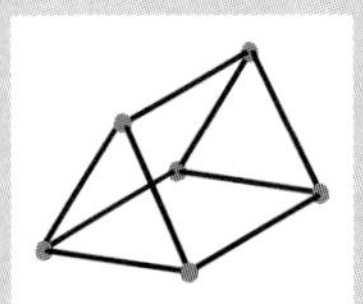

they require 9 straws of equal length

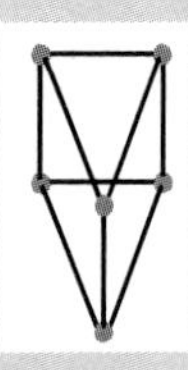

they require 5 short and 4 long straws of equal lengths.

A satisfactory way of making straws of equal length is to cut all the required number simultaneously as illustrated.

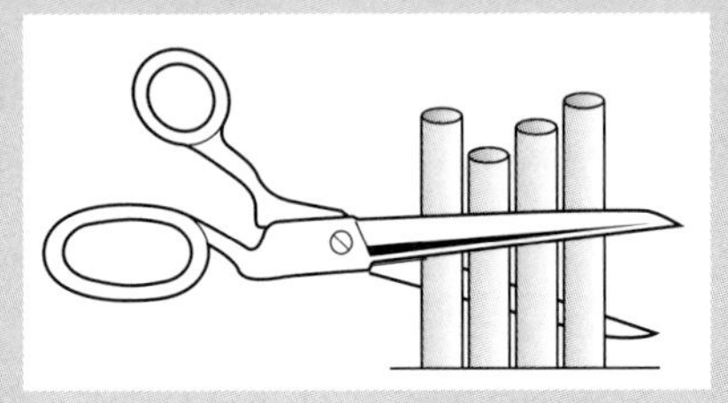

# NETS AND SKELETON MODELS: PYRAMIDS

## Introductory activities

SSM2b/4
RS/D1 RS/E4
S/4b

### 1 Triangular pyramids

Discuss different triangular pyramids, such as a pyramid made from

— 1 equilateral and 3 congruent isosceles triangles

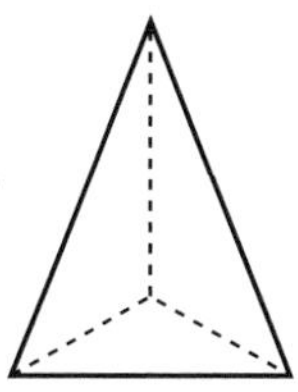

— 4 congruent equilateral triangles.

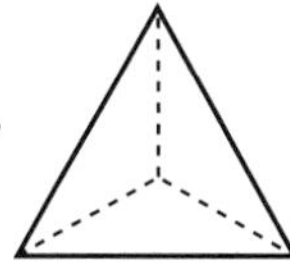

Ask the children to use the terms 'faces', 'vertices' and 'edges' to describe these triangular pyramids.

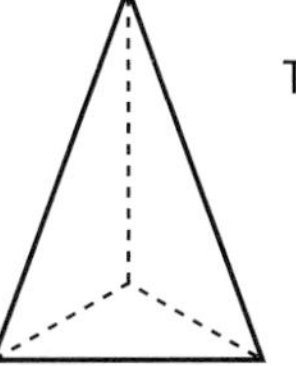

Triangular pyramid: 4 vertices
6 edges
4 triangular faces

Make sure that the children are familiar with the name **triangular pyramid.**

### 2 Finding the net of a triangular pyramid

- Open out a triangular pyramid to obtain a net, and display the net on the chalkboard.

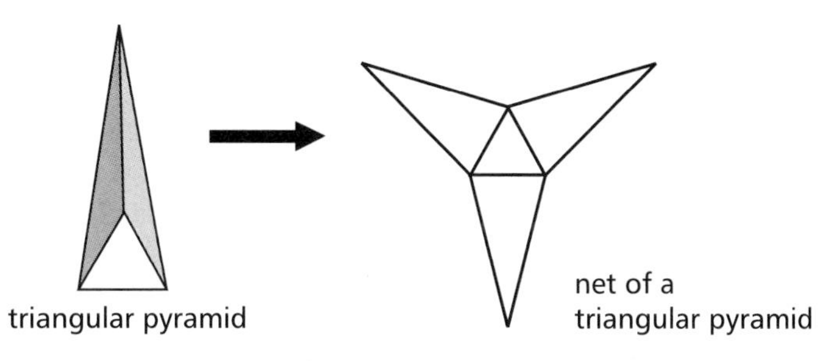


This net is made using 1 equilateral triangle and 3 congruent isosceles triangles.

- Rearrange the isosceles triangular faces of the pyramid to make a different net. Fold to show that this net makes a triangular pyramid.

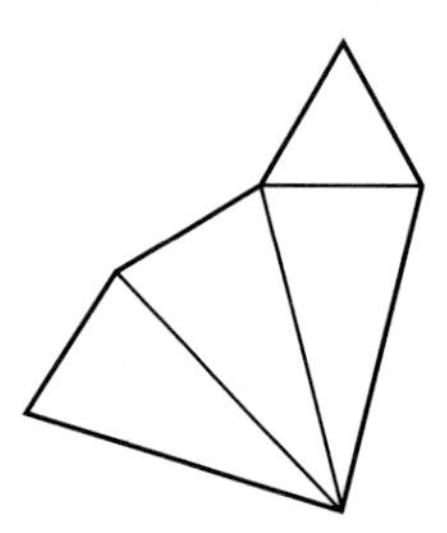

Give groups of children interlocking shapes. Ask them to find and sketch different nets for a triangular pyramid.

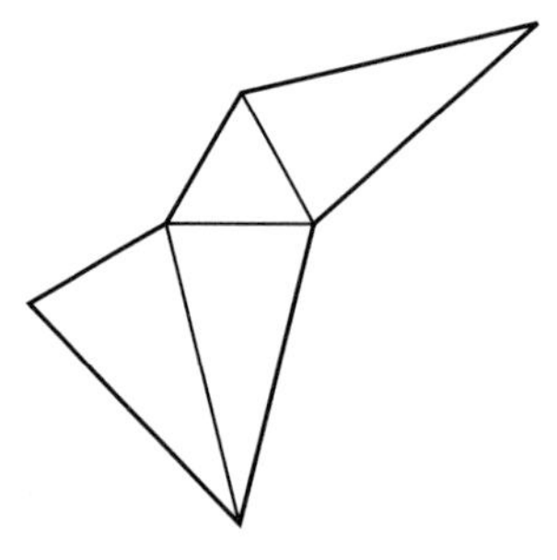

## 3 Square pyramids

Discuss different square pyramids, such as a pyramid made from

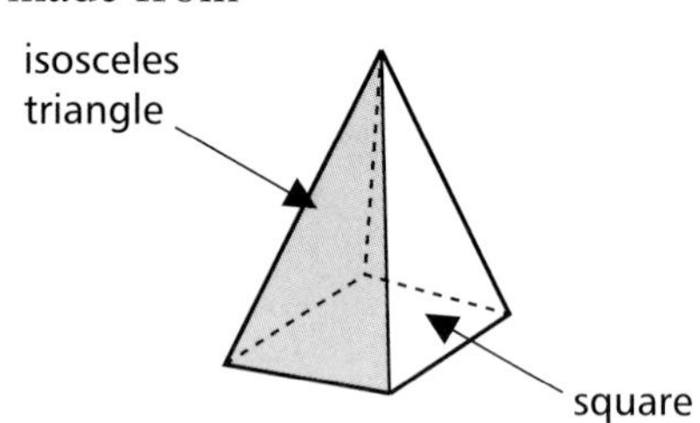


— 1 square and 4 identical isosceles triangles

equilateral
triangle

squar

— 1 square and 4 identical equilateral triangles.

Ask the children to use the terms 'faces', 'vertices' and 'edges' to describe these square pyramids.

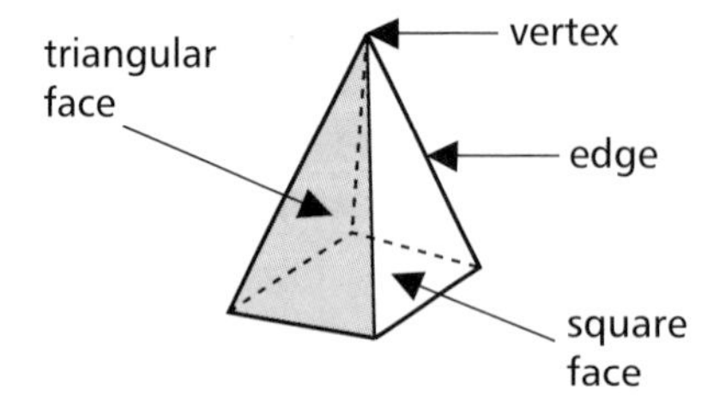


Square pyramid: 5 vertices
8 edges
1 square face
4 triangular faces

Make sure that the children are familiar with the name **square pyramid**.

## 4 Finding the net of a square pyramid

Open out a square pyramid and sketch this net, made from 1 square and 4 identical isosceles triangles, on the chalkboard.

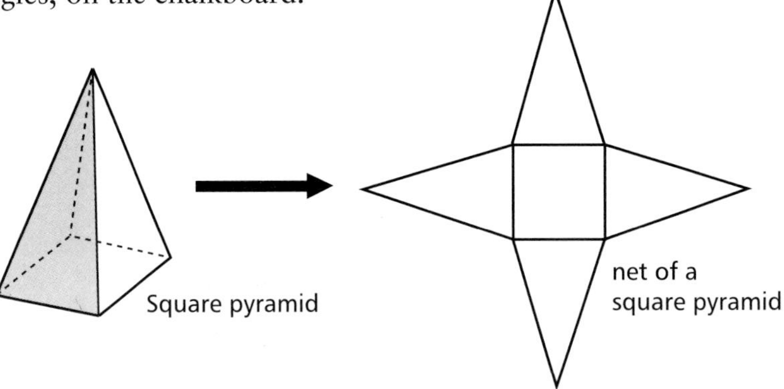


Rearrange the isosceles triangular faces of the square pyramid to make a different net. Fold to show that this net makes a square pyramid.

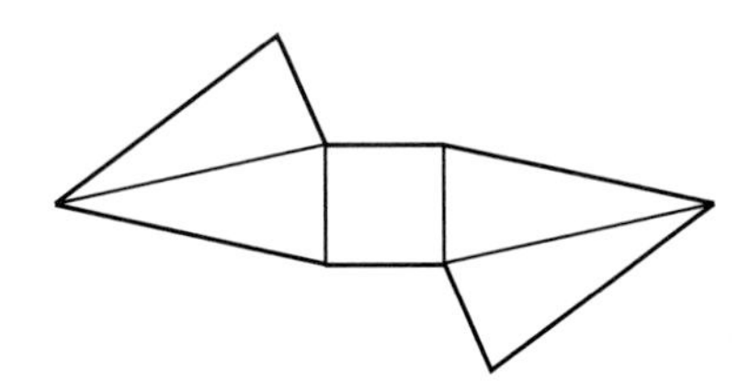

Give groups of children interlocking shapes. Ask them to find and sketch different nets for a square pyramid.

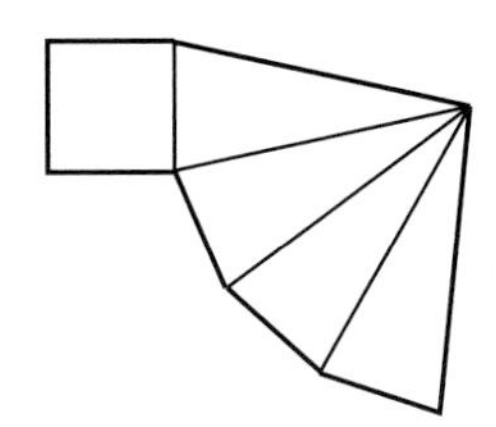

## 5 Skeleton models of triangular and square pyramids

Provide skeleton models of triangular and square pyramids.

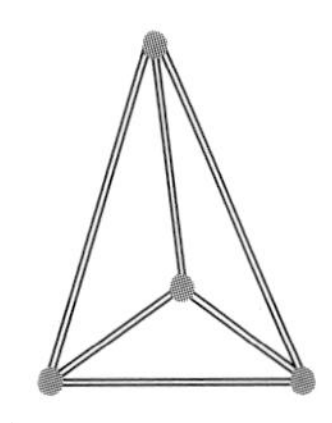

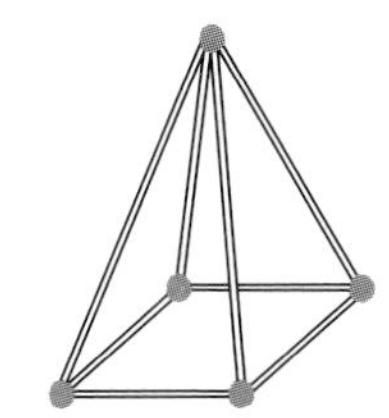

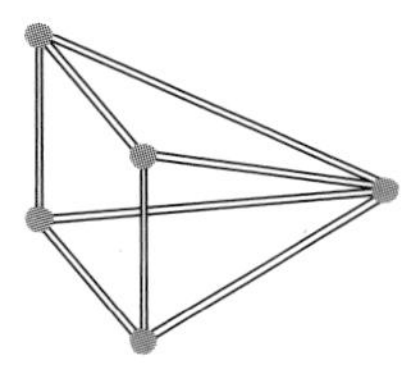

For each pyramid, discuss

— the number and length of straws required

— the number of blobs of plasticine or Blu-tack.

Emphasize that the name of a pyramid is determined by the shape of its base.

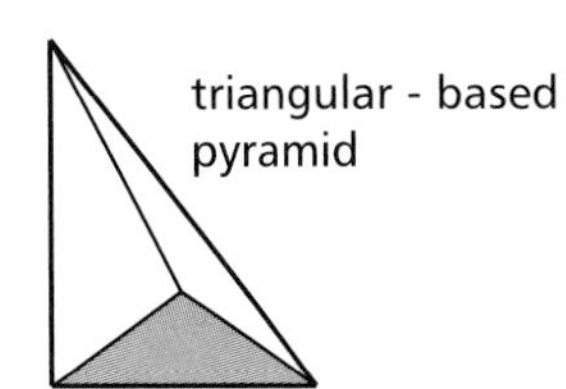


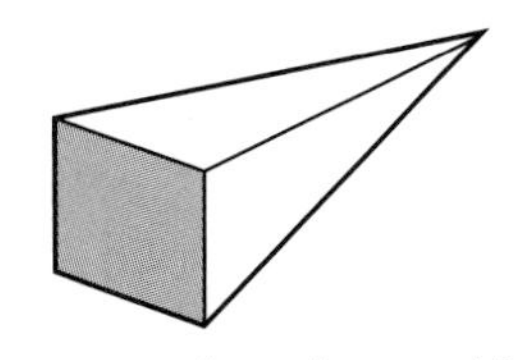

SSM2b/4
RS/D1 RS/E4
S/4b

## Textbook page 106

### *3D shape: pyramids, nets, skeleton models*

The children will need interlocking shapes, straws, plasticine or Blu-tack, and scissors to complete the questions on this page.

In questions 1, 2 and 3, they make nets of triangular and square pyramids. In questions 1 and 3, they use the terms 'faces', 'edges' and 'vertices' to describe the pyramids constructed.

In question 4, the children are asked to make skeleton models of triangular and square pyramids. They should use the method of cutting shown on page 230 of these Notes to ensure that the straws are of equal length.

SSM2c/4
PSE S/D1 S/E1
P/4abd S/3ac S/4b S/5c

## Textbook page 107 *Other activity: 2D shape*

- This activity involves the children in investigating the symmetry properties of pentominoes and in constructing shapes using some of them. It could be attempted at any time after the children have completed the 2D-shape section on Textbook pages 95 to 102.
- Possible answers are as follows.

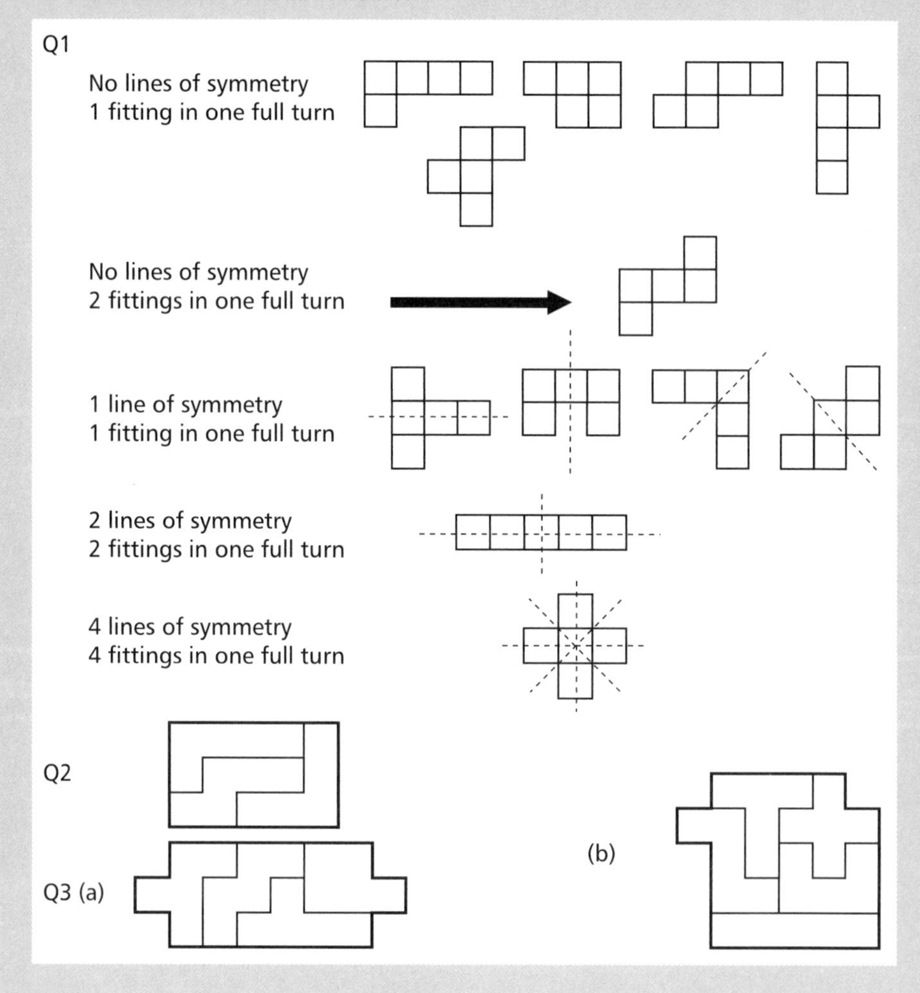

Able pupils might enjoy the challenge of fitting all 12 pieces together to make a rectangle. Some possible solutions are:

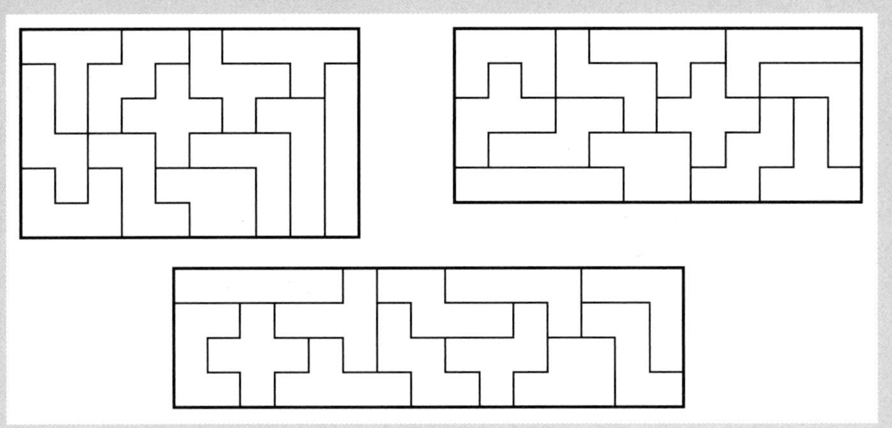

UA4b/4
SSM3c/4
SSM3c/4→5
PSE
PM/D2,4
A/C1,2,3
A/D1
P/4ab
S/4ace

# Angles

## Overview

This section

- revises and consolidates clockwise and anti-clockwise turns in degrees, and links these to the 8-point compass
- introduces measurement of angles to the nearest 5°
- introduces the terms 'acute', 'right' and 'obtuse' to describe angles.

| | Teacher's Notes | Textbook | Workbook | Reinforcement Sheets |
|---|---|---|---|---|
| **Lands Beyond: a context for area and angles** | 166 | | | |
| **Angles in degrees, turning clockwise and anti-clockwise, using the 8-point compass** | 238 | 108 | | |
| **Angles: measuring in degrees** | 240 | 109, 110 | 22 | 30 |
| **Angle activities associated with LOGO** | 242 | | | |

An Extension activity is related to the above section of work as follows:

| | Teacher's Notes | Extension Textbook |
|---|---|---|
| **Other activity: angles, investigation** | 295 | E30 |

Teaching notes for the Extension Textbook are in a separate section at the end of the Teacher's Notes.

# Resources

## Useful materials

- scissors and glue
- paper circles (preferably gummed)
- centimetre square dotty paper
- an angle measurer calibrated in 5° intervals (see Textbook page 110)
- other materials suggested within the introductory activities

## Assessment and Resources Pack

### Assessment

*Shape Check-up 3*
Textbook pages 108–10
(Angles: turns, degrees)

*Round-up 3*
Question 5

### Resources

*Resource Cards*
30, 31 Angle machine
(2D shape: angles)

# Teaching notes

Heinemann Mathematics 5 introduced the use of degrees

- for right angles (90°) and multiples (180°, 270°, 360° or 2, 3, 4 right angles)
- for half right angles (45°) and multiples (135°, 225°, 315° or $1\frac{1}{2}$, $2\frac{1}{2}$, $3\frac{1}{2}$ right angles)
- to interpret clockwise and anti-clockwise turns of these magnitudes in the setting of the 8-point compass.

The work in Heinemann Mathematics 6 begins by revising and consolidating these concepts.

The work is set in the 'Lands Beyond' context.

## ANGLES IN DEGREES, TURNING CLOCKWISE AND ANTI-CLOCKWISE, USING THE 8-POINT COMPASS

### Introductory activities

SSM3c/4
A/C1 PM/D2
S/4c

#### 1 Collect 5 *(estimating angles in degrees)*

In this game for two players, a 90° angle tester, a 45° angle tester and 10 counters are required. In addition, each player requires a ruler, a pencil and a sheet of 1 cm square dotty paper. The testers may be supplied or made by the players using paper circles.

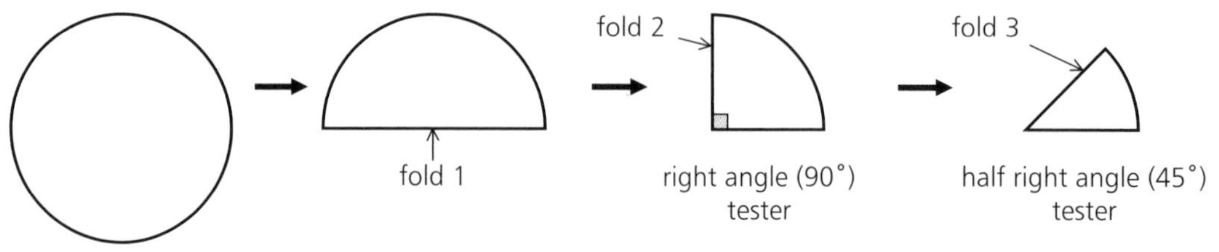


Player 1 draws an angle on their sheet.

Player 2 estimates the size of the angle using the language

'less than 45°'

'about 45°'

'between 45° and 90°'

'about 90°'

'more than 90°'.

Player 1 checks the estimate using the 45° and 90° testers.

If the estimate is correct, player 2 collects a counter.

The players take turns. The winner is the first to collect five counters.

#### 2 8-point compass *(turns in degrees)*

- Prepare a large card compass shaded in eighths and with a needle as shown.

  Revise turning clockwise and turning anti-clockwise by rotating the needle appropriately.

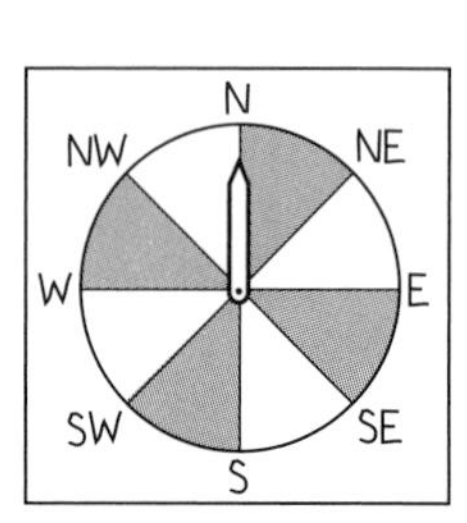

- Ask questions in which the starting direction and the turn are given, and the finishing direction has to be found. For example,

  'The needle starts at North and turns 90° anti-clockwise. In which direction is it now pointing?'

  Each answer can be confirmed using the compass card. The difficulty levels of the questions can be gradually increased:

| **Starting directions** | **Turns (clockwise and anti-clockwise)** |
|---|---|
| N, E, S and W | 90°, 180°, 270° and 360° |
| NE, SE, SW and NW | 90°, 180°, 270° and 360° |
| N, E, S and W | 45° and 135° |
| NE, SE, SW and NW | 45° and 135° |

  Turns should also be described as 1, 2, 3 and 4 right angles, and $\frac{1}{2}$ and $1\frac{1}{2}$ right angles. Turns of 225° ($2\frac{1}{2}$ right angles) and 315° ($3\frac{1}{2}$ right angles) might be included for more able children.

- Ask questions in which the starting and finishing directions are given and the turn has to be described. Confirm each answer using the compass card.

  In such examples, the turn can be given in **two** ways.

  For example,

Starting direction

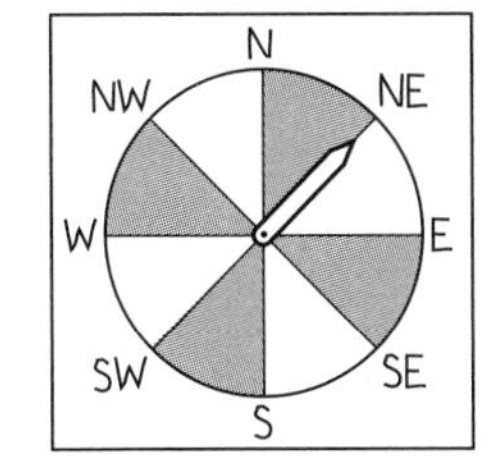


Finishing direction

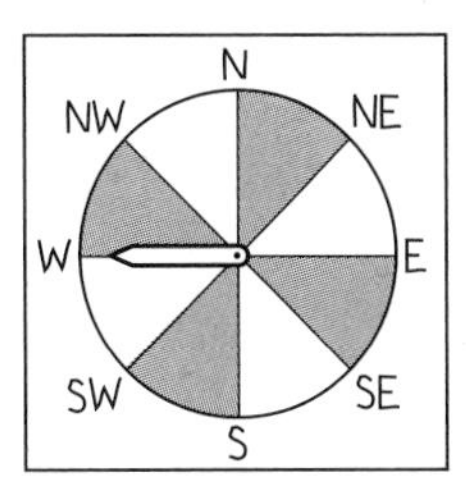


The turn is 135° anti-clockwise or 225° clockwise

Questions can be varied to cover the range of difficulties described above.

SSM3c/4
A/C1 PM/D2
S/4c

## Textbook page 108 *Angles: degrees, turns*

If the Lands Beyond context has not been met previously – it appears in the Area section – it should be discussed.

It would be helpful if the children used **gummed** paper circles to make their angle testers.

As none of the work is new, children who have carried out the introductory activities should be expected to attempt the activities with little or no further guidance.

In question 4(b), more able children might be encouraged to give **two** answers in each of lines 4, 5 and 6 in the table.

# ANGLES: MEASURING IN DEGREES

SSM3c/4→5
A/C1,2,3 A/D1
S/4a

In Heinemann Mathematics 5, the children were introduced to ordering angles by direct comparison, and to describing angles, such as a right angle and a half right angle, as 90° and 45° respectively. Multiples of 45° were introduced. The angle work is now extended to include

— the language **acute, right** and **obtuse** to describe angles

— the use of an angle measurer, calibrated in 5°, to measure given angles.

The Lands Beyond context continues with the children visiting the 'angle wall' outside one of the marquees.

## Introductory activities

### 1 Acute, right or obtuse

- Prepare a set of card angles like these:

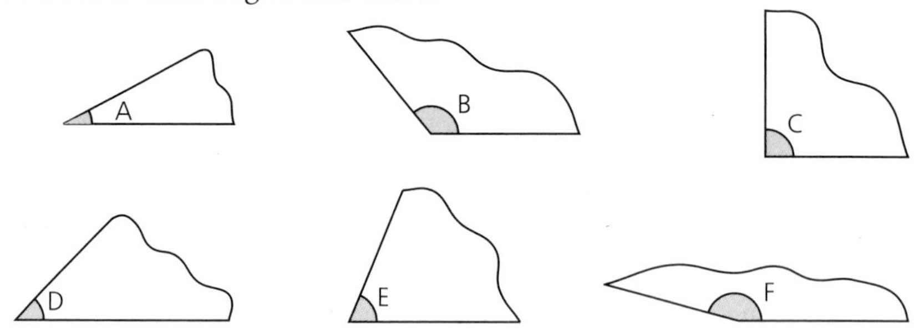


- Ask the children to
  - — compare the angles by placing one angle on top of another
  - — order the angles from smallest to largest using the direct comparison method
- They could also compare the angles with a known right angle or a right angle tester to sort them into
  - — angles greater than 90°
  - — angles equal to 90° or a right angle
  - — angles less than 90°.
- Introduce the words
  - — **acute** to describe angles less than 90°
  - — **obtuse** to describe angles between 90° and 180°.
- Ask the children to find and label

around the classroom.

## 2 Measuring angles *(5° intervals)*

- Introduce an angle measurer like this. Discuss the markings and emphasize that each division measures 5 degrees.
- Use the card angles from activity 1.

  Ask the children to estimate whether each angle is acute, right or obtuse.

  Use the angle measurer to check that angle A, for example, is acute because it measures less than 90°.
- Discuss how to place the angle on the angle measurer and how to 'read' the size of the angle, emphasizing the need to place

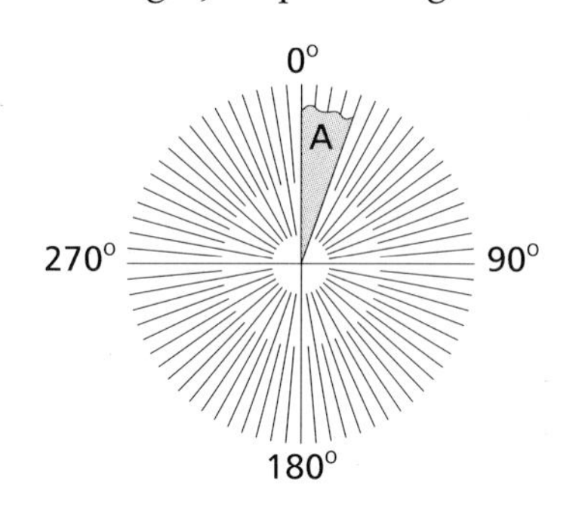


— the angle on the **right** of the 0° line with the corner at the centre of the angle measurer

— one arm of the angle along the 0° line.

Discuss

— how to count round the markings in 10° or 5° steps

— the link between the type of angle (acute, right, obtuse) and the size of the angle in degrees.

UA4b/4 SSM3c/4→5
A/C1,2,3 A/D1
S/4a

### Textbook pages 109, 110 *Angles: measuring in degrees*
### Workbook page 22

Explain to the children that the 'angle wall' is outside one of the marquees in Lands Beyond.

On Textbook page 109, in question 1, some children may need a further reminder of the meaning of acute, right and obtuse before they annotate the red angles on Workbook page 22.

In question 2(a), the children should be encouraged to cut out the angles **carefully** to ensure that they can be measured as accurately as possible. Some children may need a further demonstration of how to use the angle measurer on Textbook page 110.

In question 3, the children should sort **all** of the angles before they stick them into their jotter. The matching pairs are A and B, C and G, D and F, E and H. The children should notice that the sum of each pair of angles which make a **straight angle** is 180°.

On Textbook page 110, in question 6, it may help some children to know that the three pieces (**J, M, N**) they need to make the triangle have 'zig-zag' cuts – for example,

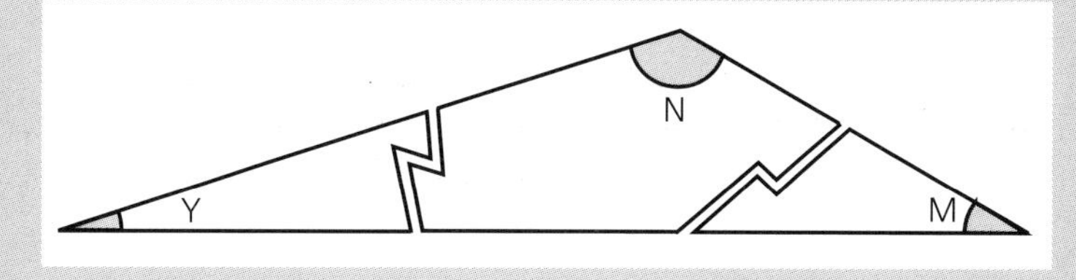

in question 7, the angle pieces (**I, K, L, D**) have 'square' cuts like this:

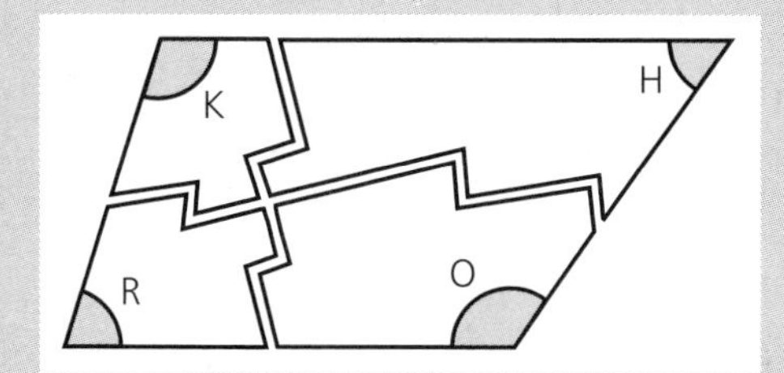


In question 6(c), children who have worked accurately enough should be able to add together the sizes of the three angles to find that the sum of the angles in the triangle is 180°.

| **M** + **N** + **J** |
|---|
| = 45° + 110° + 25° |
| =180° |

## ANGLE ACTIVITIES ASSOCIATED WITH LOGO

UA2a/4 SSM3c/4
PSE PM/D4
P/4ab S/4e

- In the Textbook and Workbook of Heinemann Mathematics 5, the children used the commands

  **FD** (forward) **BK** (back) **RT** (turn right) **LT** (turn left)

  to follow or draw pathways on squared or dotty paper grids. The alternative of carrying out these activities using a LOGO screen 'turtle' was also suggested.

- In the Assessment and Resources Pack of Heinemann Mathematics 5, Resource Cards 22 to 25 provided further activities involving pathways and turning, which were specifically designed to be implemented using LOGO on a computer screen. The notes for the cards introduced other LOGO commands such as

  **CS** or **CLEARSCREEN** **CT** or **CLEARTEXT**

  **PU** or **PENUP** **PD** or **PENDOWN** **PE** or **PENERASE**

- These notes do not attempt to provide a technical guide to using LOGO or a complete course for children. They suggest a sequence of further activities which deal with

  — using simple commands

  — using **REPEAT**

  — writing procedures

  — using procedures within procedures.

- The activities presented below are consistent with the use of LOGOTRON. They assume that the children can already use simple commands and have some experience of using LOGO. More complete programming and 'troubleshooting' information will be available in the user's manual and other documentation supplied with the relevant software.

# Using simple commands

## 1 Alphabet *(moving and turning)*

- Ask the children to use the commands **FD**, **BK**, **RT** and **LT**, along with the appropriate angle turns in degrees, to draw the capital letters below. They should list the commands used each time. The screen and text will need to be cleared, using **CS** and **CT** respectively, between letters.

LTEHKR

- At this stage it is not necessary to be prescriptive about the sizes and proportions of the letters drawn, or the number and order of commands used, as long as recognizable letters are provided.

One possible sequences of commands to draw each of the letters is shown.

L

**FD 200**
**BK 200**
**RT 90**
**FD 100**
**BK 100**

T

**FD 200**
**LT 90**
**FD 50**
**BK 100**

E

**RT 90**
**FD 100**
**BK 100**
**LT 90**
**FD 100**
**RT 90**
**FD 50**
**BK 50**
**LT 90**
**FD 100**
**RT 90**
**FD 100**

H

**FD 200**
**BK 100**
**RT 90**
**FD 100**
**LT 90**
**FD 100**
**BK 200**

K

**FD 200**
**BK 100**
**RT 45**
**FD 140**
**BK 140**
**RT 90**
**FD 140**

R

**FD 200**
**RT 90**
**FD 100**
**RT 90**
**FD 100**
**RT 90**
**FD 100**
**LT 135**
**FD 140**

- The letters **K** and **R** in particular are likely to require the use of 'trial and improvement' methods to find the angles and lengths for the diagonal lines.
- The children may feel that the appearance of the letters is spoiled by the presence of the turtle. It can be made invisible by the use of the command **HT** or **HIDETURTLE** after the sequences given above. The children should be aware that the turtle is still there, even when it is invisible, and that it can still draw (in fact, considerably faster).

The command **ST** or **SHOWTURTLE** can be used to make the turtle reappear.

# Using REPEAT

## 2 Squares *(introducing the* **REPEAT** *command)*

- Ask the children to draw a square of side 100 turtle units on the computer screen.

  Depending on the direction of turn chosen, they are likely to type something like this:

  **FD 100**
  **RT 90**
  **FD 100**
  **RT 90**
  **FD 100**
  **RT 90**
  **FD 100**
  and possibly **RT 90**

- Many children will omit the last **RT 90** shown above, since it is not strictly necessary to complete the square. However, when the starting and finishing points coincide, they should be encouraged **to return the turtle to its original position.** The importance of this will become apparent once the **REPEAT** command is introduced.

  (The command **LT 90** could be used instead of **RT 90** throughout the above program.)

- Ask the children to look at the list of commands they used to draw their square. They should notice that they have repeated **FD 100 RT 90** four times (ie the number of sides of a square).

  Tell them that there is a shorter way of doing this using LOGO, by writing the commands which are repeated inside **square** brackets.

  **[FD 100 RT 90]**

  and then using the command **REPEAT** like this:

  **REPEAT 4[FD 100 RT 90]**

  Ask the children to copy this, emphasizing the need to use the square brackets and to leave spaces as shown. Ask them to suggest how it works and why the brackets are needed.

- Initially, it is better to leave the turtle visible so that the children can follow its path. Subsequently, the turtle can be hidden as the children draw squares of different sizes. For example,

  **REPEAT 4[FD 200 RT 90]**

  **REPEAT 4[FD 273 RT 90]**

- Encourage the children to see that, while the side length can be changed, the number of repeats must always be 4 and the angle of turning must be 90° when drawing a square.

## 3 Rectangles *(using* **REPEAT***)*

- Ask the children to investigate how to draw a rectangle of breadth 100 units and length 200 units. They should try to do this first by using **FD** and **RT** or **LT** only, and then once they have been successful, by using **REPEAT**. For example,

  **REPEAT 2[FD 100 RT 90 FD 200 RT 90]**

- Some children may need help to see that an 'L' shape has to be repeated twice to make a rectangle, since it does not have four equal sides. Others may omit the second **RT 90**, the need for which should also be discussed.
- The children can then be asked to draw rectangles of other lengths/breadths.

## Writing procedures

### 4 A procedure called **SQUARE**

- LOGO is a language with a vocabulary of special words which, when used as commands, are understood and accepted. It would be very convenient if the names of basic shapes such as 'square' and 'rectangle' were part of LOGO's vocabulary.

  Ask the children to type the command

  **SQUARE**

  LOGO will respond with the message **'I DON'T KNOW HOW TO SQUARE'**. This response indicates that LOGO does not have a definition for a square and hence is unable to draw a diagram. However, one of the most powerful features of LOGO is its ability to learn new words and their definitions. We can extend LOGO's vocabulary and teach it how to draw a square.

- Ask the children to type

  **TO SQUARE** **RETURN**

  The prompt should be seen to change from a question mark (?) to

  **>**

  They should now type

  **HT** **RETURN**

  so that the turtle will not be visible in the drawing.

  Now ask the children to type the definition of a square, remembering the spaces.

  **REPEAT 4[FD 100  RT 90]** **RETURN**

  Finally, they must type

  **END** **RETURN**

  This last command tells LOGO that the definition of a square is complete. The prompt should have changed back to ? and the message '**SQUARE DEFINED**' should be on the screen.

  The children can now type the command

  **SQUARE**

  and a square, as defined, with side 100 units and the turtle automatically hidden, should be drawn.

  The children have written a LOGO **procedure** for a square which has been assigned the name **SQUARE**. This will be remembered until

  — LOGO is specifically instructed to forget the procedure with the command

  **ER "SQUARE**

  (**ER** stands for **ERASE** – note the open quotes only)

  — the **BREAK** key is pressed

  — the computer is switched off.

  In the last case, the procedure can be saved on a disk prior to switching off.

  If any children make mistakes while they are writing the procedure, they should be told to press the **ESCAPE** key and then start again.

## 5 Another square

There can only ever be **one** definition of a particular procedure name at any one time. If the children, having defined a square, try to write a procedure for a different square, say with side 200 units, they will find that typing **TO SQUARE** now produces the message '**SQUARE ALREADY EXISTS**'.

A different square must therefore be defined using a different name. For example,

**TO SQUARE200** (note – all one word) or

**TO NEWSQUARE**

would be appropriate alternatives.

## 6 RECTANGLE procedure

Now ask the children to write a procedure **RECTANGLE** which defines a rectangle of breadth 100 units and length 200 units.

# Procedures within procedures

## 7 TOWER – a procedure which uses SQUARE

- Now that the computer understands and accepts the command **SQUARE**, it can be used within other procedures. For example, ask the children to sketch what they think will be the result of the following procedure:

  **TO TOWER**

  **REPEAT 3[SQUARE FD 100] BK 300**

  **END**

- They should now check by entering the **TOWER** procedure. Typing the command **TOWER** should then produce the diagram shown here.

  Discuss the way in which this procedure uses the **SQUARE** procedure.

- Further activities like this one are included in Problem Solving Activity 25 in the Assessment and Resources Pack of Heinemann Mathematics 6.

## 8 FLAG

Ask the children to write a procedure **FLAG**, which uses the procedure **SQUARE**, to draw this shape.

Remind the children that the turtle should be returned to its original position.

Their procedure might look like this:

**TO FLAG**

**FD 100 SQUARE BK 100**

**END**

## 9 Procedures using **FLAG** *(procedures involving rotation)*

- Now ask the children to write and test procedures to draw these shapes (possible solutions are given below).

**TO 2FLAG**
**LT 90**
**REPEAT 2[FLAG RT 180]**
**RT 90**
**END**

**TO 4FLAG**
**REPEAT 4[FLAG RT 90]**
**END**

- For each 'flag' shape, ask the children to look at the number of repeats and the angle of rotation in degrees.

| Procedure | Repeats | Angle |
|---|---|---|
| **2FLAG** | 2 | 180° |
| **4FLAG** | 4 | 90° |

Remind them that a complete turn measures 360°. Discussion should lead to the idea that the angle of rotation can be found by dividing 360 by the number of repeats.

- Introduce two new designs, **3FLAG** and **8FLAG**.

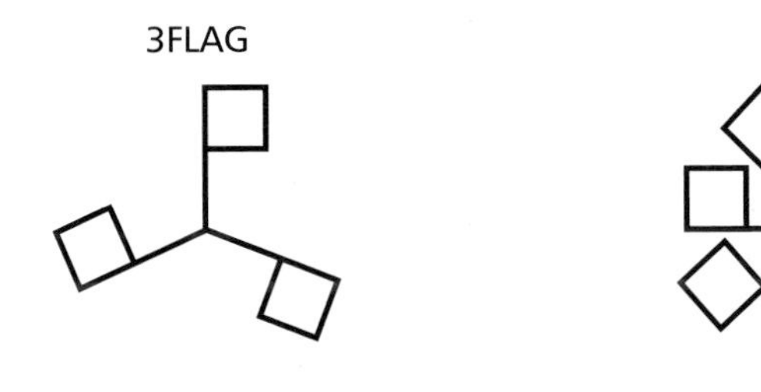


Discuss with the children the required angle of rotation for each design.

| Procedure | Repeats | Angle | |
|---|---|---|---|
| **2FLAG** | 2 | 180° | ← 360° ÷ 2 |
| **4FLAG** | 4 | 90° | ← 360° ÷ 4 |
| **3FLAG** | 3 | 120° | ← 360° ÷ 3 |
| **8FLAG** | 8 | 45° | ← 360° ÷ 8 |

- Ask the children to write and test procedures for **3FLAG** and **8FLAG**. Possible answers are as follows:

**TO 3FLAG**
**REPEAT 3[FLAG RT 120]**
**END**

**TO 8FLAG**
**REPEAT 8[FLAG RT 45]**
**END**

## 10 Further designs involving rotation

Ask the children to invent procedures for, draw and, perhaps, print out other designs involving rotation, such as

```
TO SHAPE1
REPEAT 36[SQUARE RT 10]
END
```

```
TO SHAPE2
REPEAT 10[SQUARE RT 36]
END
```

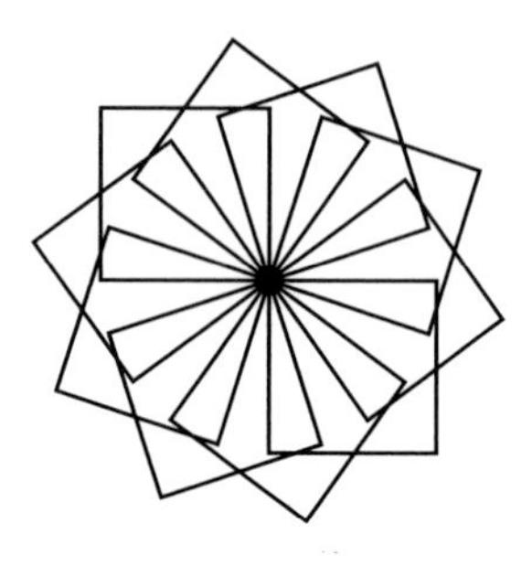

The Handling Data part of Heinemann Mathematics 6 has two sections, Handling Data and Probability. The work of the first section is presented in the context of 'Greenwatch', the title of a campaign run by the children of Topperton Primary School in an attempt to improve their environment.

# Greenwatch

## A context for Handling data

All the work on Textbook pages 111–22 and Workbook pages 34–6 is set in the context of 'Greenwatch'. The children of Class 6 in Topperton Primary School are involved in a campaign to improve the quality of the environment in and around their school.

### Introducing the context

The context could be introduced by discussing improvements the children would like to see made to

— their classroom

— the school and playground

— the area around the school.

Suggestions might include the painting of murals, the planting of shrubs and bulbs, the clearing of rubbish and improvements to the classroom and playground.

### 1 Improving our classroom

Tell the children that the Headteacher has been given a sum of money to use to improve their classroom environment. Discuss changes they would make. List these on the chalkboard, then ask the children to prioritize them.

### 2 Improving our school

Ask the children to discuss what changes they would make to the school to allow easier access by a disabled person – for example, someone in a wheelchair or a visually impaired person.

### 3 Improving our playground

Tell the children that there is going to be a competition to design a more 'child-friendly' playground. Ask them, in pairs, to submit a sketch of the playground, identifying the improvements they would make. For example,

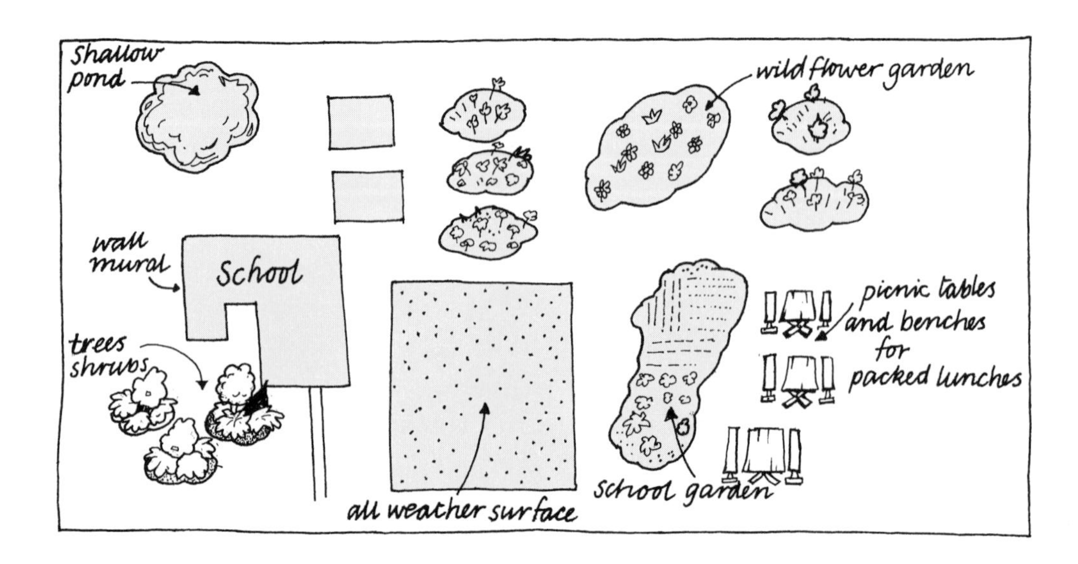

## 4 Mathematics mural

Ask the children what a mural is, and discuss any examples in the school or local area. Tell them that they are to design a **mathematics** mural for a suitable place in their school. Discuss

— what makes an attractive mural

— what sorts of image could be included

— where the mural might be painted.

Produce a 'gallery' of all the designs. The children could vote for the one they would like to see turned into an actual wall mural.

## 5 Poster competition

Explain that the local newspaper wants to encourage people to improve the environment. It is sponsoring a poster competition to highlight issues involved, such as

— litter

— ugly buildings

— lack of greenery

— effects of car pollution and abandoned wrecks

and to encourage people to respect their community. The children are invited to design an entry for the competition.

UA3bc,4d/3
UA2ab,3bc,3bc,4acd/4
UA3a/4→5
HD2abd/3
HD2abd/4
HD2c/4→5
PSE
C/C1,2
O/C1,2,3
O/D1,2
D/C1,2
D/D2
I/C1,2
I/D1
P/4ab
D/3abcd
D/4abcef
D/5a

# Handling data

## Overview

This section

- extends the children's experience of bar graphs and pictograms
- introduces bar-line and trend graphs
- develops the concept of an 'average' or 'mean' and introduces the terms 'mode' and 'median'
- introduces the organization of data in class intervals
- extends the children's experience of extracting and interpreting data from tables, and allows the use of a spreadsheet package
- includes a card database which could provide an opportunity to work with an appropriate computer database
- involves the children practically in class surveys.



An Extension activity is related to the above section work as follows:



Teaching notes for the Extension Textbook are in a separate section at the end of the Teacher's Notes.

# Resources

## Useful materials

- squared paper, including 2 mm grid
- measuring tapes marked in centimetres
- calculators
- reference books about bushes, trees, plants and shrubs
- computer database and spreadsheet packages
- other materials suggested within the introductory and additional activities

## Assessment and Resources Pack

### Assessment

*Handling data check-up 1*
Textbook pages 111–17,
Workbook page 34
(Bar line graphs: mode and mean)

*Handling data check-up 2*
Textbook page 118, Workbook pages 35–6
(Class intervals)

*Round-up 1*
Question 7

*Round-up 2*
Question 10

*Round-up 3*
Question 8, 9

### Resources

*Problem Solving Activities*
26 Absences (Trend graphs)
27 Safety first (Surveys)

# Teaching notes

## BAR GRAPHS AND PICTOGRAMS

In Heinemann Mathematics 5, the children constructed and interpreted horizontal and vertical bar graphs, using scales where one interval represented either 2 or 4 units.

The children also interpreted pictograms where a whole symbol represented 10 items and part of the symbol represented from 1 to 9 items.

This work is now revised and then extended to include a scale where one interval represents 5 units.

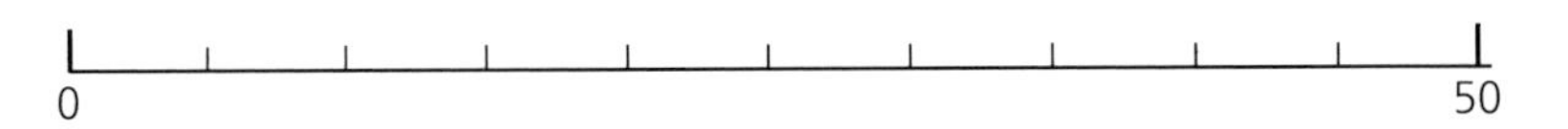


The Greenwatch campaign for environmental improvements in and around Topperton School is introduced.

UA3bc/3 HD2bd/3
D/C2 I/C1,2
D/3cd

## Introductory activities

### 1 Colour chart *(interpreting bar graphs, revision)*

- Show the children a copy of this bar graph, which displays the colours of the outside doors in the homes of the Topperton School children. Ask what each interval on the vertical axis represents, and confirm that the scale used is 1 square represents 4 doors.

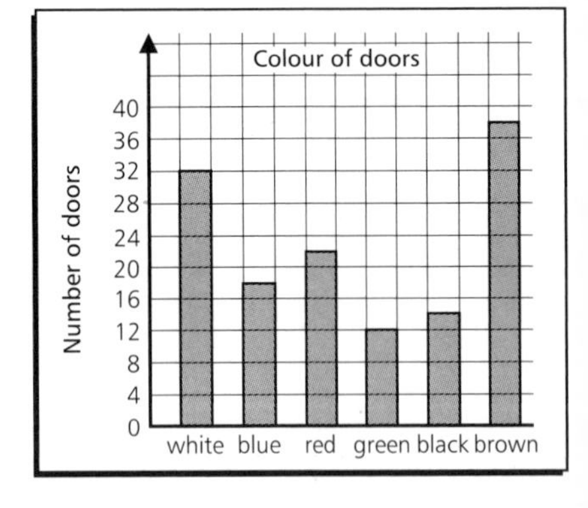


- Interpret the information in the graph by asking questions such as:

  'How many children have an outside door that is red?'

  'What is the least common door colour?'

  'How many more doors are painted brown than black?'

  'Is it true that a total of 35 children have doors painted either red or green?'

**At this point the children could try Textbook page 111.**

## 2 Anti-litter campaign *(interpreting pictograms, revision)*

- Prepare a pictogram to show the number of empty crisp packets collected in Topperton School playground last week.

| | Number of empty packets collected | Total |
|---|---|---|
| Monday | | |
| Tuesday | | |
| Wednesday | | |
| Thursday | | |
| Friday | | |

Explain that one complete symbol represents 20 packets and that a part symbol is now to be taken as half of this, ie 10 packets.

- Discuss
  - — displaying the same data as a bar graph
  - — using a new scale of 1 small interval represents 5 units (empty crisp packets).

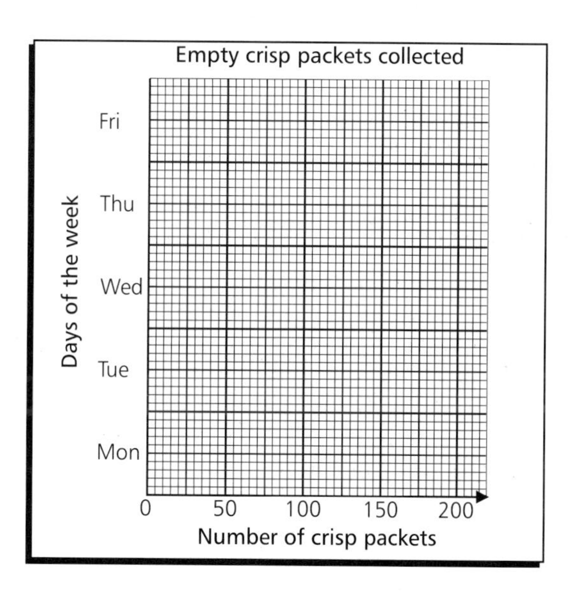


UA3b/3 HD2bd/3
D/C2 I/C1
D/3ac

### Textbook page 111 *Handling data: bar graphs*

This is the first page of the Handling data context, Greenwatch. The activities are based on classroom improvements which the children at Topperton School would like to see implemented.

In question 1, the bar graph scale is one interval represents 4 children or 4 votes, and this should be established before the questions are attempted.

In question 3(b), the children divide 248 by 4 and reach the conclusion that more than one quarter (68 votes) do want the classroom painted.

Some children may have difficulty interpreting question 4(a). For example, in the first part they must **add** the votes for any combination that contains white (44 and 34 is 78).

UA3b/3 HD2bd/4
D/C2 I/C1
D/3acd

**Workbook page 34** *Handling data: pictograms, bar graphs*

Established that the symbol [ ] represents 10 bricks and that [ ] represents 5 bricks. In question 1, the total number of bricks painted is 550. Check that the pictogram totals are correct before the children try questions 2 and 3.

In question 3, the children have to insert the missing numbers on the vertical axis and then complete the graph by using the totals from question 1.

# BAR-LINE AND TREND GRAPHS

The children are introduced to two new ideas:

— When drawing bar graphs, the information can be represented by straight lines instead of 'bars'. These graphs may be called 'bar-line' or 'spike' graphs.

— Sometimes the data represented by a bar or bar-line graph is such that the graph displays a 'trend'. This trend can be made clearer by marking only the top of each line or spike and then joining these tops using a dotted line.

The Greenwatch context continues with the children improving the school grounds by planting heathers and bulbs. A sunflower-growing competition is held in the garden, and is visited by parents and friends during an Open Week.

UA3bc/4 HD2bd/3
D/C2 I/C1,2
D/3ac D/4ef

## Introductory activities

### 1 Garden furniture *(horizontal bar-line graph)*

- Present a table of results which shows the kind of furniture for the school garden preferred by the children of one class. Draw a bar graph to show the information.

| Type of furniture | Number of children |
|---|---|
| wood | 7 |
| plastic | 10 |
| metal | 4 |
| concrete | 6 |

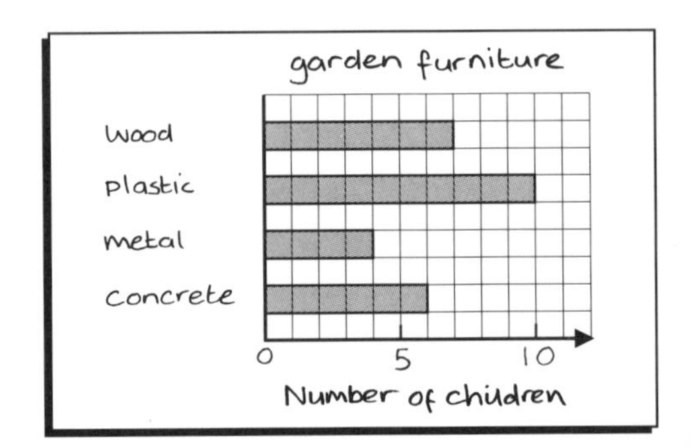


- Establish that only the **length** of each bar is needed to represent the number of children, and that a single straight line would show the same information.

  Introduce the term **bar-line graph** (or **spike graph**).

- The children could be asked to choose which type of furniture **they** would prefer for their own school garden, and to draw a bar-line graph.

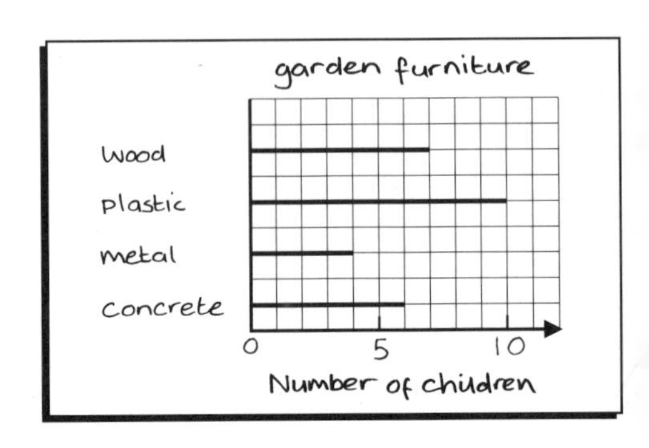

## 2 Which one?

*(vertical bar-line graph on 2 mm grid)*

- Display another set of results from the Topperton School garden survey.

  The children were asked to choose the type of garden they would prefer. The table shows how they voted.

| Type of garden | Number of votes |
|---|---|
| rockery | 60 |
| herb garden | 86 |
| wild flower garden | 24 |
| heather garden | 78 |

- Ask the children to
  - — draw and label two axes on 2 mm grid paper
  - — choose a suitable scale on the vertical axis so that at least 86 votes can be represented
  - — complete the graph by drawing straight lines.
- The children could be asked to carry out a similar survey in the school and then draw a bar-line graph.

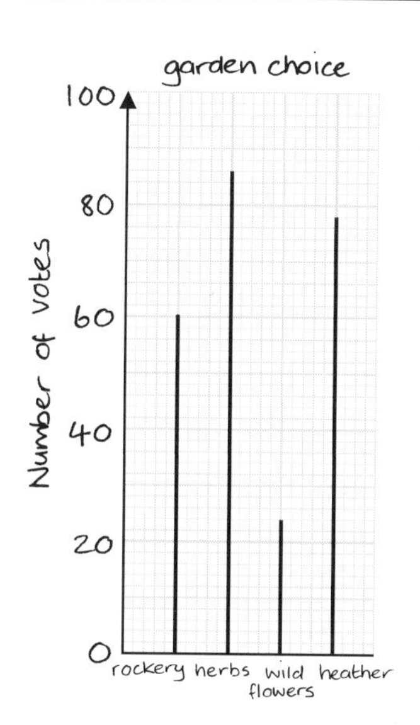


**At this point the children could try Textbook pages 112 and 113.**

## 3 Jody's sunflower graph *(trend graph)*

Sometimes the data represented on each axis are related in such a way that the lines (or spikes) display a **trend**.

Discuss Jody's sunflower graph, which appears on Textbook page 113. Establish that the height of the sunflower increased in **successive** months from February to July.

Show how, by marking only the top of each line and then joining these 'tops' using dotted lines, the trend can be shown clearly.

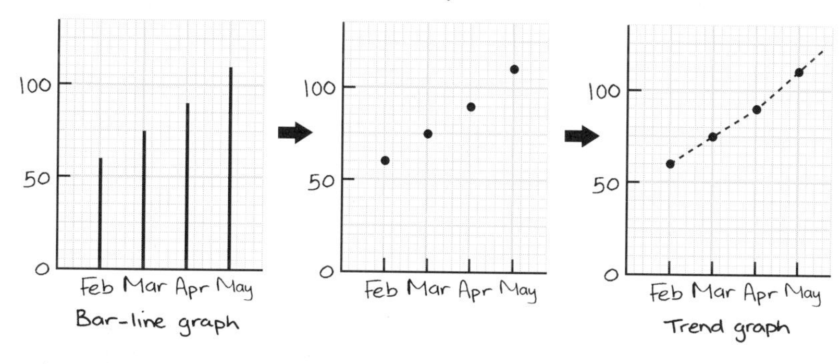


## 4 Garden log book *(trend graph)*

Present the children with data from the school garden log book, which gives the number of days each month, from July to December, when the temperature reached at least 15° C.

| Month | July | August | September | October | November | December |
|---|---|---|---|---|---|---|
| Number of days | 24 | 29 | 18 | 14 | 7 | 2 |

Ask them to draw a trend graph of this information. Discuss the graph and ask the children to write about the trend it shows.

UA3b,4d/3 HD2bd/4
D/C2 I/C1,2
D/3ac D/4e

## Textbook pages 112 and 113
### *Handling data: bar-line graphs*

The theme of making environmental improvements to the Topperton School grounds continues. The children are required to interpret and draw bar-line graphs on 2 mm grid paper. Their completed graphs should be discussed.

On Textbook page 113, the first graph has a scale where one small interval represents 5 centimetres, and this should be established with the children before they attempt question 1.

Question 2 presents a set of data from which a similar bar-line graph has to be drawn. The scale to be used is illustrated on the partial graph displayed beside the question.

UA3b,4d/3 HD2bd/4
D/C2 I/C1,2
D/3ac D/4f

## Textbook page 114 *Handling data: trend graphs*

Before questions 1 and 2 are attempted, the trend graph shown at the top of the page, its scale and the way it has been constructed should be discussed. Note that the final part of question 2(a) has two answers – 10.00 am and 2.30 pm.

In question 3(b), the drop in temperature (down 5° in a short time of only 30 minutes) could have been caused by a sudden change in outside weather conditions, such as a thunderstorm, or perhaps by a door or vent in the greenhouse being opened.

In question 4(a), the scale to be used in drawing the graph is illustrated on the partial graph displayed beside the question.

# MODE, MEDIAN, MEAN

In Heinemann Mathematics 5, the children were introduced to finding the average (mean). This work is now revised and the new terms 'mode' and 'median' are introduced.

The Greenwatch Campaign continues with the Class 6 children putting on a play, 'Greenwatch'. The children make costumes from recycled materials.

UA3a/4→5 HD2c/4→5
O/C1 I/D1
D/3c D/4ab

## Introductory activities

### 1 Sponsored swim *(introducing the mode)*

Class 6 organizes a sponsored swim to raise money for the Greenwatch Campaign.

- Display a table as shown. Remind the children that 𝍸 represents 5 children. Complete the table with the children.

| Number of lengths swum | Tally marks | Total |
|---|---|---|
| 2 | II | |
| 3 | 𝍸 | |
| 4 | 𝍸 II | |
| 5 | 𝍸 𝍸 𝍸 | |
| 6 | 𝍸 𝍸 | |
| 7 | 𝍸 III | |
| 8 | 𝍸 I | |
| 9 | IIII | |
| 10 | III | |

- Ask which number of lengths swum occurs most often (5).

  Tell the children that the number which occurs most often is called the **mode**.

  Ask how many children swam a number of lengths

  — equal to the mode (15)

  — less than the mode (14)

  — greater than the mode (31).

- Using the table, highlight that the number of lengths swum **ranges from 2 to 10**.

  (**Note**: the **actual range** is 10 – 2 = 8.
  At this stage, the children are **not** asked to calculate the actual range, but simply to indicate that the range of the number of lengths swum is **from** 2 **to** 10).

## 2 Waist sizes *(using the mode)*

Class 6 make skirts for their play, 'Greenwatch', from recycled paper. They measure the waist sizes of 45 children.

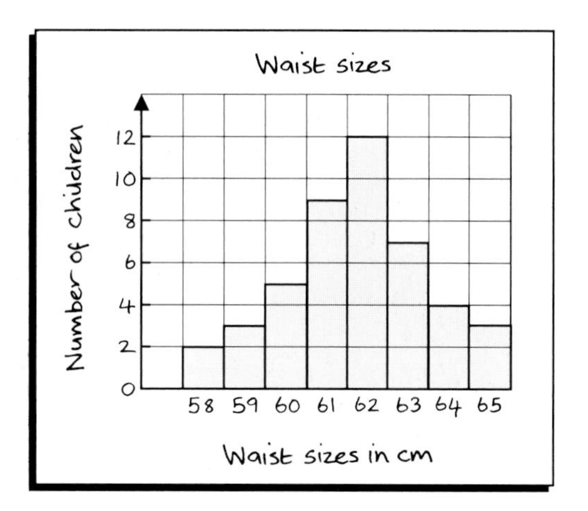


- Display a bar graph as shown, where, **for the first time, the bars are 'closed up'**, ie there is no space between adjoining bars.

  Discuss and complete the following sentence with the children.

  'The waist sizes range from ___ cm to ___ cm.'

- Question the children about the graph.

  'How many children have a waist size of 60 cm, 63 cm . . .?'

  'Which waist size occurs most often?' (62 cm).

  Ask how many children have this waist size (12).
  Remind the children that the size which occurs most often is called the **mode**.

  'Which **sizes** are • shorter than the mode • longer than the mode?'

  'How many children have a waist size

  • equal to the mode (12)

  • shorter than the mode (19)

  • longer than the mode?' (14).

**At this point the children could try Textbook pages 115 and 116.**

## 3 Heights of children *(introducing the median)*

- Select five children and arrange them in a row in order of height. The height of the child **in the middle** of the row is called **the median** height.

- Select four more children and fit them into the row, so that all nine children are standing **in order of height.**

  Ask the children to point to the child whose height is **now** the median. This may be a different child.

## 4 Ticket sales for 'Greenwatch' *(median, mean)*

The following are the number of tickets sold by seven children in Class 6 for the 'Greenwatch' play:

10, 20, 8, 20, 5, 15, 20

- Discuss how to find the number which is the median:
  - put the ticket sales in order: 5, 8, 10, 15, 20, 20, 20 (↑ median: 15)
  - pick out the number in the middle (15).
- Ask the children to find the **mean** (average) number of tickets sold. (Children are probably, at this stage, more familiar with the word 'average' than 'mean'.)

  The mean number of tickets sold by the seven children is $98 \div 7 = 14$.
- Repeat the process for another nine children in Class 6 who sold 6, 7, 10, 14, 9, 16, 20, 4 and 4 tickets.

  **Note:** In the above examples, the children have been asked to find the median in a set which contains an **odd** number of items.

  The difficulty of finding the median in a set which contains an even number of items is left until a later stage. For example, the median of the following set of numbers

  1, 2, 2, 3, 4, 5, 7, 7, 9, 10 (↑ median between 4 and 5)

  lies half way between 4 and 5,

  i.e the median is $\frac{4+5}{2} = 4\frac{1}{2}$.

UA4ad/4 HD2c/4→5
O/C1 D/C2 I/D1
D/3c D/4ab

## Textbook pages 115–17

### *Handling data: range, mode, mean, median*

On Textbook page 115, in questions 1(a) and 2(b), some children may need to be reminded to work systematically – for example, down each column – when tallying the lists of data. They should check that the total number of items tallied is 45 in question 1(a) and 60 in question 2(b).

On Textbook page 116, in question 4(a), the children could work in pairs to measure each other's headband sizes. Thereafter, it is recommended that the headband sizes are allocated and presented in a table similar to the one shown on Textbook page 116.

On Textbook page 117, in question 3, Sam has the median height, 152 cm, while Sue has the median weight, 38 kg.

Question 4 may cause some difficulty. The children have to re-order the heights and weights.

**Heights in cm**
142, 142, 143, 147, 147, 148, 152, 154, 154, 154, 156

**Weights in kg**
33, 34, 36, 36, 38, 40, 42, 44, 44, 46, 47.

Sarah now has the median height 148 cm, while Lyn has the median weight, 40 kg.

The children could also be asked to find the mean height of the 11 'Greenwatch' actors (149 cm) and the mean weight (40 kg). These values could be compared with the mean heights and weights of the seven children investigated in questions 1 and 2.

On completion of question 4, the children could be invited to find the median, mode and mean for the number of letters used in the names of the children who appear on this page. For example, for the names used in questions 1 and 2, the number of letters, in order, is

3, 3, 3, 5, 6, 7, 8

The mode is 3, the median is 5 and the mean is 5. When the other four actors' names are added to this list, the number of letters, in order, is

3, 3, 3, 4, 4, 5, 5, 6, 7, 7, 8

Again the mode is 3, the median 5 and the mean 5.

Although in this investigation the mode, the median and the mean remain unaltered when four more items are added to the list, this may not be the case for a different set of data.

# CLASS INTERVALS

The organization of data in a frequency table using class intervals and the display of these data in a bar graph are now introduced. At this stage, the children are not expected to **choose** class intervals when organizing data. Hence suitable class intervals are provided throughout the work.

The Greenwatch context develops with the Class 6 children thinking about improvements that could be made to an area of neglected ground beside the school, which has become an eyesore because of overgrown grass, bushes, weeds, rubbish and so on.

## Introductory activity

UA3bc/4 HD2abd/4
O/C1 D/D2 I/D1
D/3ac D/4a

### Age distributions *(organizing and displaying data using class intervals)*

- Discuss a survey carried out by Class 6. The children have asked people what they think should happen to the neglected ground – for example, rubbish cleared, grass cut, trees planted and so on. The people questioned were asked to give their ages.
- Display the data about ages on the chalkboard. Build up a frequency table with class intervals and then draw the bar graph. In the discussion, emphasize that
    - — a class interval includes several values:for example, the class interval 8–17 contains any of the values 8, 9, 10 . . . 17
    - — the class intervals are all of equal size: for example, 8–17, 18–27 . . .

Ages of people in survey

| | | | | | |
|---|---|---|---|---|---|
| 9 | 12 | 61 | 11 | 17 | 11 |
| 57 | 20 | 35 | 8 | 10 | 13 |
| 10 | 12 | 23 | 45 | 62 | 33 |
| 11 | 15 | 60 | 29 | 12 | 26 |
| 58 | 10 | 44 | 13 | 55 | 19 |

— the lowest value (8) is within the first class interval (8–17), and the highest value (62) is within the last class interval (58–67)

— the bar graph has no space between adjacent bars.

| Ages in years | Tally marks | Total |
|---|---|---|
| 8–17 | ~~||||~~ ~~||||~~ ~~||||~~ | 15 |
| 18–27 | \|\|\|\| | 4 |
| 28–37 | \|\|\| | 3 |
| 38–47 | \|\| | 2 |
| 48–57 | \|\| | 2 |
| 58–67 | \|\|\|\| | 4 |
| | | 30 |

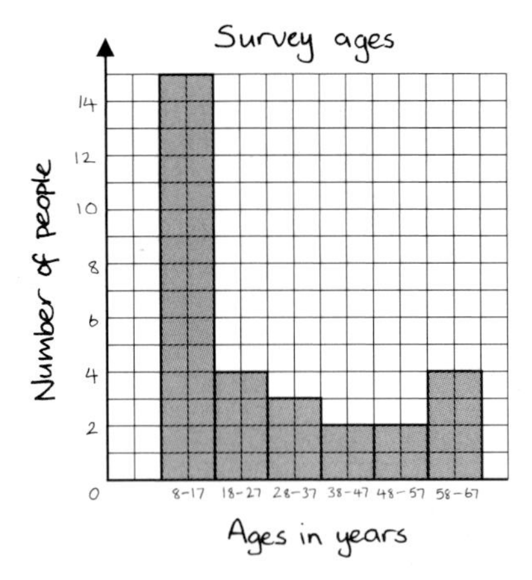


■ Repeat, if necessary, with class intervals of a different size: for example, 6–20, 21–35, 36–50, 51–65.

Note, that in this case, the lowest value is within the first class interval but is **not** the lower limit.

UA3bc/4 HD2abd/4
O/C1 D/D2 I/D1
D/3ac D/4a

## Textbook page 118 *Handling data: class intervals* Workbook pages 35 and 36

The idea of a class interval should be discussed along the lines suggested in the introductory activities, before the children are asked to attempt the work on the Textbook and Workbook pages.

On Textbook page 118, the illustration, introductory information and graphs should be discussed to set the scene.

On Workbook pages 35 and 36, some children may need an explanation that colour indicates which frequency table is associated with which graph. It may also be necessary to suggest that each piece of data should be ticked each time it is transferred to a frequency table. This should help the children to keep track of the data, and to ensure that it is all used and that no figure is used more than once for each frequency table.

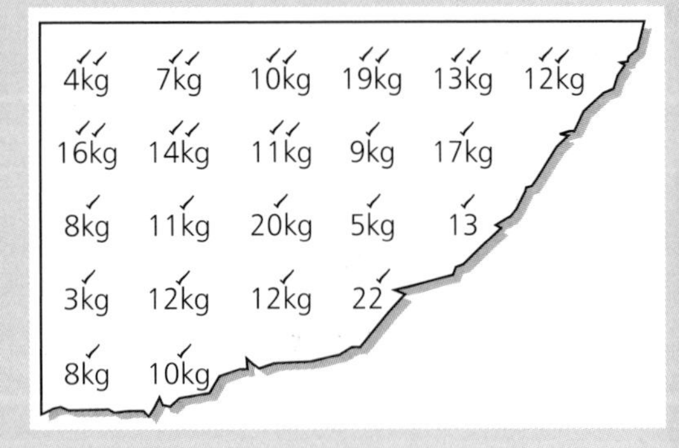


On Workbook page 35, in question 5, the following points should be discussed:

— the same data are displayed on each graph

— the class interval sizes are different on different graphs

— the lowest value (3 kg) is in the first class interval and the highest value (22 kg) is in the last class interval

— the graph with the greatest number of class intervals (blue) shows most detail

— the graph with the smallest number of class intervals (yellow) shows least detail.

# INTERPRETATION OF DATA, DATABASE AND TABLES

UA2b,3bc4 HD2ab/3
C/C1 O/C2 I/C1 I/D1
D/3ab

n Heinemann Mathematics 5, the children extracted and interpreted numerical and ext-based data from tables. In some questions more than one data 'field' had to be canned. The children also created and altered simple tables, according to nformation given.

This work is now extended to extracting symbolic data, such as the following ssociated with the growing of bushes.

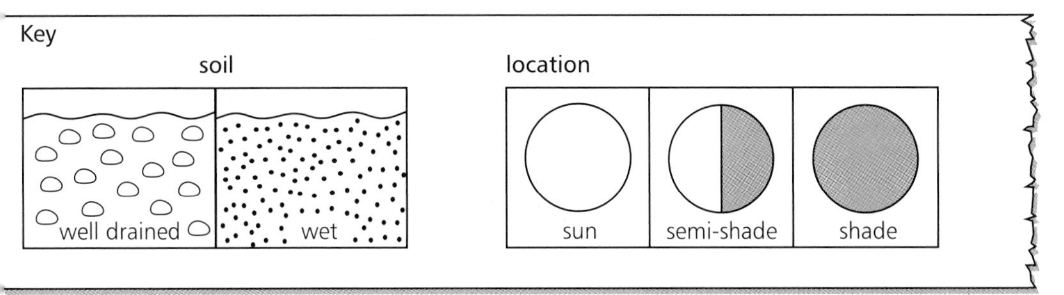


The children use one or more criteria, and create and alter more complex tables.

The information could also be used to give the children the opportunity to work with ppropriate database and spreadsheet packages.

The Greenwatch context continues with the children analysing information on data ards about the bushes in the overgrown garden in the corner of the waste land beside he school.

## ntroductory activities

### 1 Tree search

- Ask the children to find out more about some of the trees which grow in and around the school, or perhaps in nearby woodland. They could do this through their own observations or through the use of reference books, and then organize their information in a table or on a set of data cards. For example,

| Tree | Soil | Location | Height | Shape | Leaf | Interesting features |
|---|---|---|---|---|---|---|
| Birch | well-drained | sun or semi-shade | 10m - 18m | open | small | Catkins in spring |
| Beech | well-drained | sun or semi-shade | more than 18m | large-domed | oval | gold/copper leaves in autumn |

- Symbols, rather than words, could be used to represent some of the data. For example,

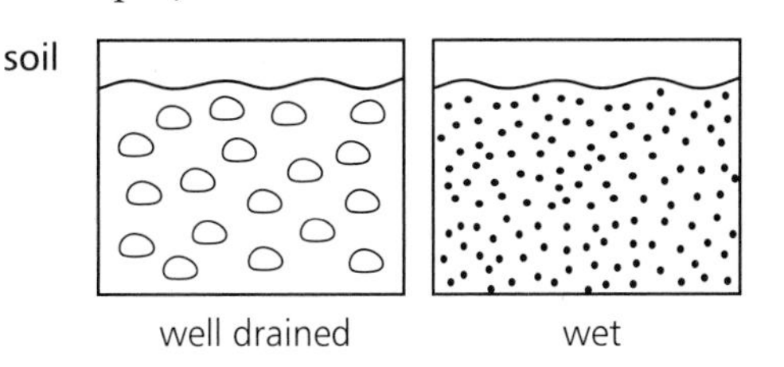


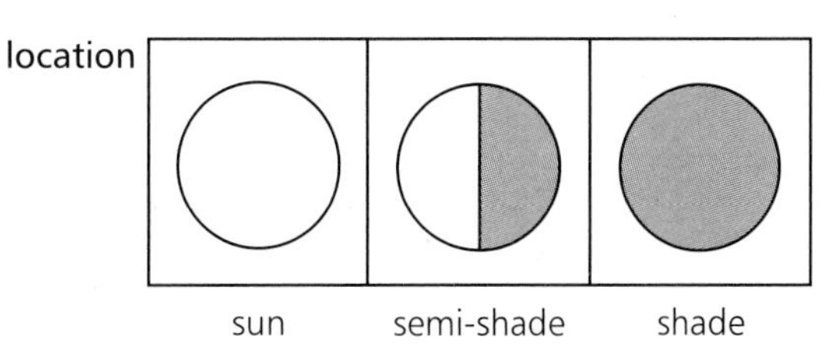

- Ask questions like these:

  'Which trees grow in well-drained soil?'

  'Which trees grow to a height of less than 10 m?'

  'Which trees grow in the sun **and** produce catkins in springtime?'

  'Which tree
  - can grow in semi-shade
  - grows taller than 18 m
  - has gold/copper leaves in autumn?'

- The data could be entered into a computer database, which would allow the children to ask a wider range of questions about the data.

## 2 Planting out

- Tell the children that they are going to plant bulbs and flowers in a part of the school grounds, or that Class 6 is going to restore the overgrown garden in the waste ground next to Topperton School.
- Ask them to find out more about suitable flowers and bulbs by using
  - — gardening books
  - — information from seed packets or bulb packs
  - — gardening catalogues

  or by talking to a gardener or someone who works at a garden centre.
- The information collected could be used to create a table or a set of data cards similar to those in introductory activity 1.

UA3bc/4 HD2ab/3
C/C1 O/C2 I/C1 I/D1
D/3b D/4c

## Textbook pages 119 and 120

### *Handling data: interpretation, database*

The data on these two pages has been simplified to make it easier for the children to use.

Before the children tackle the questions, it may be necessary to discuss the key and the meanings of the symbols used to represent soil type, location and height.

Key

soil

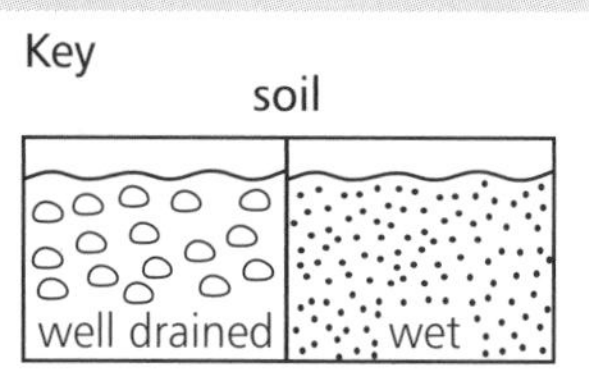


location

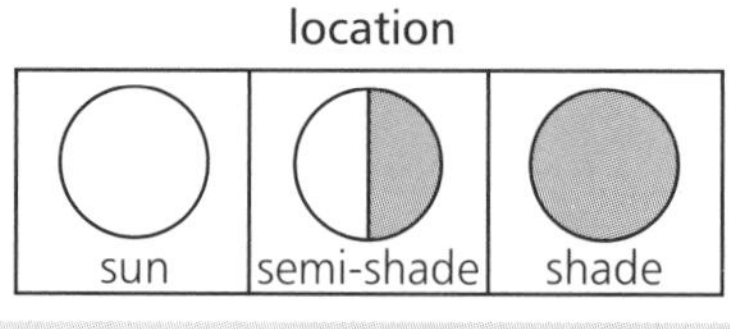


height

| shorter than 1.5m | between 1.5 and 3m | taller than 3m |
|---|---|---|

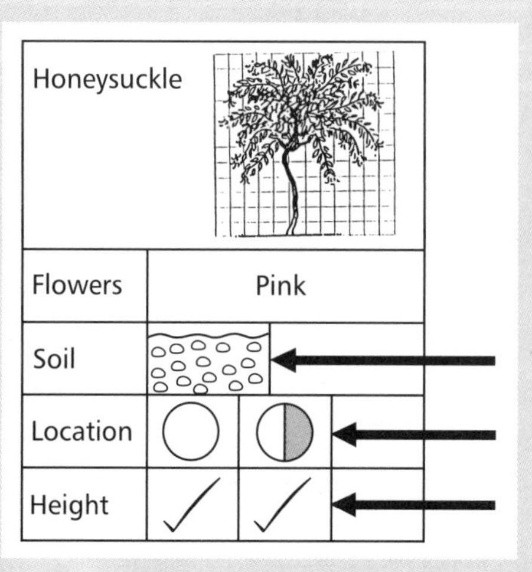


On each data card, one or more symbols may be identified for each field.

For example, the Honeysuckle grows in well-drained soil in the sun **or** semi-shade to a height of up to 3 m.

On Textbook page 120, in questions 1 and 2, the children have to look through all the cards to find the names of the appropriate bushes. In question 1(b), the children should realize that they select any bushes which grow in semi-shade, even though they may also grow in the sun or in the shade.

In question 3, the children should be encouraged to look at

— the shape of the bush

— the leaves

— the flowers or berries.

In question 4, the children are likely to require access to reference books and gardening catalogues to find data for three more cards. They should use the same fields of information as the bush cards on the page.

In question 5, the following five fields could be used for entry into a computer database.

NAME FLOWER COLOUR SOIL LOCATION HEIGHT

It is important to discuss

— the amount of data to be entered

— the way in which the data is entered: for example, in whole words or an abbreviated form.

Once the data has been entered into the database, questions similar to questions 1 and 2 can be asked.

## Textbook page 121 *Handling data: tables (spreadsheets)*

UA2a,3bc/4 HD2ab/3
PSE O/C2,3 O/D1,2 I/C1,D1
P/4ad D/3ab

On Textbook page 121, the children enter and manipulate data in a table. This information could also be entered in a spreadsheet. Suggestions about how to do this are given in the additional activities below.

In question 1, most children should be able to interpret the information from the display on the page and enter it in the table. Note that the orange group did not want any seeds.

Some children may need help to complete the **total** column at the right-hand side of the table. They need to appreciate that they should add together the **total cost** figure for each of the five columns – plants, tools, seeds, grass and slabs – to find the overall total.

In question 2, the children have to interpret and compare the data from the **two** tables they have constructed.

In question 3, some children may need to be reminded how to find an average.

Problem solving

In question 4, some children may need help to see that, in addition to changing the specific data indicated, they also need to alter the overall totals.

| group orange | plants | tools | seeds | grass ($m^3$) | slabs | |
|---|---|---|---|---|---|---|
| Number | 20 | 4 | ~~0~~ 3 | 10 | 21 | |
| Cost each | £1·75 | £5·50 | 50p | £1·50 | ~~£2~~ £2·50 | Total |
| Total cost | £35 | £22 | ~~0~~ £1·50 | £15 | ~~£42~~ £52·50 | ~~£114~~ £126 |

| group purple | plants | tools | seeds | grass ($m^3$) | slabs | |
|---|---|---|---|---|---|---|
| Number | ~~8~~ 10 | 4 | 12 | 12 | 30 | |
| Cost each | £1·75 | £5·50 | 50p | £1·50 | ~~£2~~ £2·50 | Total |
| Total cost | ~~£14~~ £17·50 | £22 | £6 | £18 | ~~£60~~ £75 | ~~£120~~ £138·50 |

UA2a,3bc/4 HD2ad/3
PSE O/E2 I/E3
D/4g

# Additional activity

## Using a spreadsheet

- A spreadsheet is a computer package which allows numbers and text to be entered into a table similar to the one on Textbook page 121.
- It gives the children the opportunity to
  - — read totals and averages
  - — change the data that has been entered
  - — create graphs
  - — explore the relationships between different rows or columns.
- At this stage, the row and column headings should be entered by the teacher, but the children should be encouraged to enter numeric data and perhaps give instructions (formulae) for calculations to be carried out. Full details about setting up the spreadsheet and entering data are usually found in the documentation accompanying each spreadsheet package.
- The data on Textbook page 121 could be used to set up a spreadsheet as follows:

**Enter labels** for rows and columns.

| | A | B | C | D | E | F | G |
|---|---|---|---|---|---|---|---|
| 1 | | Plants | Tools | Seeds | Grass | Slabs | |
| 2 | Number | | | | | | |
| 3 | Cost each (£) | | | | | | Total |
| 4 | Total cost (£) | | | | | | |

**Enter the data**
In most spreadsheet packages, data are entered in a different **mode** from labels. It is important to ensure that the data are entered in the correct mode, or the spreadsheet will not function.

At this point, only the data for the orange group should be entered. For example,

| | A | B | C | D | E | F | G |
|---|---|---|---|---|---|---|---|
| 1 | Orange | Plants | Tools | Seeds | Grass | Slabs | |
| 2 | Number | 20 | 4 | 0 | 10 | 21 | |
| 3 | Cost each (£) | 1·75 | 5·50 | 0·50 | 1·50 | 2 | Total |
| 4 | Total cost (£) | | | | | | |

**Set up instructions (formulae) for calculations**
Set up simple instructions for the appropriate cells of the spreadsheet. For example,

| | A | B | C | D | E | F | G |
|---|---|---|---|---|---|---|---|
| 1 | Orange | Plants | Tools | Seeds | Grass | Slabs | |
| 2 | Number | 20 | 4 | 0 | 10 | 21 | |
| 3 | Cost each (£) | 1·75 | 5·50 | 0·50 | 1·50 | 2 | Total |
| 4 | Total cost (£) | ↑ B2×B3 | ↑ C2×C3 | ↑ D2×D3 | ↑ E2×E3 | ↑ F2×F3 | ↑ B4+C4+D4+E4+F4 |

This should produce the following table:

| | A | B | C | D | E | F | G |
|---|---|---|---|---|---|---|---|
| 1 | Orange | Plants | Tools | Seeds | Grass | Slabs | |
| 2 | Number | 20 | 4 | 0 | 10 | 21 | |
| 3 | Cost each (£) | 1·75 | 5·50 | 0·50 | 1·50 | 2 | Total |
| 4 | Total cost (£) | 35 | 22 | 0 | 15 | 42 | 114 |

- The children can explore what happens to totals when they change some of the data entries. For example, if the orange group wants to buy 3 packets of seeds, this entry will change from zero to three.

| | A | B | C | D | E | F | G |
|---|---|---|---|---|---|---|---|
| 1 | Orange | Plants | Tools | Seeds | Grass | Slabs | |
| 2 | Number | 20 | 4 | 3 | 10 | 21 | |
| 3 | Cost each (£) | 1·75 | 5·50 | 0·50 | 1·50 | 2 | Total |
| 4 | Total cost (£) | 35 | 22 | 1·50 | 15 | 42 | 115·50 |

These two totals will automatically appear.

- Repeat for other examples:
  - — 5 fewer plants were bought
  - — seeds increased in price to 75p.
- To set up the spreadsheet for the purple group
  - — restore the prices to the ones shown on the Textbook page
  - — change the number of items to those bought by the purple group
  - — since the formulae are already set up, the totals for the purple group will appear automatically. The purple group's spreadsheet should look like this:

| | A | B | C | D | E | F | G |
|---|---|---|---|---|---|---|---|
| 1 | Purple | Plants | Tools | Seeds | Grass | Slabs | |
| 2 | Number | 8 | 4 | 12 | 12 | 30 | |
| 3 | Cost each (£) | 1·75 | 5·50 | 0·50 | 1·50 | 2 | Total |
| 4 | Total cost (£) | 14 | 22 | 6 | 18 | 60 | 120 |

- Some children may be able to work with other facilities in the spreadsheet package, such as those for finding averages or drawing graphs. Full details of how to deal with these are found in the documentation accompanying the package.

# SURVEYS

In Heinemann Mathematics 5, the children carried out a survey for which the data was collected using a yes/no questionnaire.

They now carry out a survey of the types of waste found in and around their own school, and consider ways in which to cut down waste.

## Introductory activity

### Litter survey

UA2b,3bc/4 HD2ab/3
C/C2 O/C1,2 D/C1 I/C1,2
D/5a

- Ask the children to draw a simple sketch of the playground and to record on it where litter is found.
- Ask them to analyse the type of litter found, such as paper, food containers, bottles and cans.
- The information collected could be displayed as a table, showing the quantities of litter in different playground areas.
- From the information gathered, the children should decide on the most appropriate place to locate rubbish bins in the playground.

UA2b,3bc,4acd/4 HD2ab/4
C/C2 O/C1,2 D/C1 I/C1,2
D/5a

## Textbook page 122 *Handling data: survey*

The children are asked to carry out a survey, in and around their school, of the types of waste they can see, and to consider how they might prevent or cut down this waste.

In question 1, in groups, the children have to identify different types of waste they might find around the school. To help with this, the illustration at the top of the page may have to be discussed.

In question 3(b), the intention is to subdivide the task within the group, with children working in pairs or threes to deal with separate places.

In question 3(c), each subgroup should think about realistic ways in which waste can be reduced – for example, by switching off lights in rooms where no one is working.

In question 4, the **whole** group

- — collates and discusses all the information
- — constructs a written, oral or pictorial report to explain their findings
- — makes recommendations about how they might save this waste.

UA3a,4d/4
HD3a/4
HD3b/4→5
P/4cd
D/3ef
D/4i

# Probability

## Overview

This section

- deals with language associated with likelihood
- includes ideas of 'evens' and fairness
- gives practice in listing outcomes of events.

| | Teacher's Notes | Textbook | Workbook | Reinforcement Sheets |
|---|---|---|---|---|
| **Likelihood** | 270 | 123 | | |
| **Evens, fair, unfair** | 272 | 124 | | |
| **Listing outcomes** | 275 | 125 | | |

## Resources

### Useful materials

- dice, cubes, counters, coins
- dominoes, playing cards, number cards (1–10)
- bag (or box)
- other materials suggested within the introductory activities

### Assessment and Resources Pack

#### Assessment

*Handling data Check-up 3*
Textbook pages 123–5
(Probability)

# Teaching notes

This section begins by consolidating and extending language associated with likelihood, such as 'impossible', 'certain', 'fair', 'unfair', 'evens', 'equally likely', 'unlikely' and 'probable'. The likelihood of events is represented on a probability scale ranging from 'impossible' to 'certain'.

## LIKELIHOOD

### Introductory activities

UA3a,4d/4 HD3ab/4 HD3b/4→5
P/4cd D/3e D/4i

#### Events

- Display the following words.

  impossible | unlikely | likely | certain

  Discuss events to illustrate their meaning. For example,

  'Next year will be a leap year' (impossible or certain depending on the present year).

  '*Blue Peter* will be on TV again next year' (likely).

  'There will be a school holiday on Christmas Day' (certain).

  'Great Britain will be the country to win most medals at the next Olympic Games' (unlikely).

  Provide other examples to give the children practice in describing likelihood. They could also suggest events which match descriptions such as 'unlikely', 'certain' and so on. Some children have difficulty with events such as 'I will meet a princess today', which they view as 'impossible' when they are simply 'very unlikely' but could happen.

- Consider events where a wider range of language is appropriate. These can be related to a scale like this on paper or a chalkboard.

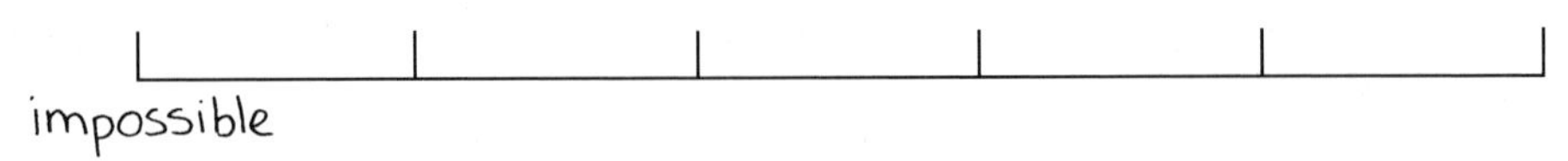


  List five other descriptions of likelihood:

  certain very unlikely likely unlikely likely

  Ask the children to decide where these descriptions fit on the scale and then discuss the positions of events such as the following:

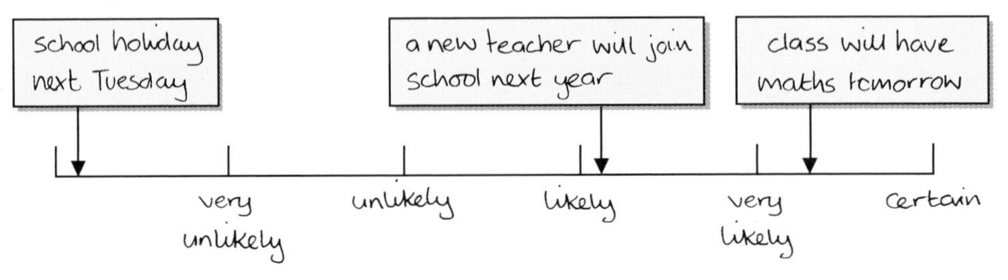

Discuss the apparent likelihood of each event with the children. For example,

'It is **almost impossible** to imagine a school holiday next Tuesday.' (There's no holiday in the school calendar, but the boiler could burst and the school be closed.)

'It is **more than likely** that there will be at least one new teacher next term.'

'You are **very, very, likely** to have maths tomorrow.' (We nearly always do.)

- Ask children to position other familiar events on the scale by thinking about how likely they are **for them.** There are no 'correct' answers, but the children should give proper reasons for the positions they choose.

Possible events are

| | |
|---|---|
| go swimming tomorrow | listen to the radio tomorrow |
| make a phone call tomorrow | receive a letter tomorrow |

## Textbook page 123 *Handling data: Probability*

UA3a,4d/4
HD3a/4 HD3b/4→5
**P**/4cd **D**/3e **D**/4i

In question 1 it is important that the children's answers are discussed. For example, it is **unlikely** that a dice throw will give a six as there are many more ways of **not** getting six, by throwing 1, 2, 3, 4 or 5. It is not certain that it will rain next week, although it is likely that it will. In part (b), it is **impossible** in real life for a frog to turn into a prince. In part (e), describing 'going to bed tonight' as **certain** is acceptable for the age group, although 'likely' could also be accepted.

Question 3 involves the likelihood of events shown on a probability scale. These could be described in a number of ways. Acceptable descriptions include

'It is **nearly impossible** that the girl in red is guilty.'

'It is **more than likely** that there will be a car chase.'

In question 4, the children may place some of the arrows in widely differing positions, due to individual differences. This is acceptable as long as their explanations demonstrate a group of ideas of likelihood and of the appropriate language. For example,

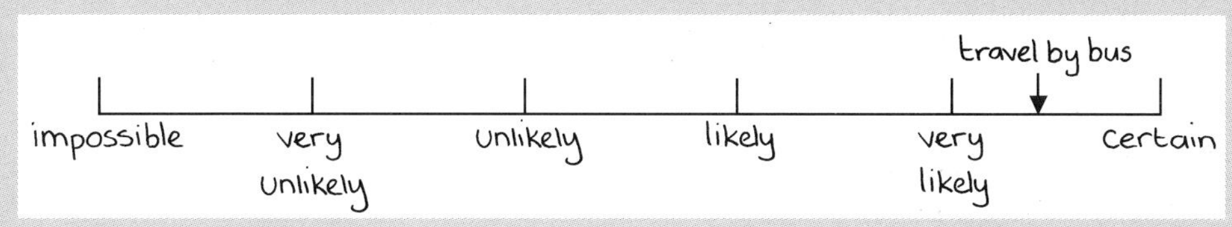


'I am very, very likely to travel by bus tomorrow because I usually go to my Gran's on a Tuesday.'

# EVENS, FAIR, UNFAIR

Simple games and activities are used to illustrate ideas of 'evens' and fairness.

## Introductory activities

UA3a,4d/4
HD3a/4 HD3b/4→5
P/4cd D/3ef

### 1 Evens, equally likely

- Revise some of the language of likelihood using a probability scale. Draw a new line in the middle of the scale and ask the children to describe how likely an event in this position would be.

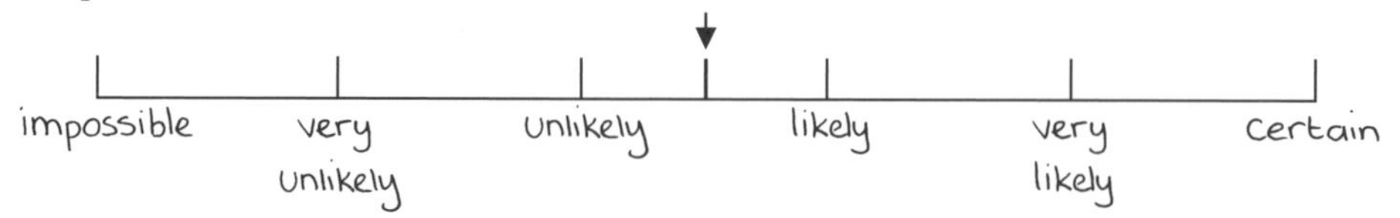


In the discussion, include phrases such as 'neither likely nor unlikely' and 'an equal chance of it happening or not happening'.

- Discuss events with two outcomes which are 'equally likely'. For example,

  'A new baby is equally likely to be a boy or a girl.'

  'A toss of a coin is equally likely to give heads or tails.'

  Introduce the idea of 'evens'.

  'There is an even chance of a head or a tail.'

  'The chance of a baby being a boy is evens.'

- Provide activities where a group of children can explore situations with two equally likely outcomes. For example,

A bag with 5 white and 5 blue cubes.

Pick a cube without looking. Record its colour.

Put it back and shake the bag. Do all this 30 times.

| white | ✓ ✓ ✓ |
|---|---|
| blue | ✓ ✓ ✓ ✓ |

Pack of playing cards.

Cut the pack. Record the colour of the card – red or black.

Put the card back and shuffle. Do all this 30 times.

| red | ✓ ✓ ✓ ✓ ✓ |
|---|---|
| black | ✓ ✓ ✓ |

Ten cards or number tiles, marked with the numbers 1 to 10. Shuffle the cards around, face down. Pick one. Record whether its number is even or odd. Put the card back. Do all this 30 times.

| even | ✓ ✓ ✓ ✓ ✓ |
|---|---|
| odd | ✓ ✓ ✓ ✓ ✓ |

- The children may, for example, pick 16 blue and 14 white cubes. They may need persuading that the ticks for blue and white would tend to even up as they went along and that, if they were to continue for 100 or 1000 trials, the numbers of blue and white would remain very close. There is an even chance of blue or white, although one cannot expect 'real life' to work out exactly every time.

## 2 It's not fair

- Put 30 pieces of folded paper in a bag – 1 blank piece and 29 marked with a cross (✗).

  Play a game with the children where a child picks one piece of folded paper from the bag. If there is a cross on it, the child has to perform a forfeit, chosen by the teacher. If it is blank, the child chooses a forfeit for the teacher.

  After 3 or 4 turns, there is likely to be some feeling that the game is unfair. Allow the children to examine the pieces of paper. They could count the crosses (29) and compare them with the blanks (1).

  Discuss why this game is unfair – a cross and a blank are not equally likely to be picked. A cross is very, very likely; the chance of choosing it is more than evens. A blank is very unlikely; the chance of choosing it is less than evens. The children should suggest an equal number of crosses and blanks to make the game 'fair'.

- Remind the children of the cube, card and number activities from activity 1 above and discuss why these were 'fair'. They should realize that there were an **equal number** of white and blue cubes/red and black cards/odd and even numbers, from which to choose.

- Ask them if the forfeit game would be fair if they used playing cards and the teacher paid a forfeit when they chose a face card (King, Queen or Jack). A good way of showing that this is still unfair is to separate the pack into two piles.

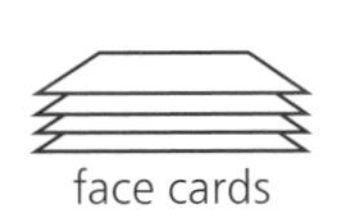

  There is a less than even chance of the child choosing a face card.

## 3 A number game

- Consider a game where there are more than two possible outcomes. Ask each child to choose a whole number from 1 to 20 and write it down. Now reveal a large diagram like this:

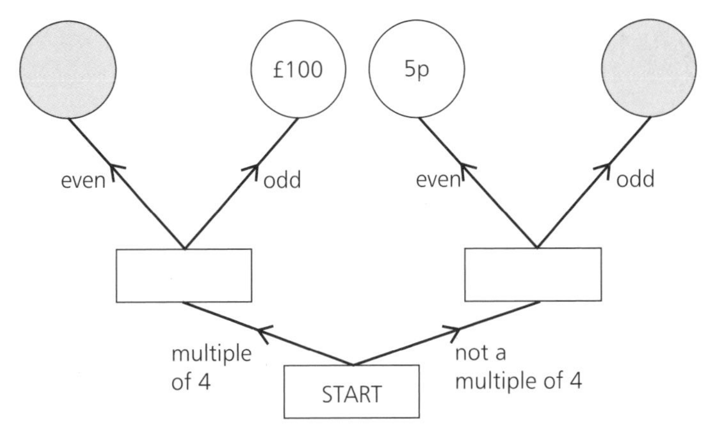

Handling data
Probability

- Discuss the paths followed by a number like 19 (not a multiple of 4/odd) which end up at the circle on the right. Ask the children where their numbers lead and discuss the results.
    - — About half of the children should end up at the circle on the right. This is the 'most likely' outcome.
    - — Some should finish at the 5p circle and some at the circle on the left.
    - — No one should reach the £100 circle.

  Discuss the £100 route. Only the numbers 4, 8, 12, 16 and 20 can follow the first path. None of these is odd, so it is impossible to reach the £100.
- Make clear that the four outcomes are **not** equally likely. Ask the children to suggest changes to make them equally likely. This is a difficult task, but some children may suggest changing the first path to '10 or less'/'greater than 10'.

UA3a,4d/4 HD3a/4
P/4cd D/3ef

## Textbook page 124 *Handling data: Probability*

In question 1, some children may need help with game C. Laura is more likely to win because she has four winning outcomes (the factors of 6 are 1, 2, 3, 6) and only two numbers (4 and 5) give Jamie a win.

The children should read the rules for the Planets Game in question 2 very carefully. It is important that they do **not** replace the first cube each time. They should find that

- — a win at Jupiter is impossible
- — Mars is a likely destination (the most likely of the four)
- — Saturn and Venus are unlikely finishing points.

The theoretical chances (for teachers only) of finishing at each planet are

| Saturn | Jupiter | Mars | Venus |
|---|---|---|---|
| 20% | nil | 60% | 20% |

The game is very unfair. Replacing the first cube would improve it, as would having equal numbers of yellow and red cubes. Doing both of these would make the four outcomes equally likely.

**Problem solving**

Question 3 is challenging. One method of investigation is to consider systematically all the possible outcomes. For example, for card (a),

| | | |
|---|---|---|
| $3 \times 6 = 18$ | $4 \times 6 = 24$ | $8 \times 6 = 48$ |
| $3 \times 7 = 21$ | $4 \times 7 = 28$ | $8 \times 7 = 56$ |
| $3 \times 10 = 30$ | $4 \times 10 = 40$ | $8 \times 10 = 80$ |

Only one of the nine outcomes (21) is an odd number, so a win is unlikely with this card. Some children may realize that, for card (b), all the numbers in the top row are even, the products must be even, and therefore a win is impossible.

# LISTING OUTCOMES

Some practice is now provided in listing all the possible outcomes of an event.

## Introductory activities

UA3a/4 HD3a/4
P/4cd

- Discuss possible outcomes of simple events with the children, using material where necessary to help them think about what could happen. For example,

  Throw a die ⟶ 1, 2, 3, 4, 5, 6

  Toss a coin ⟶ head, tail

  Pick a card. Find which suit ⟶ Heart, Club, Spade, Diamond

  Drop a drawing pin ⟶ land point up, land point down

  Pick a domino and count the spots ⟶ 0, 1, 2, 3, . . . 12

  Sit a driving test ⟶ pass, fail

- Consider a raffle with tickets numbered 1 to 50. Ask the children to suggest possible numbers if the winning ticket

  is a multiple of 10 ⟶ 10, 20, 30, 40, 50

  is between 18 and 22 ⟶ 19, 20, 21

  has two equal digits ⟶ 11, 22, 33, 44

- Discuss a situation where two choices are combined – for example, when choosing a flavour of crisps **and** a type of drink.

| Crisps | Drink |
|---|---|
| Prawn | Cola |
| Onion | Orange |
| Plain | |

There are six possibilities:

| | |
|---|---|
| Prawn crisps and Cola | Prawn crisps and Orange |
| Onion crisps and Cola | Onion crisps and Orange |
| Plain crisps and Cola | Plain crisps and Orange |

Emphasize the usefulness of listing systematically.

UA3a/4 HD3a/4
P/4cd

## Textbook page 125 *Handling data: Probability*

In question 2(e), the children could think of possible units digits and add 2 to find the tens digits, giving

20, 31, 42, 53, 64, 75, 86, 97

In 2(f), numbers with two digits which add to 7 show a certain symmetry:

16, 25, 34, 43, 52, 61 and also 70

Their digits show patterns when listed vertically.

16
25
34
43
52
61
70

Question 4 is fairly challenging if the children are to find all the possible totals. The three dominoes should be made available. The totals are:

One domino: 8, 9, 11

Two dominoes: 17, 19, 20

Three dominoes: 28

The teaching notes follow for the thirty pages of activities in the Heinemann Mathematics 6 Extension Textbook. The answers are provided separately in the Answerbook.

Extension

## Contents

# Extension

SSM3a/4
RS/D5
S/4b

## Extension Textbook page E1 *Other activity: translation*

- The activities on this page can be attempted at any time. They involve the children in
  - — making patterns by translating a shape horizontally or vertically
  - — creating a design on a square tile and translating this to generate a repeating pattern.
- In question 1, the children should trace the shape to make a card version. It may be worth
  - — discussing how each of the patterns is made
  - — introducing the term 'translate' as another word for 'slide'.

Children may find it helpful to draw a feint guideline to slide their shape along.

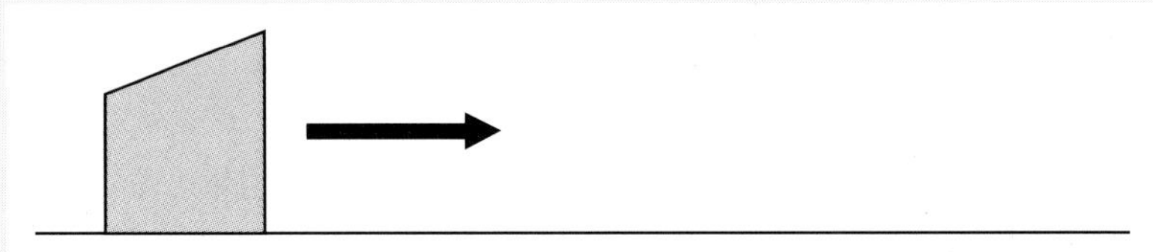

- In question 2, the children might describe their own patterns using phrases such as 'sliding the shape to make a pattern of shapes touching side to side' or 'shapes overlapping'.
- In question 4, the children could experiment by changing the shape or size of the tile used. For example,

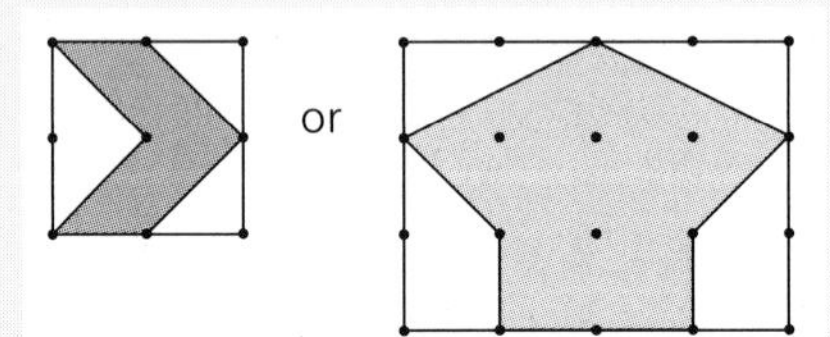


UA2bc/4 N2a,3d/4
PSE RTN/D1 AS/D2
P/4ab N/4abk

## Extension Textbook pages E2 and E3
### *Place value: addition and subtraction*

- The work on this page may be attempted at any time, but is probably best done after the children have completed Textbook pages 1–5 and Workbook pages 1–3.
- On page E2, in question 1(b), the children are expected to notice that
  - — each pair of numbers has the same digits, but the second number has them in reverse order
  - — in each answer, all the digits are the same.

Discuss the children's descriptions to ensure that these features are included. In question 1(c), the children should realize that the 'rule for the total' breaks down when the added digits have a sum of ten or more: that is, when there is carrying from one column to another.

- In question 2(a), many children will restrict their solutions to numbers where there is no decomposition for the units and tens, and consider the hundreds and thousands as a two-digit number. For example,

$$\begin{array}{r} 1079 \\ -630 \\ \hline 449 \\ \hline \end{array} \quad \begin{array}{r} 1189 \\ -740 \\ \hline 449 \\ \hline \end{array} \quad \begin{array}{r} 1299 \\ -850 \\ \hline 449 \\ \hline \end{array} \quad \text{rather than} \quad \begin{array}{r} 1128 \\ -679 \\ \hline 449 \\ \hline \end{array}$$

Some children may choose a three-digit number such as

$$\begin{array}{r} \square\square\square\square \\ -738 \\ \hline 449 \\ \hline \end{array} \quad \text{chosen number}$$

and then **add** to discover the remaining number: 1187, in this instance. Any number in the range 551 to 999 can be subtracted.

Similarly, in question 2(b), there are many solutions. Any number in the range 100 to 999 could be used as the three-digit number. In this instance, carrying must be involved. For example,

$$\begin{array}{r} 1110 \\ +891 \\ \hline 2001 \\ \hline \end{array} \quad \begin{array}{r} 1891 \\ +110 \\ \hline 2001 \\ \hline \end{array} \quad \begin{array}{r} 1687 \\ +314 \\ \hline 2001 \\ \hline \end{array}$$

The children could choose a three-digit number and then **subtract** from the answer to find the top number.

The children may like to collect and display their different solutions for the answers to both (a) and (b), and then discuss how they found these.

- In question 3(e), the children meet the challenge of changing two digits: that is, changing 30 to 28, which can be done by subtracting 2.

In question 3(f), the expected calculation is +70, so for question 3(g) some children might realize that the answer sought is one less than the answer to the previous example, calculating +69 for that reason. Others will go back to first principles and simply see 30 being changed to 99, adding 69. It could be interesting to discuss the children's methods with them.

- On page E3, in question 1, the children should realize that they require two single-digit numbers where the sum is 10 or more. Possible solutions include

$$\begin{array}{r} 2 \\ +8 \\ \hline 10 \\ \hline \end{array} \quad \begin{array}{r} 3 \\ +7 \\ \hline 10 \\ \hline \end{array} \quad \begin{array}{r} 4 \\ +6 \\ \hline 10 \\ \hline \end{array} \quad \begin{array}{r} 6 \\ +4 \\ \hline 10 \\ \hline \end{array} \quad \begin{array}{r} 7 \\ +3 \\ \hline 10 \\ \hline \end{array} \quad \begin{array}{r} 8 \\ +2 \\ \hline 10 \\ \hline \end{array}$$

$$\begin{array}{r} 3 \\ +9 \\ \hline 12 \\ \hline \end{array} \quad \begin{array}{r} 4 \\ +8 \\ \hline 12 \\ \hline \end{array} \quad \begin{array}{r} 5 \\ +7 \\ \hline 12 \\ \hline \end{array} \quad \begin{array}{r} 7 \\ +5 \\ \hline 12 \\ \hline \end{array} \quad \text{and so on up to} \quad \begin{array}{r} 9 \\ +8 \\ \hline 17 \\ \hline \end{array}$$

They should also find that the maximum total possible is 17, and that totals such as 11 have the same digit repeated and so cannot be used. The digit in the red box must be 1, and 0 can only occur in the units answer position.

# Extension

- In questions 2 and 3, some children might find it useful to use the digits written on separate pieces of paper or card. This allows the children to move the digits easily to other positions and to avoid using the same digit twice. The given digits reduce the number of possible answers and give the children a start for their solutions.

  In question 3, there are several possible solutions, including

| 632 | 652 | 657 |
|---|---|---|
| +457 | +437 | +432 |
| 1089 | 1089 | 1089 |

- Some children may realize that a solution for question 3 gives them possible numbers for the subtraction in question 4. For example:

UA2c/4 UA2d/5 N2a/4
PSE RN/D1
P/4a N/3h

## Extension Textbook page E4 *Place value: rounding*

- The work on this page involves rounding to the nearest 10, 100 and 1000. It can be attempted any time after the children have completed Textbook page 8.
- In question 1, the children should use a process of elimination to determine where each person was looking for shells and pebbles. By comparing the information in the speech bubbles with the numbers given in the table (when these are rounded to the nearest 10), the following answers should be obtained:

  Dorothy – *Pool*  Sean – *Hut*  Roy – *Cave*  Elaine – *Boat*

  The lagoon, which is the only place remaining, must have been where Simon was looking. In part (b), a possible speech bubble for Simon is

- In question 2, the children should use a similar process of elimination to determine the town in which each person lives. They are required to round the number of people in the table to the nearest 1000, and the number of houses to the nearest 100. The information in the speech bubble for Dorothy produces only one answer, i.e. Corbett. Once this connection is made, the towns where the other people live can be found. These are:

  Roy – *Aberdon*  Sean – *Bixton*  Elaine – *Denby*

## Extension Textbook page E5 *Other activity: temperature*

N2b/5 HD2b/4
RTN/E1 D/D1 I/D1
N/4e

- On this page, the children interpret data presented on a map and as a graph. Comparison of negative numbers is involved when answering the questions. The work on this page may be attempted at any time.
- In question 4(d), there are two pairs of months with equal temperatures:

  May and October both both have a temperature of 6°C

  March and December both have a temperature of $^-3$°C.
- In question 5, the children are expected to use the same scales on the axes as in the given graph for question 4.

## Extension Textbook page E6
### *Multiplication: multiples, digital roots*

UA3a,4bc/4 N3c/4
PSE MD/C1,3 PS/C1
P/4cd N/4gi A/3a A/4c

- The work on this page could be attempted at any time after the children have completed Textbook page 12 and Workbook pages 4 and 5, which deal with multiples, patterns and digital roots linked to multiplication tables.
- In questions 1 and 2, some children may need help to interpret the illustration – an ambulance may be kept in any station where the ambulance's number belongs to the set of multiples displayed on the station's roof.

  In question 1, the given ambulance numbers **each** belong to only **one** of the three sets of multiples.

  In question 2(a), the children are asked to make up a number which is a multiple of 2 **and** a multiple of 3 – for example, any multiple of 6.

  In question 2(c), the number to be made up must belong to **all three** sets of multiples – for example, any multiple of 30.
- In question 3(b), the next three rows are

  $(5 \times 4) - 3 = 17 = (4 \times 4) + 1$ $\quad$ $3 \times (152 + 37) = 3 \times 189 = 567$

  $(4 \times 3) - 2 = 10 = (3 \times 3) + 1$ $\quad$ $3 \times (189 + 37) = 3 \times 226 = 678$

  $(3 \times 2) - 1 = 5 = (2 \times 2) + 1$ $\quad$ $3 \times (226 + 37) = 3 \times 263 = 789$
- For questions 4 and 5, some children may need to be reminded about the meaning of the term 'digital root'.

  In question 5, the digital roots of the stations of the 4 and 5 times tables are linked in a diagram to produce the same pattern in each case.

| 4 times table | |
|---|---|
| multiple | digital root |
| 4 | 4 |
| 8 | 8 |
| 12 | 3 |
| 16 | 7 |
| 20 | 2 |
| 24 | 6 |
| 28 | 1 |
| 32 | 5 |
| 36 | 9 |
| 40 | 4 |

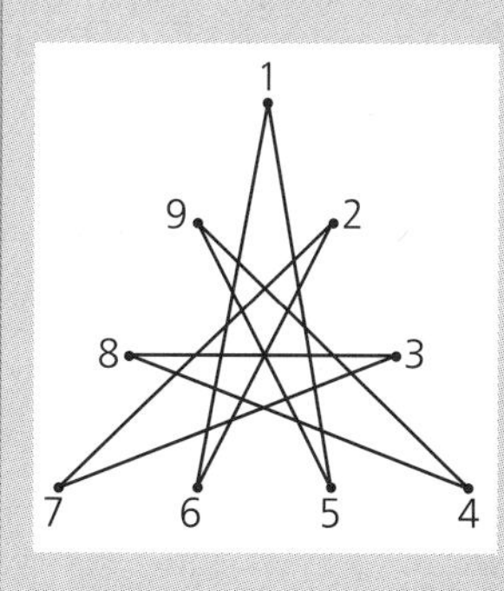


| 5 times table | |
|---|---|
| multiple | digital root |
| 5 | 5 |
| 10 | 1 |
| 15 | 6 |
| 20 | 2 |
| 25 | 7 |
| 30 | 3 |
| 35 | 8 |
| 40 | 4 |
| 45 | 9 |
| 50 | 5 |

Extension

UA3a/4 UA4bc/4→5 NC3a/4
PSE MD/D3
P/4cd N/4l U/4bc

## Extension Textbook page E7 *Division: divisibility by 9*

- This page provides
  - — activities involving division of a whole number with up to four digits by 9
  - — an investigation of one method of checking for divisibility by 9.

The work can be attempted at any time after completion of Textbook pages 24 and 25, and Workbook page 6 in the Division section.

- It may be necessary to discuss the language 'is a multiple of', 'divides exactly by' and 'sum of the digits', either before the children begin or after they have completed question 1.
- The children's answers to question 1 should be checked before they are used in question 2.
- In question 2(b), it is intended that the numbers in the yellow column be described as 'multiples of 9'. Some children may write 'divide exactly by 9'. Hence, in question 2(c), the test might be 'if the sum of the digits is a multiple of 9 (or divides exactly by 9) then the number divides exactly by 9'.
- In question 4, the children should notice that the 'remainder when divided by 9' and the 'sum of digits' are the same.
- The investigation can be extended by asking the children to find remainders for numbers such as 164, 378, 4352, 6091 and so on, where the sum of the digits is greater than 9 and so cannot be equal to the remainder. However, if the sum of the digits is in turn divided by 9 then the remainder may be found. For example,

$$6091 \div 9 = 676 \text{ (R7)}$$

$$6091 \xrightarrow{\text{sum of digits}} 16 \qquad 16 \div 9 = 1 \text{ (R7)}$$

Alternatively, if the sum of the digits is repeatedly added until a single-digit total is obtained, this too gives the remainder. For example,

$$6091 \xrightarrow{\text{sum of digits}} 16 \xrightarrow{\text{sum of digits}} 7 \text{ (R7)}$$

$$7487 \longrightarrow 26 \longrightarrow 8 \text{ (R8)}$$

N4a/4
AS/D3 MD/D4
N/4jl N/5j

## Extension Textbook page E8 *Money: calculator*

- On this page, the children use a calculator to find costs related to holidays. The activities can be attempted at any time.
- Essential information is given in the price table and in the 'child reductions' flash at the top of the page. The children should be encouraged to read these carefully before answering the questions.
- In question 1(a), the children should appreciate that a child reduction discount will **only** be available to the Greens, as they are the only family to satisfy the qualifying condition of having at least 2 adults travelling with the children.
- After answering questions 1 and 2, the children could be asked to suggest reasons for the price variations shown for the different weeks in July and August.

- In question 3, the Greens are assumed to have taken the option, introduced in question 2, of going on holiday for two weeks rather than one.

  The children need to be aware that the extra charge for a room with a sea view applies to **every** family member.

- In question 4, some children may look for prices in the table which, when multiplied by 4, give £1500. A more efficient method is to divide £1500 by 4, giving £375, and then to look for this price in the table. A basic 7-night holiday for 4 adults at the Hotel Splendide, during the first half of August, costs £1500.

N2c/4 N3g/5
RTN/D3 FPR/D1
N/4g N/5b

## Extension Textbook page E9 *Fractions: simplification*

- This page extends the work on simplification by using numbers which may require more than one step to give the fraction in its simplest form. It can be attempted at any time after the completion of the Fraction section on Textbook pages 30–35 and Workbook pages 8 and 9.
- In question 1(a), the children have to count the number of equal parts which make up the diagram representing the whole pie. This gives the initial denominator for each fraction as 20. Simplifying gives the fractions $\frac{1}{4}$, $\frac{1}{5}$, $\frac{3}{20}$, $\frac{1}{10}$, $\frac{3}{10}$.
- For the last part of question 2, because of the way the information is given, the children have to calculate the total for themselves before finding the fractions.

  Good $\frac{42}{63} = \frac{6}{9} = \frac{2}{3}$

  Bad $\frac{21}{63} = \frac{3}{9} = \frac{1}{3}$

potatoes
42 are good
21 are bad

UA2bcd,3a/4 N2b/4
PSE RTN/D4 AS/D3
P/4bcd N/4dk

## Extension Textbook page E10 *Decimals: place value*

- This page provides further place value work involving numbers with at most two decimal places. It can be attempted at any time after the completion of Textbook page 47 and Workbook page 11.
- In the first problem in question 1, a good clue to start with is 'the units digit is four less than the tens digit', leading to a consideration of the following possibilities:

  95·□□, 84·□□, 73·□□,
  62·□□, 51·□□, 40·□□

  Using the first two clues, the possibilities become

  95·59, 84·48, 73·37,
  62·26, 51·15, 40·04

  Since, from the final clue, the sum of **all** the digits is 20, the answer must be 73·37 metres.

**Extension**

- In the second problem in question 1, the first two clues give the following possibilities.

  2 1 · 1 □, 4 2 · 2 □, 6 3 · 3 □ 8 4 · 4 □

  Encourage the children to write down these possibilities. The third clue, however, reduces the possibilities to

  4 2 · 2 1, 8 4 · 4 2

- In question 2, the first two clues give the following possibilities.

  1 □ · 3 4, 2 □ · 6 5, 3 □ · 9 6,

  The third clue reduces these possibilities to

  2 5 · 6 5 and 3 0 · 9 6

  The fourth clue enables the answer of 30·96 metres to be found.

- In question 3, the children should compare the number they must 'enter' in the calculator display with the number they have to 'make'. For example, in question 3(a),

  They should then decide whether to add or subtract, and which digits have to be changed (in this case, units and tenths). They should then add 2·30 (or 2·3) to 5·08 to obtain 7·38.

UA2acd/4→5 N2b,3fg/4
P/4abc N/4dk

## Extension Textbook page E11

### *Decimals: sequences, addition, subtraction*

- This page provides further work in adding and subtracting decimals with and without the use of a calculator. It can be attempted at any time after Textbook page 51 has been completed.

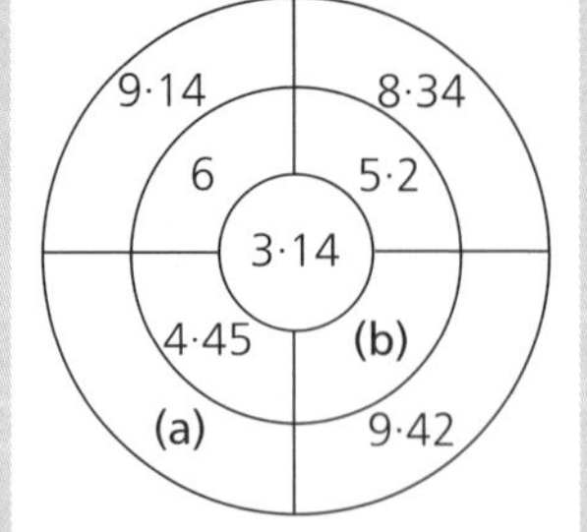


- In question 2, in the first target, the children have to notice that the 'bullseye' number 3·14 added to 6 gives 9·14, and that 3·14 added to 5·2 gives 8·34. Hence the answer to **(a)** is 7·59 (3·14 + 4·45) while the answer to **(b)** is 6·28 (9·42 – 3·14).

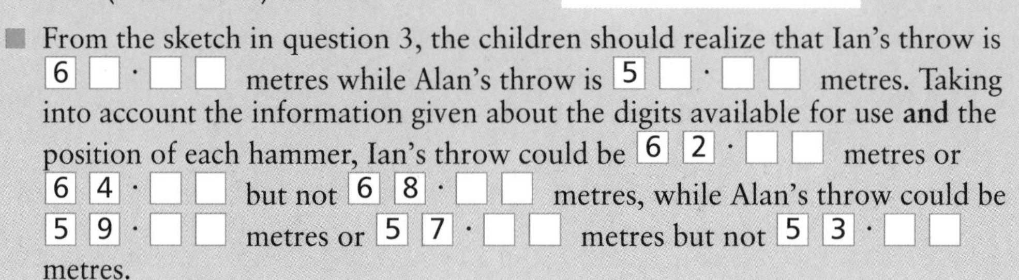

- From the sketch in question 3, the children should realize that Ian's throw is 6 □ · □ □ metres while Alan's throw is 5 □ · □ □ metres. Taking into account the information given about the digits available for use **and** the position of each hammer, Ian's throw could be 6 2 · □ □ metres or 6 4 · □ □ but not 6 8 · □ □ metres, while Alan's throw could be 5 9 · □ □ metres or 5 7 · □ □ metres but not 5 3 · □ □ metres.

Noting that the difference between the two throws is 5·09 metres and using a trial and improvement strategy, the children should find the following answers:

Ian's throw is 64·82 metres, Alan's throw is 59·73 metres **or**
Ian's throw is 62·48 metres, Alan's throw is 57·39 metres.



UA3a/4 SSM2b/4
PSE ME/C3 ME/D4 RS/D1,2,3
P/4bd M/4c S/4b

## Extension Textbook page E12 *Other activity: 2D shape*

- The work on this page involves making various shapes on a 16-pin nailboard. The activities can be attempted at any time after the completion of the 2D shape work on Textbook pages 95–102.
- The children should make the shapes on the nailboard and then record each shape on centimetre squared dotty paper. If 16-pin nailboards are not available, larger boards could be used with sixteen pins marked off.
- Possible answers for question 1 are:

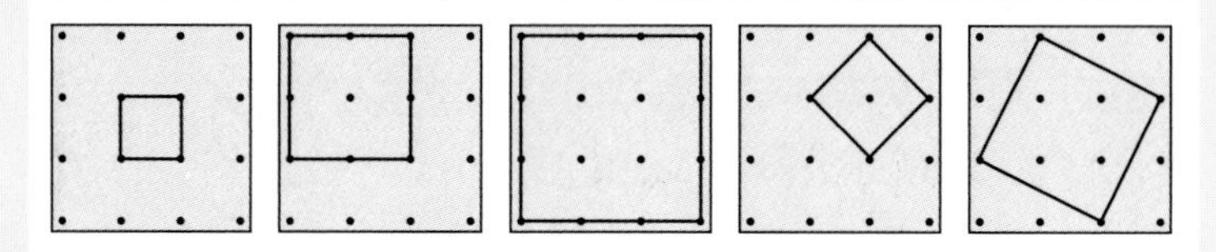

- In question 2, it is possible to make a shape with 16 sides.

- Possible answers for question 3 are:

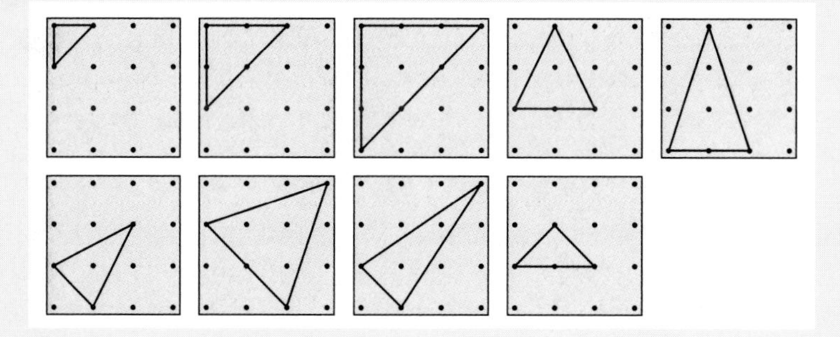

- Possible answers for question 4 are:

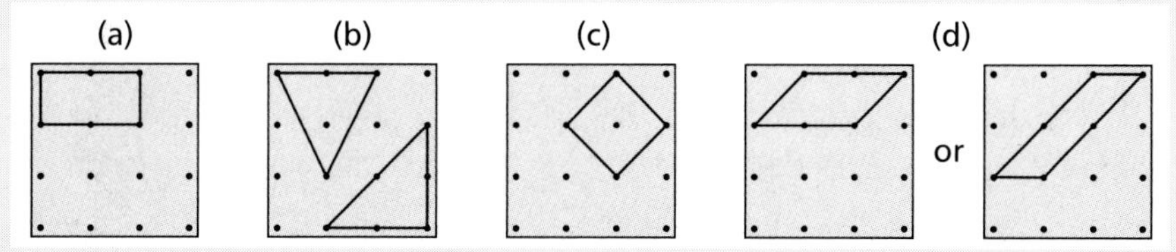


- In question 5, many hexagons can be made. Four different areas are shown.

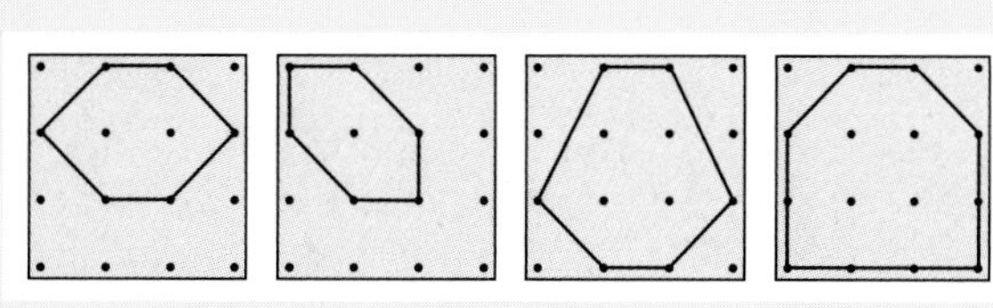

N2c/4
RTN/D4
N/4f N/5c

## Extension Textbook page E13

### *Percentages: fractions as percentages*

- This page extends the work on percentages on Textbook pages 52 and 53 and Workbook pages 12–14. It can be attempted at any time after the completion of the Percentages section.
- The examples involve expressing one number as a fraction of another, simplifying the fraction where appropriate, and then changing it to a percentage. The percentages, with one exception, are limited to 50%, 25%, 75% and 10%, which the children should remember as being equivalent to $\frac{1}{2}$, $\frac{1}{4}$, $\frac{3}{4}$, and $\frac{1}{10}$ respectively.
- Remind the children of the four fraction/percentage equivalents given at the top of the page, and discuss the worked example. The phrase 'out of', as in '7 out of 14', is included as one which is in common use and which should help children to find a fraction initially.
- The last part of question 3 leads to $\frac{4}{20}$ as the fraction of women who work on the farm. Simplifying gives $\frac{1}{5}$ which is not equivalent to any of the percentages on which the other examples are based. The children could think of $\frac{1}{5}$.
  - — as $\frac{2}{10}$, leading to doubling of 10% to give 20% **or**
  - — as $\frac{20}{100}$, leading to 20% directly.

UA2abcd/4 N3a/4
PSE PS/D1 PS/E1,2 FE/D1
P/4ad A/3a A/4d

## Extension Textbook pages E14 and E15

### *Pattern: problem solving*

- This problem solving activity extends the work on pattern. It can be attempted at any time after the children have completed Textbook pages 55–8 and Workbook pages 15 and 16.
- The context of the problem could be discussed with the children. Controller 'S', the head of MI9, knows that one of the four agents shown on page E14 is a double agent. To discover his identity the children must first make up a key to the secret code. Each missing number in a sequence corresponds to a letter. The key is made by collecting together the numbers and letters. For example, in question 1, the letter B corresponds to the number 54.
- In question 1, the letters do not appear in alphabetical order. It may be helpful for the children to list all the letters of the alphabet sequentially before they start, and to fill in the key as they answer each part of the question. For example,

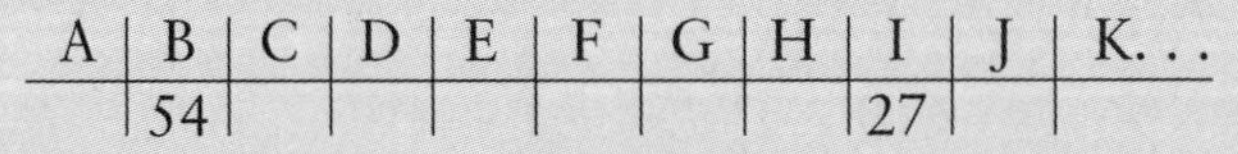

| A | B | C | D | E | F | G | H | I | J | K... |
|---|---|---|---|---|---|---|---|---|---|---|
| | 54 | | | | | | | 27 | | |

  Not all the letters of the alphabet are coded.
- In question 2, decoding the print-out reveals that the agent has brown hair, wears glasses and has no beard, so he must be Orion.

## Extension Textbook page E16 *Multiplication: other methods*

N2a/4 N3d/4→5
AS/C3 MD/E1
N/4I

- This page introduces and provides practice in further alternative methods of multiplying by a two-digit number. The activities can be attempted after the multiplication work on Textbook pages 60–64 has been completed.
- When the worked example for Russian multiplication is discussed, particular attention should be given to
  - the fact that any remainders are ignored in the numbers which are halved
  - the need to score out **both** numbers whenever the **halved** number is even.

The remaining first numbers should be totalled using a pencil-and-paper method. A calculator should be used only to check the final answer.

- In question 2, the final two examples are more challenging as the numbers cannot be halved or quartered so easily. For example, in 2(g), the children need to see that $23 \times 50$ is equivalent to $11\frac{1}{2}$ hundreds or 1150.

## Extension Textbook page E17 *Length: scale*

SSM/4
MD/D1 PFS/E3
N/4i M/5a

- The work on this page involves finding and comparing true lengths of objects drawn to different scales. The activities can be attempted at any time after completion of the Length section on Textbook pages 69–74 and Workbook page 17.
- The children will need a ruler marked in centimetres and half centimetres. A piece of string or pipecleaner about 20 cm long can be bent to follow the curves of the python, and then straightened out and measured. The length of the python should be about 17 cm.
- In question 1, true lengths could be given in centimetres or, where more appropriate, in metres and centimetres.
- In question 2, the dinosaur bone is too long. The doors would not close.

  The pony will fit easily.

  The tyre will fit if standing upright.

  The python, although much longer than the trailer, can curl up to fit.

Extension

UA2a,3a,4d/4 SSM2c,4c/4
PSE ME/C3,7 AS/D2
P/4abd M/4c

## Extension Textbook page E18 *Area: the square metre, $m^2$*

- This page provides several activities involving the square metre which may be best carried out in the school hall or playground. The work can be attempted after the children have met the terms 'parallelogram', 'isosceles triangle' and 'congruent' on Textbook pages 95–8 and Workbook page 33 of the 2D shape section, and have completed the Area section on Textbook pages 79 and 80 and Workbook pages 25–7.
- In question 1, the large sheets of paper could be cut from newspaper pages, provided that these have both length and breadth longer than 50 cm. The answer to part (c), one square metre or $1\,m^2$, should be checked before the children continue.
- In question 2, it may be helpful to draw attention to the illustration, which suggests how a rectangle may be marked out.
- In question 3, the children should reason that, for example, if there are 6 squares in one row and the total area is 30 squares then there must be 5 rows (what times 6 gives 30?). Alternatively, they could draw the rectangles on centimetre squared paper, building them up one row at a time until the required total area is drawn.
- In question 4, if necessary, remind the children of the meaning of 'congruent'. The solutions to part (d) should look like this:

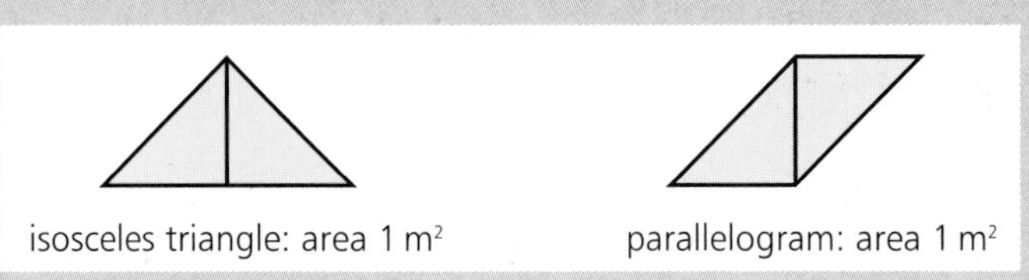


- In question 5, the square and two triangles could be used to mark out the triangle in the same way as the two squares in question 2. The children should be encouraged to include a sketch in their answer to part (b).

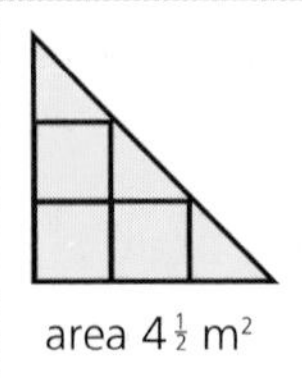


Some children might mark out a square of side 3 m, then form the triangle by inserting a diagonal.

## Extension Textbook page E19 *Volume: cubic centimetre, $cm^3$*

- The work on this page deals with finding the volumes of prisms which are not cuboids, using a method involving layers. The activities can be attempted after the children have completed the work on Textbook page 83.

- In question 1, the shapes have a single layer and the children can find their volumes by counting cubes.

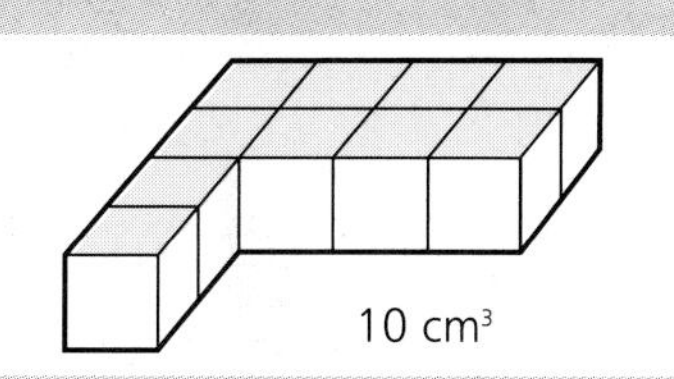


- In questions 2 and 3, the number of cubic centimetres in one layer can be found by counting the number of cubes in the top layer of the shape. The volume can then be found by multiplying this number by the number of layers.

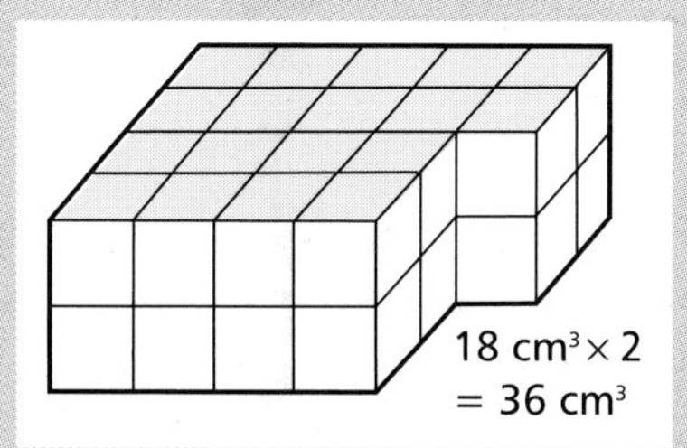


Extension

SSM4c/4
ME/D3
M/4c

## Extension Textbook pages E20 and E21
### *Measure: Imperial units*

N2c/4 SSM4a/5
ME/D9 FPR/C1
M/5b N/3c

- The work on these two pages can be attempted at any time. It deals with the Imperial units – inches, feet, pounds and pints – and attempts to relate each of these units to the equivalent metric units, with which the children are familiar.

  The rough metric equivalents used are

  — one inch (1 in) is about $2\frac{1}{2}$ centimetres

  — one foot (1 ft) is about 30 centimetres

  — one pound (1 lb) is about $\frac{1}{2}$ kilogram

  — one pint is just more than $\frac{1}{2}$ litre.

- The activities on **page E20** investigate the inch and the foot. There should be some introductory discussion of these units, their lengths and where they are likely to occur.

  The children are **not** expected to change centimetres into inches or feet, but rather to take measurements given in inches and feet and then find their equivalents in centimetres. For example,

  A door is 3 feet wide. How many centimetres is this? (about 90 cm)

  A man is 6 feet tall. How many centimetres is this? (about 180 cm)

  In question 2, the content relates to clothes sizes, where inches are often still used. The children take each of Superhero's measurements, convert it to centimetres and then choose the correct size of helmet, shirt, jacket and trousers.

  In questions 3 and 4, the children will have to find or know their heights and collar sizes in centimetres.

- The activities on **page E21** investigate the pound and the pint. These Imperial units are often used in shopping contexts, and some discussion of them would be worthwhile before the children begin.

## Extension

In question 3, the weights of the items on the shopping list have to be changed from pounds to kilograms and then compared with the weights of the items purchased by Mog.

Personal scales may need to be available for question 4 as the children have to find or know their weight in kilograms.

In question 5(a), the children should calculate that the jug holds just more than $1\frac{1}{2}$ litres, so at least six $\frac{1}{4}$-litre glasses can be filled. In part (b), they should calculate that the four guests drink a total of 6 pints of punch, which is just more than 3 litres.

UA2a,3c/4 SSM4a/4
PSE T/D1,2
P/4a M/4e

## Extension Textbook pages E22 and E23

### *Time: 24-hour clock, durations*

- The problem solving activity on these pages extends the work on 24-hour notation and can be attempted at any time after the children have completed the work on Textbook pages 89 and 90 and Workbook page 29 in the Time section.
- Discuss the context with the children. They have to work out the identity of a mystery person who handed the Channel 6TV doorman a cheque for £100 000 as a donation to the Hospital Appeal. To identify the mystery person, the children have to interpret information from the illustrations and from the statements made by the characters who appear on these two pages.
- It may be worth emphasizing that the donation was made between 15.00 and 17.00.
- In question 1, the children should work out that three of the people could have handed in the cheque:
  - **I. M. Broke**, who left Sound Sense, a shop next to the studios, at 16.15
  - **Stan**, who was in the vicinity of the studios between 15.30 and 15.40
  - **B. Packer**, who was exploring the town between 14.30 and 16.45.
- In question 2, the children need to work out that the only times the cheque could have been handed to the doorman were
  - between 14.00 and 15.20
  - between 15.45 and 16.15
  - between 16.30 and 18.00

when he was welcoming people to the studios.

- By a process of elimination the children should realize that B. Packer must have been the mystery donor.

  I. M. Broke and Stan are ruled out because
  - between 4.15 pm, when I. M. Broke paid £25.30 for an item in Sound Sense, and 4.30 pm, when she arrived home for the evening, the doorman was answering the telephone.
  - between 15.30 and 15.40, when Stan was in the vicinity of the studios, the doorman was having a teabreak.

B. Packer, while 'exploring the town', is the only person who could have been close to the studios when the doorman was welcoming people to Channel 6TV.



UA3ab/4 SSM3ab/4
PSE PM/D3
P/4d A/4f

## Extension Textbook page E24 *Co-ordinates: transformations*

- The activities on this page allow the children to explore simple transformations of a shape on a co-ordinate grid. The work can be attempted at any time after the section on co-ordinates on Textbook page 94 and Workbook pages 31 and 32 has been completed.
- The intention is that the children in a group share the work and then discuss and describe the effect on the 'head' shape of changing the co-ordinates in various ways.
- In question 2(a), the children should be clear that the given points

  (7, 12) (7, 8) (8, 8) . . .

  are the **new** points derived by adding 6 to the second co-ordinates only of the points given for the original head shape in question 1(b). The effect is to slide the head up 6 units. For some children, this slide could be described as a **translation.**

  In 2(b), subtracting 5 from the first co-ordinate of each point slides the head 5 units to the left.

  Adding 4 to each co-ordinate, in 2(c), slides the head to the right and upwards.

  There will be similar translations in questions 2(d) and (e), depending on the numbers added or subtracted by the children. To avoid negative co-ordinates, they should subtract no more than 7 from the first co-ordinate or 2 from the second one.
- In question 3, multiplying **one** of the co-ordinates translates the shape and also 'stretches' it in one direction only. For example, in 3(a), multiplying the first co-ordinate by 2 has the following effect:

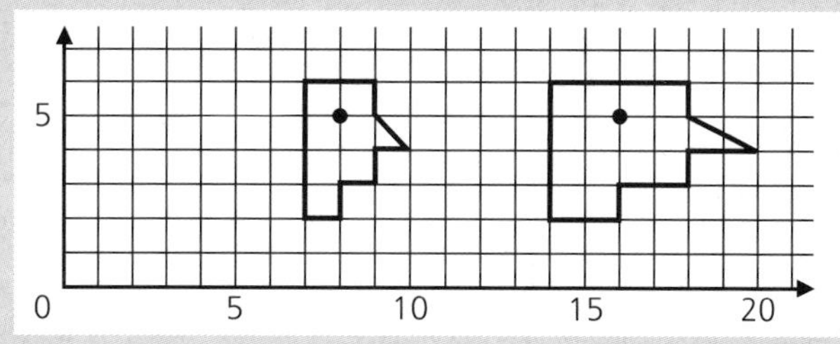


  In part (b), multiplying the second co-ordinate by 3 slides the head up and stretches it vertically.
- In question 4, the head is flipped over or reflected (about the dashed line) when its co-ordinates are reversed.

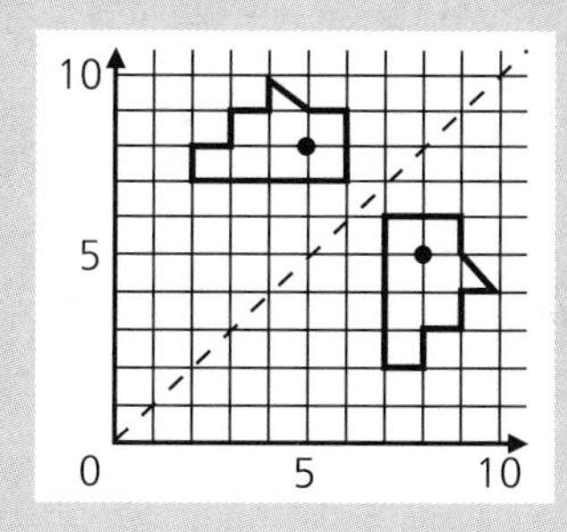

N3b/4
FE/D1 PM/D3
A/4bf

## Extension Textbook page E25

### *Other activity: co-ordinate patterns*

- This page deals with patterns in the co-ordinates of points which lie in a straight line. The activities are challenging and should only be attempted after the work on co-ordinates on Textbook page 94 and Workbook pages 31 and 32 has been completed. It would also be desirable for the work on pattern on Textbook pages 55 and 56 and Workbook pages 15 and 16 to have been attempted.
- In question 1(b), the children should realize that the number of brains is three times the number of Nards. Put in another way, each Nard has 3 brains. In 1(c), the points (5, 15) and (11, 33) should be marked as each has its second co-ordinate three times the first.
- The graph representing Nard hearts in question 2(a) should show points lying in a straight line. As the second co-ordinate (representing hearts) has to be twice the first co-ordinate (representing Nards), the correct points in 2(b) are (8, 16), (10, 20), (14, 28) and (25, 50). There are many possibilities for the final point, but in each the second co-ordinate should be twice the first.
- In question 3(a), among other possibilities, the children should notice the pattern 'the second co-ordinate is half the first' or 'the first co-ordinate is double the second'. This should lead them to realize that (12, 6) and (18, 9) would be 'red' points in question 2(c).

SSM2bc/4
RS/D1,2,3 RS/E1 S/D1
S/3c S/4a S/5a

## Extension Textbook page E26 *2D Shape: properties*

- The activities on this page extend the children's knowledge and experience of properties of 2D shapes. They can be attempted after completion of the work on 2D shape on Textbook pages 95–8 and the Angle section on Textbook pages 108–10.
- Specific properties investigated on the page are
  - — right, acute and obtuse angles
  - — equal angles
  - — equal sides
  - — congruent shapes
  - — lines of symmetry
  - — diagonals.
- The shapes are

| | |
|---|---|
| parallelogram (P) | pentagon (Y) |
| kite (T) | equilateral triangle (W) |
| rhombus (V) | right-angled isosceles triangle (S) |
| trapezium (Q and X) | isosceles triangle (U) |
| | obtuse-angled scalene triangle (R). |

The children are **not** expected to identify a trapezium or an obtuse-angled scalene triangle by name.

Extension

- Throughout the work, the children should pay particular attention to the use of the words 'only' and 'exactly' when looking, for example, for shapes with
  - '**only** 1 obtuse angle' or
  - '**exactly** 2 lines of symmetry'.
- In question 2, some children may try to find equal sides by measuring with a ruler. Others may rely, in the first instance, on visual perception. In the latter case, the children should be encouraged to check by measuring or tracing and fitting.
- In question 5, lines of symmetry can be identified by inspection, turning the page around, where appropriate, so that possible axes appear 'vertical'. Alternatively, mirrors could be used.

UA2b,3a,4bc/4 N3a/4
PSE FE/D1
P/4cd A/4bd

## Extension Textbook page E27

### *Other activity: pattern, area, perimeter*

- This page provides further pattern work, which involves expressing simple relationships in words. It can be attempted after the completion of the pattern work on Workbook page 16 and Textbook page 56, but may be left until the work on area and perimeter in Workbook page 27 and the introduction of the term 'congruent' on Textbook page 98 have been experienced.
- Many children are likely to find the answer to question 1(c) by noticing that the perimeter increases by one unit each time. By extending the table, the perimeter for an area of 10 triangles is found to be 12 units. In part (d), the children must read **across** the rows of the table to compare the perimeter of each strip with its area. They should notice that the perimeter of each strip is always 2 greater than its area.
- In question 2(b), the children should notice that the perimeter increases by 2 units each time. Using this fact, the answers to part (c) can be found by extending the table. However, some children may realize that, for example, for a strip of area

  3 squares ☐☐☐, the perimeter is $2 \times 3 + 2$ units

  4 squares ☐☐☐☐, the perimeter is $2 \times 4 + 2$ units

  5 squares ☐☐☐☐☐, the perimeter is $2 \times 5 + 2$ units.

  This provides a way of answering question 2(d). The perimeter of each strip can be found by multiplying its area by 2 and then adding 2.

**Extension**

UA2ab,3b/4→5 HD2b/3
PSE O/C3 I/E1 O/D1
P/4cd D/3b

## Extension Textbook pages E28 and E29

### *Handling data: interpretation, database*

- The activities on these pages can be tackled any time after completion of the work on interpreting data which appears on Textbook pages 119 and 120.
- Some children may require guidance in interpreting the illustrations and the symbols which provide information about eight types of fungus. The symbols used are:

denotes fungi which are not safe to eat – they are **poisonous**

denotes fungi which are safe to eat – they are **edible**

denotes fungi which are safe to eat but have an unpleasant taste – they are **inedible**

denotes fungi which are found in **broad-leaved** woodlands

denotes fungi which are found in **pine** woodlands

denotes fungi which are found in **mixed** woodlands

The other symbol used is the calendar, which shows when each fungus can be seen. For example, Destroying Angels can be seen from August to October.

- On **page E28**, in question 1, the children should complete the table by writing information or placing ticks under the appropriate symbols.

| Name | cm | cm | | | | | | | |
|---|---|---|---|---|---|---|---|---|---|
| Destroying Angel | 6-10 | up to 15 | Aug-Oct | | | ✓ | | | ✓ |

They then use the information to help them answer questions 2 to 6.

- In question 2, some children may find it helpful to list the range of months, May to December, then place ticks under the months when each fungus appears. The months with eight ticks are those when it is possible to see all the fungi.

Extension

- In questions 3 and 4, the children have to look for the names of the fungi which have **both** properties. Some children may find it helpful to list all of the fungi that have the **first property** – for example, for question 3(b),

  poisonous fungi – Destroying Angel, Fly Agaric, Spring Bolete

  and then to look **only** at the illustrations of these fungi to find which one has the shortest cap (6–10 cm, Destroying Angel).
- On **page E29**, in questions 5 and 6, a similar process of elimination could be used. The children should be encouraged to read the clues in order, from top to bottom, to locate a fungus which has
  - **all three** listed properties (question 5)
  - **all four** listed properties (question 6).

UA2b,3a/4 SM2b,3c/4
PSE A/C1,2
P/4bcd S/3b S/4a

## Extension Textbook page E30

### *Other activity: angles investigation*

- This investigation involves making a variety of triangles on a 16-pin nailboard, using a 90° angle tester to check for acute, right and obtuse angles, and using a 45° angle tester to check the sizes of some of the acute angles. The work can be attempted at any time after the children have completed the Angles section on Textbook pages 108–10 and Workbook page 22.
- If 16-pin nailboards are not available, alternatives are to mark off 16 pins on larger boards or provide a specially prepared grid. The distance between 'pins' should be at least 2 cm. Angle testers can be made by folding circles of diameter at least 10 cm.

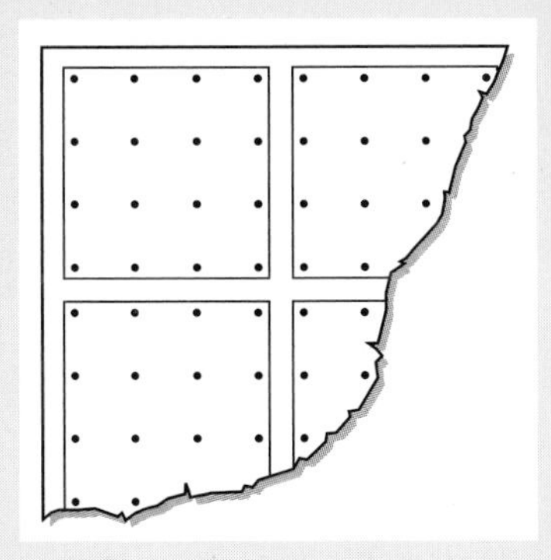

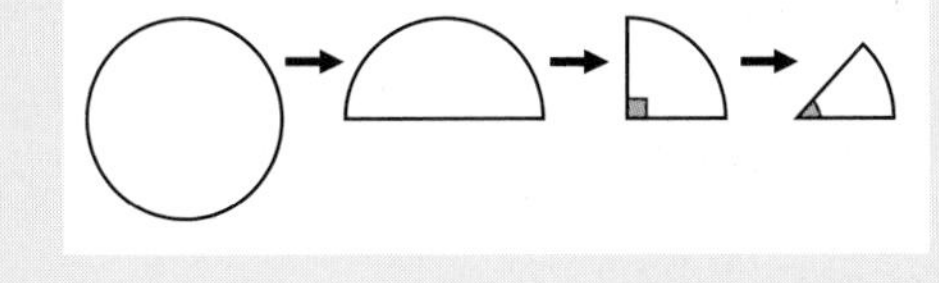

- In question 1, if necessary, explain that each triangle should be made on the nailboard **before** estimating the types of angle.
- In question 4, the children must realize that the line joining two 'diagonally opposite' pins is a line of symmetry for the square it passes through – it divides opposite corners of the square into two angles of 45°.

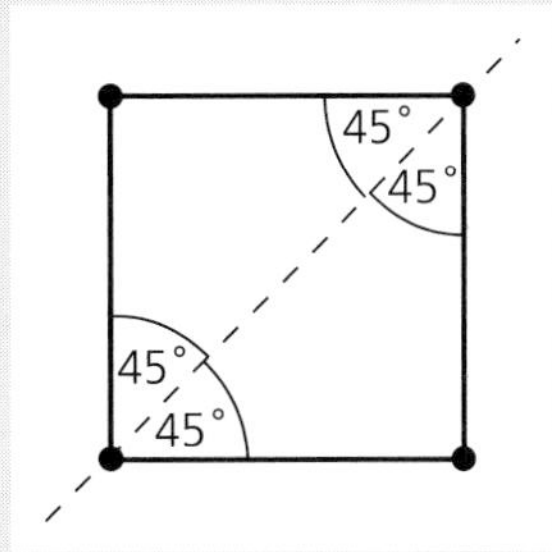

# Heinemann Mathematics 6: Comparison Chart for National Curriculum (England and Wales) Programme of Study for Key Stage 2*

| | Programme of Study | Teacher's Notes | Textbook | Workbook | Reinf't Sheets | Extension Textbook | Assessment | Problem Solving Activities | Resource Cards |
|---|---|---|---|---|---|---|---|---|---|
| **USING AND APPLYING MATHEMATICS** | **1. Pupils should be given opportunities to:**<br>**a** use and apply mathematics in practical tasks, in real-life problems and within mathematics itself<br>**b** take increasing responsibility for organizing and extending tasks<br>**c** devise and refine their own ways of recording<br>**d** ask questions and follow alternative suggestions to support the development of reasoning | Although presented as a separate section of the Programme of Study, 'Using and Applying Mathematics' is integral to all sections of work. The philosophy underpinning the whole of Heinemann Mathematics is that mathematical ideas should be developed through activity. By using suggestions for practical work given in the Teacher's Notes, Assessment and Resources Pack and children's materials, teachers will therefore be giving children opportunities to use mathematical equipment and materials. The Teacher's Notes also contain ideas for activities which can play an important role in developing both mathematical language and communication and mathematical reasoning. | | | | | | | |
| | **2. Making and monitoring decisions to solve problems** | pp. 26–30, 44–5, 95, 97, 105, 108–11, 119, 137, 151, 157, 163–5, 170–7, 182, 191, 202–3, 242–8, 263–4, 266–8, | pp. 4–5, 9–11, 17–18, 20–1, 23–6, 37–8, 41, 45–9, 54, 64, 70, 76–80, 92–3, 121, 122 | pp. 2–3, 6, 8, 9, 11, 17, 24–5, 27 | R3, R6, R7, R14, R22, R24 | E2–E4, E10, E11, E14–E15, E18, E22–E23, E27–E30 | **Round-up** 3 | 2–12, 14–16, 18, 19, 22, 23, 25–8, 30–2 | **23, 24** (True or false cards)<br>**26** (Tangrams) |
| | **3. Developing mathematical language and forms of communication** | pp. 31–3, 38–9, 42–3, 52–4, 56–7, 62–5, 96–108, 114–19, 124–7, 151, 170–7, 190–202, 254–64, 266–8, 270–5 | pp. 3, 12, 17–21, 25, 39–40, 52, 56–8, 79–80, 85–6, 89–91, 95, 96, 98, 103, 111–14, 118–25 | pp. 2–6, 10, 12–14, 16–17, 18, 20, 24–5, 27, 29–30, 33–6, | R2, R6, R10, R13–15, R17, R22, E29 | E6, E7, E10, E12, E18, E22–E24, E27–E30 | **Round-ups** 2, 3 | 6–8, 13–22, 24–9, 31 | **26** (Tangrams)<br>**(26, 30, 31)** (Strip patterns) |
| | **4. Developing mathematical reasoning** | pp. 38–9, 52–4, *119*, 124–8, 161, 164, 171, 177, 198, 270–4 | pp. 12, 29, *54*, 56, 59, 75, 78, 80, 89–90, 109–10, 112–17, 123, 124 | pp. 4, 5, 16, 22, 24–5, 27 | R17, R24, R28 | E6, E7, E18, E27 | **Round-up** 1 | 1, 4, 5, 7, 9, 17, 18, 25, 30–2 | |
| **NUMBER** | **1. Pupils should be given opportunities to:**<br>**a** develop flexible and effective methods of computation and recording, and use them with understanding<br>**b** use calculators, computers and a range of other resources as tools for exploring number structure and to enable work with realistic data<br>**c** develop the skills needed for accurate and appropriate use of equipment | In Heinemann Mathematics children are provided with material in a variety of contexts and formats. Opportunities are provided for them to develop confidence in different aspects of Number work. This includes introducing strategies in mental calculation.<br>Calculators are built in from Heinemann Mathematics 2 onwards and are an increasingly important tool in the later stages of the course.<br>Through using the variety of resources provided within the course children have a wide experience of different recording techniques. | | | | | | | |
| | **2. Developing an understanding of place value and extending the number system**<br>**a** read, write and order whole numbers, understanding that the position of a digit signifies its value; use their understanding of place value to develop methods of computation to approximate numbers to the nearest 10 or 100, and to multiply and divide | pp. 26–33, 38–43, 55–6, 66–7 | pp. 1–3, 6–8, 13, 22 | pp. 1–3 | R1, R2, R4 | E2–E4, E16 | **Number** 1, 2<br>**Round-up** 1 | | **4–6** (Escape from the Island) |

NUMBER

| | | | | | | | |
|---|---|---|---|---|---|---|---|
| system to negative numbers in context, and decimals with no more than two decimal places in the context of measurement and money | | | | | **Round-up** 3 | | **13–17** (Castle attack) |
| **c** understand and use, in context, fractions and percentages to estimate, describe and compare proportions of a whole | pp. 78–86, 289 | pp. 31–5 | pp. 8, 9 | R11 | E9, E13, E20, E21 | **Number** 7, 12<br>**Round-ups** 2, 3 | 7, 8 | **7, 8** (Equivalent fractions game) |
| **3. Understanding relationships between numbers and developing methods of computation** | | | | | | | | |
| **a** explore number sequences *eg count in different sizes of step, doubling and halving, using a multiplication square*, explaining patterns and using simple relationships; progress to interpreting, generalizing and using simple mappings, *eg C=15n for the cost of n articles at 15p*, relating to numerical, spatial or practical situations, expressed initially in words and then using letters as symbols | pp. 52–4, 121–8, 143 | pp. 12, 55–9, 68 | pp. 4, 5, 15, 16 | R17 | E7, E14–E15, E27 | **Round-up** 3 | 1, 6, 9, 17, 32 | |
| **b** recognize the number relationship between co-ordinates in the first quadrant of related points on a line or shape, *eg vertices of a rectangle, a graph of the multiples of 3* | | | | | E25 | | | |
| **c** consolidate knowledge of addition and subtraction facts to 20; know the multiplication facts to 10x10; develop a range of mental methods for finding quickly from known facts those that they **cannot** recall; use some properties of numbers, including multiples, factors and squares, extending to primes, cubes and square roots | pp. 26–30, 52–4, 56–7, 62–8, 70–3, 127 | pp. 1–2, 11, 12, 14–28, 57–8 | pp. 1, 4–7 | R1, R5–R9 | E6 | **Round-up** 1 | 2, 6, 11, 13 | |
| **d** develop a variety of mental methods of computation with whole numbers up to 100, and explain patterns used; extend mental methods to develop a range of non-calculator methods of computation that involve addition and subtraction of whole numbers, progressing to methods for multiplication and division of up to three-digit by two-digit whole numbers | pp. 26–30, 35–7, 40–1, 52–6, 62–7, 70–3, *132–43* | pp. 1–2, 4–7, 11–13, 17–28, *60–7* | p. 1, 4–7 | R1, R3, R4, R6–R9, *R18–21* | E2–E3, E16 | **Number** 2, 13 | 2, 4, 6 | **1–3** (All in the mind)<br>**4–6** (Escape from the island) |
| **e** understand multiplication as repeated addition, and division as sharing and repeated subtraction; use associated language and recognize situations to which the operations apply | pp. 64–8, 96–7 | pp. 19–25, 39–41, 48–9 | p. 6 | R6, R7 | | **Number** 13 | | |
| **f** understand and use the relationships between the four operations, including inverses | | | | | E11 | | | |
| **g** extend methods of computation, to include addition and subtraction with negative numbers, all four operations with decimals, and calculating fractions and percentages of quantities, using a calculator where appropriate | pp. *84*, 94–5, 97–108, *118* | pp. *34*, 37–8, 41, 45–9, *53* | | R14, *R16* | E9, E11 | **Number** 5, 7, 8, 10, 12<br>**Round-ups** 2, 3 | | **12** (Five plus fifty)<br>**13, 14** (Tartan Tours)<br>**15–17** (Castle attack) |
| **h** understand and use the features of basic calculator, interpreting the display in the context of the problem, including rounding and remainders | pp. 44–5, 70–3, 109–11 | pp. 9–10, 26, 50–1 | p. 7 | R6–R8 | | **Number** 6<br>**Round-up** 2 | 12 | |

* References in italics are to related work at Level 4→5 or Level 5.

# Heinemann Mathematics 6: Comparison Chart for National Curriculum (England and Wales) Programme of Study for Key Stage 2* *(Cont.)*

| Strand | Programme of Study | Teacher's Notes | Textbook | Workbook | Reinf't Sheets | Extension Textbook | Assessment | Problem Solving Activities | Resource Cards |
|---|---|---|---|---|---|---|---|---|---|
| NUMBER | **4. Solving numerical problems** | | | | | | | | |
| | **a** develop their use of the four operations to solve problems, including those involving money and measures, using a calculator where appropriate | pp. 35–7, 95, 97–108, 143, 151–7, 160–5, 191–9 | pp. 17, 18, 23–4, 27–9, 37–8, 41, 45–9, 69–75, 78, 86 | pp. 17, 29 | R7, R9, R14, R22 | E8 | **Number** 4, 6, 10, 11<br>**Measure** 5<br>**Round-ups** 1, 3 | 1, 3, 10–12, 15 | **13, 14** (Tartan Tours) |
| | **b** choose sequences of methods of computation appropriate to a problem, adapt them and apply them accurately | pp. 70–3 | | | | | | 12 | |
| | **c** check results by different methods, including repeating the operations in a different order or using inverse operations; gain a sense of the size of a solution, and estimate and approximate solutions to problems | pp. 44–5, 56–6, 70–3 | pp. 9–10, 13, 26 | | | | **Number** 11 | | |
| SHAPE, SPACE AND MEASURES | **1. Pupils should be given opportunities to:** | | | | | | | | |
| | **a** use geometrical properties and relationships in the solution of problems | | | | | | | | |
| | **b** extend their practical experience using a wide range of materials | | | | | | | | |
| | **c** use computers to create and transform shapes | | | | | | | | |
| | **d** consider a wide range of patterns, including some drawn from different cultural traditions | | | | | | | | |
| | **e** apply their measuring skills in a range of purposeful contexts | | | | | | | | |
| | **2. Understanding and using properties of shape** | | | | | | | | |
| | **a** visualize and describe shapes and movements, developing precision in using related geometrical language | pp. 212–15, 217–18 | pp. 96, 97 103 | pp. 18, 20, 33 | R29 | | | 21, 24, 25 | |
| | **b** make 2-D and 3-D shapes and patterns with increasing accuracy, recognize their geometrical features and properties, and use them to classify shapes and solve problems | pp. 217–22, 227–33 | pp. 95–100, 104–6 | pp. 18, 20, 33 | R29 | E12, E26, E30 | **Shape** 2<br>**Round-ups** 2, 3 | 14, 20, 21, 23–5 | **26** (Tangrams)<br>**32** (Shape template) |
| | **c** understand the congruence of simple shapes; recognize reflective symmetries of 2-D and 3-D shapes and rotational symmetries of 2-D shapes | pp. 207–8, 212–16, 219–21, 223–4 | pp. 96–8, 101–3, 107 | pp. 18, 20, 32 | | E18 | **Shape** 1, 2<br>**Round-ups** 1, 2 | | **27–9** (Shape link)<br>**32** (Shape template) |
| | **3. Understanding and using properties of position and movement** | | | | | | | | |
| | **a** transform 2-D shapes by translation, reflection and rotation, and visualize movements and simple transformations to create and describe patterns | | | | | E1, E24 | | 25 | **(26, 30, 31)** (Strip patterns) |
| | **b** use co-ordinates to specify location, *eg map references, representation of 2-D shapes* | pp. 207–8, 217–18 | p. 94 | pp. 31–3 | R29 | E24 | **Shape** 1<br>**Round-up** 1 | 22, 23 | **25** (Archaeological dig) |
| | **c** use right angles, fractions of a turn and, later, degrees, to measure rotation, and use the | pp. 238–9, 242–8 | p. 108 | | | E30 | **Shape** 3<br>**Round-up** 3 | 25 | **30, 31** (Angle machine) |

As for all the sections of the Programme of Study, support for a practical teaching approach can be found in the Teacher's Notes. For each maths topic within Measure and Shape there are opportunities for children to gain valuable hands-on experience within a meaningful context.

| Section | | Programme of Study | | | | | | | | |
|---|---|---|---|---|---|---|---|---|---|---|
| **SHAPE, SPACE AND MEASURES** | **a** | choose appropriate standard units of length, mass, capacity and time, and make sensible estimates with them in everyday situations | pp. 151–7, 160–5, 190–202 | pp. *69*, *73–4*, 75–7, 85–91 | pp. 24, 28–30 | R23, R26–8 | E20–E23 | **Measure** 1, 2, 5<br>**Round-ups** 2, 3 | 19 | **18** (APEC plan)<br>**19–22** (Channel 6 TV) |
| | | extend their understanding of the relationship between units; convert one metric unit to another, know the rough metric equivalents of Imperial units still in daily use | pp. *150–7*, 182, *184–5* | pp. *78*, 81–2, *84* | p. 28 | R25 | | | | |
| | **b** | choose and use appropriate measuring instruments, interpret numbers and read scales to an increasing degree of accuracy | pp. 92–3, 98–103, 161–3, 180–2 | pp. 36, 44, 75–7 | p. 24 | R13 | | **Measure** 4<br>**Round-up** 3 | 18 | **23, 24** (True of false cards) |
| | **c** | find perimeters of simple shapes; find practically the circumference of circles, being introduced to the ratio '$\pi$'; find areas and volumes by counting methods, leading to the use of other practical methods, *eg dissection* | pp. 150–1, 170–7, 183–5 | pp. 70, 79–80, 83–4 | pp. 17, 24–5, 27 | *R22*, R24 | E18, E19 | **Measure** 1, 3, 4<br>**Round-up** 1 | 14, 16, 17, 24 | **18** (APEC plan) |
| **HANDLING DATA** | **1.** | **Pupils should be given opportunities to:** | | | | | | | | |
| | **a** | formulate questions about an issue of their choice, and consider them using statistical methods | As for all the sections of the Programme of Study, support for a practical teaching approach can be found in the Teacher's Notes. There are opportunities for children to gain valuable hands-on experience of gathering, interpreting and presenting data within a context that is relevant to them. Advice is also given in the Teacher's Notes for using software packages currently available. | | | | | | | |
| | **b** | access and collect data through undertaking purposeful enquiries | | | | | | | | |
| | **c** | use computers as a source of interesting data, and as a tool for representing data | | | | | | | | |
| | **2.** | **Collecting, representing and interpreting data** | | | | | | | | |
| | **a** | interpret tables used in everyday life; interpret and create frequency tables, including those for grouped discrete data | pp. 261–4, 266–8 | pp. 118–21 | pp. 35–6 | | | **Handling data** 1, 2<br>**Round-ups** 2, 3 | | |
| | **b** | collect and represent discrete data appropriately using graphs and diagrams, including block graphs, pictograms and line graphs; interpret a wider range of graphs and diagrams that represent data, including pie charts, using a computer where appropriate | pp. 254–7, 261–4, 268 | pp. 111–14, 118–21 | pp. 34–6 | | E5, E28–E29 | **Handling data** 1, 2<br>**Round-ups** 1, 2 | 26, 27 | |
| | **c** | understand and use measures of average, leading towards the mode, the median and the mean in relevant contexts, and the range as a measure of spread | pp. 258–60 | pp. 115–17 | | | | **Handling data** 1<br>**Round-ups** 2, 3 | | |
| | **d** | draw conclusions from statistics and graphs, and recognize why some conclusions can be uncertain or misleading | pp. 254–7, 261–2, 266–7 | pp. 111–14, 118 | 34–6 | | | | 26 | |
| | **3.** | **Understanding and using probability** | | | | | | | | |
| | **a** | develop understanding of probability, through experience as well as experiment and theory, and discuss events and simple experiments, using a vocabulary that includes the words "evens", "fair", "unfair", "certain", "likely", "probably" and "equally likely" | pp. 270–5 | pp. 123–5 | | | | **Handling data** 3 | | |
| | **b** | understand that the probability of any event lies between impossibility and certainty, leading to the use of the scale from 0 to1 | pp. 270–4 | p. 123 | | | | **Handling data** 3 | | |
| | **c** | recognize situations where probabilities can be based on equally likely outcomes, and others where estimates must be based on experimental evidence; make or approximate these estimates | This work is introduced from Heinemann Mathematics 7. | | | | | | | |

* References in italics are to related work at Level 4→5 or Level 5.

| AT | Strand | Statement of Attainment | Teacher's Notes | Textbook | Workbo |
|---|---|---|---|---|---|
| INFORMATION HANDLING | PROBLEM SOLVING AND ENQUIRY | | pp. 215, 242–8, 266–7 | pp. 4, 5, 17–21, 23–6, 29, 37, 38, 41, 45–51, 54, 57–8, 70, 73–4, 78, 92, 93, 98, 103, 107, 121 | pp. 2, 3, 6, 18, 33 |
| | COLLECT | By obtaining information for a task from a variety of given sources, including a simple questionnaire with yes/no type responses. | pp. 263–4 | pp. 119–20 | |
| | | By conducting a survey which extends beyond the class. | p. 268 | p. 122 | |
| | ORGANIZE | By using a tally sheet with grouped tallies. | pp. 258–62, 268 | pp. 115–18, 122 | pp. 35–6 |
| | | By entering data in a table using row and column headings. | pp. 263–4, 268 | pp. 119–22 | |
| | | By using a database where the teacher defines the headings or fields. | | p. 121 | |
| | DISPLAY | By constructing a table or chart. | p. 268 | p. 122 | |
| | | By constructing a bar graph with axes graduated in multiple units and discrete categories of information. | pp. 254–7 | pp. 111–17 | p. 34 |
| | INTERPRET | From displays and databases by retrieving specific records. | pp. 254–7, 263–4, 268 | pp. 111–22 | |
| | | From displays and databases by identifying the most and least frequent items. | pp. 254–7, 268 | pp. 112–14, 122 | |
| NUMBER, MONEY AND MEASUREMENT | RANGE AND TYPE OF NUMBERS | Work with whole numbers up to 10,000 (count, order, read/write). | | | |
| | | Work with thirds, fifths, eighths, tenths and simple equivalences such as one half = two quarters (practical applications only). | | | |
| | | Work with decimals to 2 places when reading/recording money, and using calculator displays. | | | |
| | MONEY | Use coins/notes to £5 worth or more, including exchange. | | | |
| | ADD AND SUBTRACT | Add and subtract mentally for one digit to or from whole numbers up to 3 digits; beyond in some cases involving multiples of 10. | pp. 26–30, 40–1 | pp. 1–2, 6 | p. 1 |
| | | Add and subtract mentally for subtraction by 'adding on'. | pp. 26–30 | pp. 1–2, 6 | p. 1 |
| | | Add and subtract, without a calculator, for whole numbers with 2 digits, added to or subtracted from 3 digits. | pp. 132–6, 139–42, 150–1 | pp. 61–70 | p. 17 |
| | | Add and subtract, with a calculator, for 3-digit whole numbers. | | | |
| | MULTIPLY AND DIVIDE | Multiply and divide mentally within the confines of all tables to 10. | pp. 52–4, 62–7 | pp. 12, 19–22 | pp. 4, 5 |
| | | Multiply and divide mentally for any 2- or 3-digit whole number by 10. | pp. 132, 133, 140–2 | pp. 60, 61, 66–8 | |
| | | Multiply and divide, without a calculator, for 2-digit whole numbers by any single-digit whole number. | pp. 132, 133 | pp. 60, 61 | |
| | | Multiply and divide, with a calculator, for 2- or 3-digit whole numbers by a whole number with with 1 or 2 digits. | pp. 70–3 | pp. 17–18, 26–8 | p. 7 |

* References in italics are to related work at Level B.

## hart for Mathematics 5–14, Level C*

| Reinforcement Sheets | Extension Textbook | Assessment | Problem Solving Activities | Resource Cards |
|---|---|---|---|---|
| R3, R6, R7, R14, R22, R29 | E2–E4, E6, E7, E10–E12, E14–E15, E18, E22–E24, E27–E30 | **Round-ups** 1–3 | All | **23, 24** (True or false cards)<br>**26** (Tangrams)<br>**(26, 30, 31)** (Strip patterns) |
| | | **Handling data** 1, 2 | | |
| | | | 27 | |
| | | **Handling data** 1, 2<br>**Round-ups** 2, 3 | 27 | |
| | | | | |
| | E28–E29 | | | |
| | | | | |
| | | **Round-up** 2 | | |
| | | | | |
| | | | 27 | |
| | | | | |
| | | | | |
| | | | | |
| | | | 10 | |
| R1 | | **Number** 2 | *(Level B, 13)*<br>2, 11 | |
| R1 | | | 6 | |
| R18–R20 | E16 | | 3 | |
| | | | | |
| R6 | E6 | **Round-up** 1 | 6 | |
| R20, R21 | | **Number** 13 | | |
| | E6 | | | |
| R8, R9 | | **Number** 6 | 1 | |

# Heinemann Mathematics 6: Compariso

| AT | Strand | Statement of Attainment | Teacher's Notes | Textbook | Workbo |
|---|---|---|---|---|---|
| NUMBER, MONEY AND MEASUREMENT | ROUND NUMBERS | Round 3-digit whole numbers to the nearest ten (e.g. when estimating). | pp. 38–40 | | p. 7 |
| | FRACTIONS, PERCENTAGES AND RATIO | Find simple fractions ($\frac{1}{3}$, $\frac{1}{5}$, $\frac{1}{10}$) of quantities involving 1- or 2-digit numbers. | pp. 82–3, 289 | p. 19 | |
| | PATTERNS AND SEQUENCES | Work with patterns and relationships within and among multiplication tables. | pp. 52–4, 121–6 | pp. 12, 55, 56 | pp. 4, 5, 15, |
| | FUNCTIONS AND EQUATIONS | Use a simple 'function machine' for operations involving doubling, halving, adding and subtracting. | pp. 121–3 | | |
| | MEASURE AND ESTIMATE | Measure in standard units weight: accuracy extended to include 20g weights; 1 kg = 100 g. | pp. 160–1, 164 | pp. 75, 78 | |
| | | Measure in standard units volume: litre, $\frac{1}{2}$ litre, $\frac{1}{4}$ litre. | | | |
| | | Measure in standard units area: shapes composed of rectangles/squares or irregular shapes using tiles or grids in square centimetres and metres. | pp. 170–6 | pp. 79, 80 | pp. 25–7 |
| | | Estimate length and height in easily handled standard units: m, $\frac{1}{2}$m, $\frac{1}{10}$m, cm. | | | |
| | | Select appropriate measuring devices and units for length. | pp. 150–1 | p. 69 | |
| | | Read scales on measuring devices to the nearest graduation where the value of an intermediate graduation may need to be deduced. | pp. 92–3, 101–3, 161–3, 180–2 | pp. 36, 44, 76–7, 81–2 | pp. 24, 28 |
| | | Realize that weight and area can be conserved when shape changes. | | | |
| | TIME | Use 12-hour times for simple timetables. | pp. 190–4 | pp. 85–8 | |
| | | Conventions for recording time. | pp. 190–4 | pp. 85–8 | |
| | | Work with hours, minutes. | pp. 191–4 | pp. 86–8 | |
| | | Use calendars. | | pp. 92, 93 | |
| SHAPE, POSITION AND MOVEMENT | RANGE OF SHAPES | Identify 2D shapes within 3D shapes. | pp. 215, 227–8 | p. 104 | |
| | | Draw circles using a variety of methods. | pp. 215, 222 | pp. 99–100 | |
| | | Recognize 3D shapes from 2D drawings. | | | |
| | POSITION AND MOVEMENT | Describe the main features of a familiar journey or route. | | | |
| | | Create paths on squared paper described by instructions such as 'forward 5, right 90, forward 7, left 90'. | | | |
| | SYMMETRY | Find lines of symmetry of shapes drawn on squared grids. | p. 215<br>(level E, 223–4) | (level E, 101–2) | |
| | | Complete the missing half of a simple symmetrical shape or pattern on a squared grid. | | | |
| | ANGLE | Know that a right angle is 90°. | pp. 238–41 | pp. 108–10 | p. 22 |
| | | Use 'right', 'acute' 'obtuse' to describe angles. | pp. 240–1 | pp. 109–10 | p. 22 |
| | | Know that a straight angle is 180° | pp. 240–1 | pp. 109–10 | p. 22 |

* References in italics are to related work at Level B.

## hart for Mathematics 5–14, Level C* *(Cont.)*

| Reinforcement Sheets | Extension Textbook | Assessment | Problem Solving Activities | Resource Cards |
|---|---|---|---|---|
| R8 | | | | |
| | E20–E21 | | | |
| R17 | E6 | | 6 | |
| | | | | |
| | | **Round-up** 3 | | |
| | | | | **23, 24** (True or false cards) |
| R24 | E12, E18 | **Measure** 3<br>**Round-up** 1 | 16 | **18** (APEC plan) |
| | | | *13* | |
| | | | | **23, 24** (True or false cards) |
| R13, R23, R25 | | **Number** 9<br>**Measure** 2, 4<br>**Round-up** 3 | 18 | |
| | E18 | | | |
| R26, R27 | | **Measure** 5 | 19 | |
| R26, R27 | | **Measure** 5<br>**Round-up** 2 | 19 | **19–22** (Channel 6 TV) |
| R26, R27 | | | 19 | **23, 24** (True or false cards) |
| | | | | **23, 24** (True or false cards) |
| | | | | |
| | | | | |
| | | | | |
| | | | | |
| | | | 25 | |
| | | | | |
| | | | | |
| | E30 | **Shape** 3 | | **30, 31** (Angle machine) |
| | E30 | **Shape** 3<br>**Round-up** 3 | | **30, 31** (Angle machine) |
| | | | | **30, 31** (Angle machine) |

| | Strand | Statement of Attainment | Teacher's Notes | Textbook | Workbo |
|---|---|---|---|---|---|
| INFORMATION HANDLING | COLLECT | By selecting sources of information for tasks, including a questionnaire which allows several responses to each question. | | | |
| | ORGANIZE | By using diagrams or tables. | | p. 121 | |
| | | By using a database or spreadsheet table with up to three fields defined by pupils. | pp. *266–7* | pp. 121, 126 | |
| | DISPLAY | By constructing graphs (bar, line, frequency polygon) and pie charts involving simple fractions or decimals. | | | |
| | | By constructing graphs (bar, line, frequency polygon) and pie charts involving continuous data which has been grouped. | pp. 261–2 | p. 118 | pp. 35–6 |
| | INTERPRET | From a range of displays and databases by retrieving information subject to one condition. | pp. 258–64, *266–7* | p. 115–21 | pp. 35–6 |
| NUMBER, MONEY AND MEASUREMENT | RANGE AND TYPE OF NUMBERS | Work with whole numbers up to 100,000 (count, order, read/write). | pp. 31–3 | p. 3 | pp. 2–3 |
| | | Work with whole numbers up to a million (read/write only). | pp. 31–3, 44–5 | pp. 3, 9–10 | pp. 2–3 |
| | | Work with fractions (all previous plus twentieths, fiftieths, hundredths) and equivalences among these and decimals (in applications). | pp. 78–81, 83, 85–6, 92–4, 98–103 | pp. 30, 32–3, 35–6, 42–4, 54 | pp. 9, 10 |
| | | Percentages, decimals to 2 places and equivalences among these in applications in money and measurement. | pp. 98–103, 114–19 | pp. 42–4, 52–4 | pp. 10–14 |
| | MONEY | Use all UK coins/notes to £20 worth or more, including exchange. | | | |
| | ADD AND SUBTRACT | Add and subtract mentally for 2-digit whole numbers, beyond in some cases, involving multiples of 10 or 100. | PP. 26–30, 40–1 | pp. 1–2, 6, 7, 11, 73, 74 | p. 1 |
| | | Add and subtract, without a calculator, for 4 digits with at most 2 decimal places (easy examples only). | pp. 35–7, 93–4, 104–7, 154–5, 160–1 | pp. 4–5, 37, 38, 41, 45–9, 75 | |
| | | Add and subtract, with a calculator, for 4 digits with at most 2 decimal places. | pp. 44–5, 108–10 | pp. 9–10, 50–1 | |
| | MULTIPLY AND DIVIDE | Multiply and divide mentally for whole numbers by single digits (easy examples only). | pp. 55–6, 152–3 | pp. 13, 71–2 | |
| | | Multiply and divide mentally for 4-digit numbers including decimals by 10 or 100. | pp. 55–6, 66–7, 95–6, 134–6 | pp. 13, 22, 39, 40, 62–4 | |
| | | Multiply and divide, without a calculator, for 4 digits with at most 2 decimal places by a single digit. | pp. 56–7, 64–5, 68, 95–6, 134–6, 164 | pp. 14–16, 20–1, 23–5, 39–41, 48–9, 62–4, 78 | p. 6 |
| | | Multiply and divide, with a calculator, for 4 digits with at most 2 decimal places by a whole number with 2 digits. | pp. 70–3, 108–10 | pp. 17–18, 27–8, *29*, 50–1 | |
| | ROUND NUMBERS | Round any number to the nearest appropriate whole number, ten or hundred. | pp. 38–43, 70–3, 108–10 | pp. 6–8, 25, 27–8, 50–1 | p. 6 |
| | FRACTIONS, PERCENTAGES AND RATIO | Find simple fractions (1/7, 3/4, 3/5, 60/100) of quantities involving at most 4 digits (easy examples only). | pp. 64–5, 84, 118–19 | pp. 20–4, 34, 53 | |
| | PATTERNS AND SEQUENCES | Continue and describe more complex sequences. | | pp. 57–9 | |
| | FUNCTIONS AND EQUATIONS | Recognize and explain simple relationships between two sets of numbers or objects. | pp. 121–6 | pp. 55, 56, 59, 70 | pp. 15–17 |

* References in italics are to related work at Level E.

## hart for Mathematics 5–14, Level D*

| Reinforcement Sheets | Extension Textbook | Assessment | Problem Solving Activities | Resource Cards |
|---|---|---|---|---|
| | | | | |
| | E28–E29 | | | |
| | | | | |
| | E5 | **Round-up** 1 | | |
| | | **Handling data** 2 | | |
| | E5, *E28–E29* | **Handling data** 1, 2<br>**Round-up** 2 | 26 | |
| R2 | E2–E3, *E5* | **Number** 3 | | |
| R2 | | **Number** 1<br>**Round-up** 1 | | |
| R11, R12 | E9 | **Number** 7–9<br>**Round-ups** 2, 3 | 7, 8 | **7, 8** (Equivalent fractions game)<br>**9–11** (Go-karting) |
| R12, R13, R15, R16 | E10, E11, E13 | **Number** 9, 12<br>**Round-up** 3 | 8 | **13, 14** (Tartan Tours) |
| | | | | |
| R1, R4 | | **Number** 2, 3 | 2, 4 | **1–3** (All in the mind)<br>**4–6** (Escape from the island) |
| R3, R14 | E2–E3, E11, E18 | **Number** 2, 8, 10<br>**Measure** 1<br>**Round-ups** 1–3 | | **12** (Five plus fifty) |
| | E8, E10, E11 | **Number** 11 | | **13, 14** (Tartan Tours) |
| | *E16*, E17 | **Measure** 1 | 15 | |
| R18, R19 | | **Number** 8 | | |
| R5–R7, R18, R19 | E7 | **Number** 4, 5, 10<br>**Round-ups** 1, 2 | 15 | **12** (Five plus fifty)<br>**23, 24** (True or false cards) |
| R9 | E8 | **Number** 6, 11<br>**Round-up** 2 | *12*, 15 | **13, 14** (Tartan Tours) |
| R4, R9 | E4 | **Number** 3 | | **4–6** (Escape from the island) |
| R6, R7, R16 | E9 | **Number** 5, 7, 12<br>**Round-up** 2 | | **7, 8** (Equivalent fractions game)<br>**15–17** (Castle attack) |
| | E14–E15 | | 9, 17 | |
| R17, R22 | E14–E15, E25, E27 | **Round-up** 3 | 9, 32 | |

# Heinemann Mathematics 6: Compariso

| | Strand | Statement of Attainment | Teacher's Notes | Textbook | Workbo |
|---|---|---|---|---|---|
| NUMBER, MONEY AND MEASUREMENT | MEASURE AND ESTIMATE | Measure in standard units length: small lengths in millimetres; large lengths like buildings in metres. | | pp. 73, 74 | |
| | | Measure in standard units weight: extended range of articles, for example own weight. | pp. 161–3 | p. 78 | |
| | | Measure in standard units volume: accuracy extended to small containers in millimetres; 1l = 1000 ml. | pp. 180–4 | pp. 81–4 | |
| | | Measure in standards units area: right-angled triangles on cm squared grids. | | | |
| | | Measure in standard units temperature. | | | |
| | | Estimate small weights, small areas, small volumes in easily handled standard units. | | p. 78 | p. 28 |
| | | Recognize when kilometres are appropriate. | p. 154–5 | pp. 73, 74 | |
| | | Select appropriate measuring devices and units for weight. | pp. 161–3 | pp. 76–8 | p. 24 |
| | | Be aware of common Imperial units in appropriate practical applications. | p. 289 | | |
| | TIME | Use 24-hour times and equate with 12-hour times. | pp. 195–9 | pp. 89–90 | |
| | | Calculate duration in hours/minutes, mentally if possible. | pp. 191–4, 198–9 | pp. 86–8, 93 | |
| | | Time activities in seconds with a stopwatch. | pp. 199–201 | p. 91 | p. 30 |
| | | Calculate speeds (practical activities only). | | | |
| | PERIMETER, FORMULAE, SCALES | Calculate perimeter of simple straight-sided shapes by adding lengths. | pp. 150–1, *152–3* | pp. 70, *71–2*, 80 | pp. 17, 27 |
| SHAPE, POSITION AND MOVEMENT | RANGE OF SHAPES | Discuss 3D and 2D shapes referring to faces, edges, vertices, diagonals, sides angles. | pp. 212–14, 217–21, 227–33 | pp. 95–8, 103–6 | pp. 18, 20, 3 |
| | | Recognize pentagon, hexagon. | pp. 212–14 | | |
| | | Identify and name equilateral and isosceles triangles. | pp. 212–14, 217–18 | pp. 97, 98, 103 | pp. 20, 33 |
| | | Extend shape vocabulary to radius, diameter, circumference. | pp. 222, *229–33* | pp. 99–100, *106*, *107* | |
| | | Create or copy a tiling using a shape template. | pp. 219–21 | | |
| | | Make 3D models, solid or skeletal, including using nets: cube and cuboid only. | pp. 227–8 | p. 104 | |
| | | Use the rigidity property of triangles in model making. | | | |
| | POSITION AND MOVEMENT | Give directions for a route or journey. | | | |
| | | Use an 8-point compass rose. | pp. 238–9 | p. 108 | |
| | | Use a coordinate system to locate a point on a grid. | pp. 207–8, 217–18 | pp. 94, 98 | pp. 18, 31–3 |
| | | Create patterns by rotating a shape. | pp. 242–8 | | |
| | SYMMETRY | Identify and draw lines of symmetry, generally up to 4. | pp. 207–8, 212–14, *223–4* | pp. 96, 97, 101–2, 103, 107 | pp. 18, 20, 3 |
| | | Create symmetrical shapes. | pp. 207–8 | p. 103 | p. 32 |
| | ANGLE | Draw, copy and measure angles accurately within 5 degrees. | pp. 240–1 | pp. 109–10 | p. 22 |
| | | Use standard notation, 060°, 150°, 300°, to express bearings. | | | |

* References in italics are to related work at Level E.

## hart for Mathematics 5–14, Level D* *(Cont.)*

| Reinforcement Sheets | Extension Textbook | Assessment | Problem Solving Activities | Resource Cards |
|---|---|---|---|---|
| | | | | |
| | | | | **23, 24** (True or false cards) |
| R25 | E19 | **Measure** 4 | 18, 24 | |
| | E12 | | 17 | |
| | | | | |
| | | **Measure** 2 | | |
| | | | | |
| R23 | | | | **23, 24** (True or false cards) |
| | E20–E21 | | | |
| R28 | E22–E23 | **Round-up** 2 | | **19–22** (Channel 6 TV) |
| R26, R27 | E22–E23 | **Measure** 5<br>**Round-up** 2 | 19 | |
| | | | | **23, 24** (True or false cards |
| | | | | |
| R22, R24 | *E17* | **Measure** 1, 3<br>***Round-up** 1* | 14, 15 | **18** (APEC plan) |
| R29 | E12, E26 | **Shape** 1<br>***Round-ups** 2, 3* | 14, 21–3 | **26** (Tangrams)<br>**32** (Shape template) |
| | E12, E16 | **Shape** 1 | 21 | **26** (Tangrams)<br>**32** (Shape template) |
| R29 | E12, E26 | **Shape** 1<br>**Round-up** 2 | 22 | **26** (Tangrams)<br>**32** (Shape template) |
| | | | 20 | |
| | E1 | | | |
| | | | 24 | |
| | | | | |
| | | | | |
| | | **Shape** 3 | | |
| R29 | E24, E25 | **Shape** 1<br>**Round-up** 1 | 22, 23 | **25** (Archaeological dig) |
| | | | 25 | |
| | E26 | **Shape** 2 **Round-up** 1<br>***Round-up** 2* | | **32** (Shape template)<br>***27–9** (Shape link)* |
| | | **Shape** 1 | | |
| | | **Shape** 3 | | **30, 31** (Angle machine) |
| | | | | |

# Heinemann Mathematics 6: Comparison Chart fo

| AT | Statement of Attainment | Teacher's Notes | Textbook | Workbo |
|---|---|---|---|---|
| PROCESSES | **a** select the materials required for a task. | | pp. 17–18 | |
| | **b** explain the current work and describe the results. | | pp. 17–18 | |
| | **c** obtain, collect or generate some information. | | | |
| NUMBER | **a** read, write and order whole numbers to at least 1000; use the knowledge that the position of a digit indicates its value. | pp. 38–41 | pp. 6, 7 | |
| | **b** use and understand the conventional way of recording in money. | | | |
| | **c** recognize and understand simple everyday fractions and their notation. | pp. 78, 92–4, 289 | pp. 30, 36 | |
| | **d** have quick recall and be able to use addition and subtraction number facts to 20. | | | |
| | **e** add and subtract money expressed in conventional notation. | | | |
| | **f** have quick recall and be able to use multiplication facts up to 5 x 5, and all those in 2, 5 and 10 multiplication tables. | | | |
| | **g** solve problems involving multiplication or division of whole numbers or money, using a calculator where necessary. | | | |
| | **h** approximate to the nearest 10 or 100. | pp. 38–41 | pp. 6, 7 | |
| | **i** understand 'remainders' given the context of calculation, and know whether to round up or down. | | | |
| ALGEBRA | **a** explain number patterns and predict subsequent numbers where appropriate. | | | |
| | **b** find the number patterns and equivalent forms of 2-digits numbers. | | | |
| | **c** recognize whole numbers which are exactly divisible by 2, 5 and 10. | | | |
| | **d** understand and work with simple function machines. | pp. 121–3 | p. 55 | p. 15 |
| MEASURES | **a** use a wider range of metric units. | | | |
| | **b** choose and use appropriate units and instruments in a variety of situations, interpreting numbers on a range of measuring instruments. | pp. 92–3, 161–3, 190–4 | pp. 36, 76–7, 85–8, 92, 93 | p. 24 |
| | **c** make estimates based on familiar units. | | | |
| SHAPE AND SPACE | **a** recognize squares, rectangles, circles, triangles, hexagons, pentagons, cubes, rectangular boxes (cuboids), cylinders, spheres, and describe them. | pp. 212–15, 217–21 | pp. 95–8, 103, 107 | pp. 18, 20, |
| | **b** recognize right-angled corners in 2D and 3D shapes. | pp. 212–15 | pp. 95, 96 | pp. 18, 20 |
| | **c** recognize reflective symmetry in a variety of shapes in two dimensions. | pp. 207–8, 212–15 | pp. 96, 97, 103, 107 | pp. 18, 20, |
| HANDLING DATA | **a** extract specific pieces of information from tables and lists. | pp. 256–7, 261–4 | pp. 111–14, 118, 121 | pp. 34–6 |
| | **b** enter and access information in a simple database, which could be a computer database. | pp. 263–4 | pp. 119–21 | |
| | **c** construct and interpret bar charts. | pp. 254–62 | pp. 111–18 | pp. 34–6 |
| | **d** create and interpret graphs (pictograms) where the symbol represents a group of units. | pp. 254–5 | | p. 34 |
| | **e** place events in order of 'likelihood' and use appropriate words to identify the chance. | pp. 270–4 | pp. 123, 124 | |
| | **f** understand and use the idea of 'fifty-fifty' or 'evens' and say whether events are more or less likely than this. | | p. 124 | |

## orthern Ireland Common Curriculum, Level 3

| einforcement heets | Extension Textbook | Assessment | Problem Solving Activities | Resource Cards |
|---|---|---|---|---|
| | | | | |
| | | | | |
| | | | | |
| 4 | | | | |
| | | | 10 | |
| | E20–E21 | | | |
| | | | 2, 13 | |
| | | | 10 | |
| | | | | |
| | | | | **13, 14** (Tartan Tours) |
| 4 | E4 | **Number** 3 | | **4–6** (Escape from the island) |
| | | | | |
| | E6, E14–E15 | | | |
| | | | | |
| | | | | |
| | | | | |
| | | | | |
| 23, R26, R27 | | **Measure** 2, 4, 5<br>**Round-ups** 2, 3 | 18 | |
| | | | | |
| 29 | | **Shape** 2 | 14, 20, 22 | **32** (Shape template) |
| | E30 | **Shape** 2 | | **32** (Shape template) |
| | E26 | **Shape** 1, 2<br>**Round-up** 1 | | **32** (Shape template) |
| | | | | |
| | E28–E29 | | | |
| | | **Round-up** 2 | | |
| | | | | |
| | | **Handling data** 3 | | |
| | | **Handling data** 3 | | |

# Heinemann Mathematics 6: Comparison Chart fo

| AT | Statement of Attainment | Teacher's Notes | Textbook | Workbo |
|---|---|---|---|---|
| PROCESSES | **a** select the materials and mathematics required. | pp. 242–8 | pp. 4, 5, 19–21, 23–6, 37–8, 41, 45–51, 54, 57–8, 73–4, 78, 92, 93, 98, 103, 107, 127 | pp. 2–3, 6, 1 33 |
| | **b** use trial and improvement methods. | pp. 215, 242–8 | pp. 19–21, 23–4, 37–8, 41, 45–51, 54, 70, 73–4, 107 | p. 17 |
| | **c** produce some information relevant to the problem and use this to make some predictions with prompting from the teacher. | pp. 270–5 | pp. 23–4, 29, 57–8, 123–5 | |
| | **d** record findings and present them in oral, written or visual forms. | pp. 270–5 | pp. 23–4, 29, 37–8, 70, 107, 121, 123–5 | p. 17 |
| NUMBER | **a** read, write and order whole numbers. | pp. 31–3, 42–5 | pp. 3, 8–10 | pp. 2–3 |
| | **b** understand and use the relationship between place values in whole numbers. | pp. 31–3, 42–5, *79–81*, 83, *85–6* | pp. 3, 8–10, *32, 33, 35, 54* | pp. 2–3, *9* |
| | **c** understand the effect of multiplying a whole number by 10 or 100. | pp. 55–6, 66–7, *117–19*, 140–2 | pp. 13, 22, *53, 54,* 66–8 | p. *14* |
| | **d** use, with understanding, decimal notation to 2 decimal places, in the context of measurement. | pp. 98–103, 108–10 | pp. 42–4, 50–1 | pp. 10, 11 |
| | **e** understand the meaning of negative whole numbers in familiar contexts only. | | | |
| | **f** recognize and understand simple percentages. | pp. 114–19 | pp. 52, 53 | pp. 12–14 |
| | **g** have quick recall of multiplication facts up to 10 × 10 and use them in multiplication and division problems. | pp. 52–6, 62–8, 84 | pp. 12, 13, 19–25, 34 | pp. 4–6 |
| | **h** (using whole numbers) add or subtract mentally two 2-digit numbers; add mentally several single-digit numbers. | pp. 26–30, *118–19* | pp. 1–2, 11, *53* | p. 1 |
| | **i** without a calculator add and subtract two 3-digit numbers, multiply a 2-digit number by a single digit number and divide a 2-digit number by a single digit number. | pp. 132–6, 139–42, 150–3 | pp. 60–72 | p. 17 |
| | **j** be able to use a calculator to add, subtract, multiply and divide whole numbers, including the precedence of operations. | pp. 44–5, 70–3 | pp. 9–10, 17–18, 26–9 | p. 7 |
| | **k** solve addition or subtraction problems using numbers with no more then 2 decimal places. | pp. 35–7, 93–4, 104–10, 154–5, 160–1 | pp. 4–5, 37–8, 41, 45–51, 73–5 | |
| | **l** solve multiplication or division problems, including remainders. | pp. 56–7, 62–8, 70–3, 95–6, 108–10, 132–6, 139–42, 164 | pp. 14–28, 39–41, 48–51, 60–8, 78 | p. 6 |
| | **m** estimate in addition and subtraction calculations to obtain approximate answers. | pp. 40–1 | pp. 6, 7 | |
| | **n** read a calculator display to the nearest whole number. | pp. 70–3, 108–10 | pp. 27–8, 50–1 | p. 7 |
| | **o** know how to interpret calculator results which have rounding errors. | | | |
| ALGEBRA | **a** apply strategies, such as doubling and halving, to explore properties of numbers. | | pp. *57–8* | |
| | **b** generalize, mainly in words, patterns which arise in various situations. | pp. 52–4, 121–6 | pp. 55, 56, *57–9* | pp. 5, 15, 16 |
| | **c** understand and use multiples and factors. | pp. 52–4, 62–5 | pp. 12, 19–21, 25, *57–8*, 59 | pp. 4–6 |
| | **d** understand and use simple formulae expressed in words. | pp. 121–6 | pp. 55, 56, 59, 70 | pp. 15–17 |
| | **e** understand that multiplication and division are inverse operations and use this to check calculations, using a calculator where necessary. | pp. 70–3 | pp. 26, 29 | |
| | **f** know the conventions of the coordinate representation of points; work with co-ordinates in the first quadrant. | | | |

* References in italics are to related work at Level 5.

## rthern Ireland Common Curriculum, Level 4*

| inforcement eets | Extension Textbook | Assessment | Problem Solving Activities | Resource Cards |
|---|---|---|---|---|
| R6, R7, R13, R29 | E2–E4, E11, E14–E15, E18, E22–E23 | **Round-ups** 1, 3 | 3, 5–7, 9–12, 14–16, 22–8 | **23, 24** (True of false cards)<br>**26** (Tangrams) |
| R7, R13, R22 | E2–E3, E10–E12, E18, E30 | **Round-ups** 2, 3 | 2, 4–6, 8, 10, 11, 13, 16, 18–21, 24, 25, 30, 31 | |
| | E6, E7, E10, E11, E27–E30 | | 1, 4, 5, 7, 9, 12, 14, 15, 17, 25–7 | **23, 24** (True or false cards) |
| R22 | E6, E7, E10, E12, E14–E15, E18, E24, E27–E30 | | 1, 3–5, 7, 9, 10, 12–26, 28–32 | **23, 24** (True or false cards)<br>**26** (Tangrams)<br>**(26, 30, 31)** (Strip patterns) |
| | E2–E3 | **Number** 1<br>**Round-up** 1 | | |
| *R11* | E2–E3, *E9* | **Number** 1, 3, *7*<br>***Round-up*** *2* | *7, 8* | |
| , R20, R21 | *E13* | **Number** *12,* 13 | *8* | ***7, 8*** *(Equivalent fractions game)*<br>***15–17*** *(Castle attack)* |
| , R13 | E10, E11 | **Number** 9<br>**Round-up** 3 | | **9–11** (Go-karting) |
| | | | | |
| , R16 | E13 | **Number** 12<br>**Round-up** 3 | 8 | **15–17** (Castle attack) |
| R7 | E6, E9 | **Number** 7<br>**Round-up** 1 | 6 | |
| *R16* | | **Number** 2, *12* | 2, 4, 11 | **1–3** (All in the mind)<br>**4–6** (Escape from the island) |
| -R22 | E6, E17 | | 6, 15 | |
| R9 | E8 | **Number** 6, 11 | 1, 3, 12, 15 | |
| R13 | E2–E3, E10, E11 | **Number** 2, 8, 10, 11<br>**Measure** 1<br>**Round-ups** 1, 2 | 3 | **12** (Five plus fifty)<br>**13, 14** (Tartan Tours)<br>**23, 24** (True or false cards) |
| R7, R9, R18–R21 | E7, E8, E16 | **Number** 4–6, 8, 10, 11, 13<br>**Round ups** 1, 3 | | **12** (Five plus fifty)<br>**13, 14** (Tartan Tours)<br>**23, 24** (True or false cards) |
| | | **Number** 3 | | |
| R9 | | **Number** 6 | | |
| | | **Round-up** 2 | | |
| | | | | |
| | E7, E25, E27 | **Round-up** 3 | 1, 9, 17, 32 | |
| | E6, E7 | | 16, 17 | |
| , R22 | E14–E15, E27 | | 19 | |
| | | **Number** 11 | | |
| | E24, E25 | | | **25** (Archaeological dig) |

# Heinemann Mathematics 6: Comparison Chart f

| AT | Statement of Attainment | Teacher's Notes | Textbook | Workb |
|---|---|---|---|---|
| MEASURES | **a** understand the relationship between units. | pp. 108–10, 150–1, *152–3*, 154–5, 161–4, 180–2, 190–4, 199–201 | pp. 50–1, 69, 70, *71–2*, 73–4 76–8, 81–2, 85–8, 91, 92 | pp. 11–17, 28, 30 |
| | **b** understand the concept of perimeter. | pp. 150–1, 160–1, *289* | pp. 75, 80 | p. 27 |
| | **c** find areas by counting squares, and volumes by counting cubes, using whole numbers. | pp. 170–6, 183–4 | pp. 79, 80, 83, 84 | pp. 25–7 |
| | **d** make sensible estimates of a range of measures in relation to everyday objects or events. | pp. 199–201 | p. 78 | |
| | **e** understand and use the 24-hour clock. | pp. 195–9 | pp. 89–90, 93 | |
| SHAPE AND SPACE | **a** understand and use language associated with angle. | pp. *219–21*, 240–1 | pp. *98*, 109–10 | pp. *18*, 22 |
| | **b** make simple 2D and 3D shapes from given information and know associated language. | pp. 212–14, 217–18, 222, 227–33 | pp. 95–7, 99–100, 103–7 | pp. 18, 20, |
| | **c** understand eight points of the compass; use clockwise and anticlockwise appropriately. | pp. *223–4*, 238–9 | pp. *101–2*, *107*, 108 | |
| | **d** specify location by means of co-ordinates (in first quadrant). | pp. 207–8, 217–18 | pp. 94, 98 | pp. 18, 31- |
| | **e** use a computer package to investigate position in terms of angle and distance. | pp. 242–8 | | |
| | **f** recognize reflective symmetry in a variety of shapes in three dimensions. | | | |
| HANDLING DATA | **a** specify an issue for which data are needed; collect, group and order discrete data using tallying methods with given equal class intervals and create a frequency table for grouped data. | pp. 258–62, *268* | pp. 115–18, *122* | pp. 35–6 |
| | **b** understand, calculate and use the mean and range of a set of discrete data. | pp. 258–60 | pp. 115–17 | |
| | **c** interrogate data in a computer database. | | pp. 119–20 | |
| | **d** use a decision tree diagram by posing questions to sort and identify a collection of objects. | | | |
| | **e** construct, read and interpret a bar-line graph for a discrete variable (where the length of the bar-line represents the frequency). | pp. 256–7 | pp. 112–13 | |
| | **f** construct and interpret a line graph and know that the intermediate values may or may not have a meaning. | pp. 256–7 | p. 114 | |
| | **g** use computer packages to produce various forms of representation of data. | pp. 266–7 | | |
| | **h** understand and use 0 and 1, as the limits of the probability scale. | | | |
| | **i** give and justify subjective estimates of probabilities in a variety of situations. | pp. 270–4 | p. 123 | |

* References in italics are to related work at Level 5.

# rthern Ireland Common Curriculum, Level 4* *(Cont.)*

| nforcement ets | Extension Textbook | Assessment | Problem Solving Activities | Resource Cards |
|---|---|---|---|---|
| R23, R25–7 | *E17* | **Measure** 1, 2, 5<br>**Round-ups** *1*, 2, 3 | *15* | **19–22** (Channel 6 TV)<br>***18*** (APEC plan)<br>**23–24** (True or false cards) |
| | E20–E21 | **Measure** 1, 3 | 14 | **19–22** (Channel 6 TV) |
| | E12, E18, E19 | **Measure** 3, 4<br>**Round-up** 1 | 16, 17, 24 | **18** (APEC plan) |
| | | | 15 | **23, 24** (True or false cards) |
| | E22, E23 | ***Measure*** 3<br>**Round-up** 2 | | **19–22** (Channel 6 TV)<br>***23, 24*** (True or false cards) |
| | E26, E30 | **Shape** 1, *2*, 3<br>**Round-ups** *2*, 3 | | **30, 31** (Angle machine) |
| | E1, E12 | **Round-ups** 2, 3 | 20, 21, 23, 24 | **26** (Tangrams)<br>**32** (Shape template) |
| | | **Shape** 3<br>***Round-up*** *2* | | ***27–9*** *(Shape link)* |
| | | **Shape** 1<br>**Round-up** 1 | 22, 23 | **25** (Archaeological dig) |
| | | | 25 | |
| | | | | |
| | | **Handling data** 2<br>**Round-up** 3 | 27 | |
| | | **Handling data** 1<br>**Round-up** 3 | | |
| | | | | |
| | | | | |
| | | **Handling data** 1<br>**Round-up** 1 | | |
| | | | 26 | |
| | | | | |
| | | | | |
| | | **Handling data** 3 | | |

# Checklist of materials needed

| 100 cm flats (plastic or wooden) | | |
|---|---|---|
| angle tester (see Textbook page 110) | | |
| balances (2-pan, scales, spring) | | |
| Blu-tack | | |
| calculators | | |
| centimetre cubes (interlocking) | | |
| circles, gummed paper | | |
| compasses | | |
| computer database and spreadsheet packages | | |
| construction kits ('Clixi', 'Polydron', etc) | | |
| containers holding less than 1 litre | | |
| counters (for pattern work) | | |
| cubes, dice (for time work) | | |
| glue sticks, sticky tape | | |
| hard boiled egg, sand, plastic bag (for work on weight) | | |
| litre jug, cup and teaspoon (for work on measure) | | |
| marker cones or bean bags (for work on length) | | |
| measuring jars | | |
| metre stick (calibrated in hundredths – cm) | | |
| metric tape, long (10 m, 20 m or 25 m) | | |
| number cards | | |
| number lines (0–100) (decimal – tenths) | | |
| number square (0–100) | | |
| paper, card (coloured) for shape work | | |
| paper, dotty (cm squared) | | |
| place value cards | | |
| plastic cube container holding 1 litre | | |
| plasticine | | |
| reference books about trees and shrubs (for data handling) | | |
| road atlas or similar showing distance between towns (for length) | | |
| scissors | | |
| sets of 6 card or plastic squares | | |
| shapes, 2D (paper, card, plastic) | | |
| shapes, 3D (cubes, cuboids, triangular prisms) | | |
| shapes 3D (square and triangular pyramids) | | |
| squared paper (2 mm, $\frac{1}{2}$ cm and 1 cm) | | |
| stopwatches (analogue and digital) | | |
| strips (plastic, wood, metal) and fasteners, for making shapes | | |
| trundle wheel | | |
| TV programme guide (for work on time) | | |
| weights, sets of (inc. 20 g 10 g and 5 g) | | |

# Index